Emerging Technologies in Brachytherapy

T0252561

Series in Medical Physics and Biomedical Engineering

Series Editors: John G. Webster, E. Russell Ritenour, Slavik Tabakov, and Kwan-Hoong Ng

Other recent books in the series:

Emerging Technologies in Brachytherapy
William Y. Song, Kari Tanderup, and Bradley Pieters (Eds)

Environmental Radioactivity and Emergency Preparedness
Mats Isaksson and Christopher L. Rääf

The Practice of Internal Dosimetry in Nuclear Medicine
Michael G. Stabin

Radiation Protection in Medical Imaging and Radiation Oncology
Richard J. Vetter and Magdalena S. Stoeva (Eds)

Graphics Processing Unit-Based High Performance Computing in Radiation Therapy
Xun Jia and Steve B. Jiang (Eds)

Statistical Computing in Nuclear Imaging
Arkadiusz Sitek

The Physiological Measurement Handbook
John G. Webster (Ed)

Radiosensitizers and Radiochemotherapy in the Treatment of Cancer
Shirley Lehnert

Diagnostic Endoscopy
Haishan Zeng (Ed)

Medical Equipment Management
Keith Willson, Keith Ison, and Slavik Tabakov

Targeted Muscle Reinnervation: A Neural Interface for Artificial Limbs
Todd A. Kuiken; Aimee E. Schultz Feuser; Ann K. Barlow (Eds)

Quantifying Morphology and Physiology of the Human Body Using MRI
L. Tugan Muftuler (Ed)

Monte Carlo Calculations in Nuclear Medicine, Second Edition: Applications in Diagnostic Imaging
Michael Ljungberg, Sven-Erik Strand, and Michael A. King (Eds)

Vibrational Spectroscopy for Tissue Analysis
Ihtesham ur Rehman, Zanyar Movasaghi, and Shazza Rehman

Emerging Technologies in Brachytherapy

Edited by

William Y. Song
Sunnybrook Health Sciences Centre
University of Toronto

Kari Tanderup
Aarhus University Hospital
Aarhus University

Bradley R. Pieters
Academic Medical Center
University of Amsterdam

CRC Press
Taylor & Francis Group
Boca Raton London New York

CRC Press is an imprint of the
Taylor & Francis Group, an **informa** business

CRC Press
Taylor & Francis Group
6000 Broken Sound Parkway NW, Suite 300
Boca Raton, FL 33487-2742

First issued in paperback 2019

© 2017 by Taylor & Francis Group, LLC
CRC Press is an imprint of Taylor & Francis Group, an Informa business

No claim to original U.S. Government works

ISBN-13: 978-1-4987-3652-7 (hbk)
ISBN-13: 978-0-367-87406-3 (pbk)

Library of Congress Cataloging-in-Publication Data

Names: Song, William Y., 1977- author. | Tanderup, Kari, author. | Pieters, Bradley, 1965- author.
Title: Emerging technologies in brachytherapy / William Y. Song, Kari Tanderup, and Bradley Pieters.
Other titles: Series in medical physics and biomedical engineering.
Description: Boca Raton, FL : CRC Press, Taylor & Francis Group, [2017] | Series: Series in medical physics and biomedical engineering | Includes bibliographical references and index.
Identifiers: LCCN 2016037962| ISBN 9781498736527 (hardback ; alk. paper) | ISBN 1498736521 (hardback ; alk. paper) | ISBN 9781498736541 (E-book) | ISBN 1498736548 (E-book)
Subjects: LCSH: Radioisotope brachytherapy--Technological innovations.
Classification: LCC RC271.R27 S66 2017 | DDC 615.8/424--dc23
LC record available at https://lccn.loc.gov/2016037962

**Visit the Taylor & Francis Web site at
http://www.taylorandfrancis.com**

**and the CRC Press Web site at
http://www.crcpress.com**

We dedicate this book to cancer patients worldwide.

Contents

SECTION II **Imaging for Brachytherapy Guidance**

SECTION III **Brachytherapy Suites**

Acknowledgments

THE EDITORS OF THIS book express their gratitude to all the contributors. The authors have made a tremendous effort to provide us with fully up-to-date, state-of-the-art, and emerging technologies and opinions in brachytherapy.

The editors also wish to record their gratitude and thanks for the skilful support and encouragement of the Taylor & Francis team, in particular Francesca McGowan, Emily Wells, and Rebecca Davies.

Finally, the editors wish to record their gratitude and thanks for the support, encouragement, and forbearance shown by their families and colleagues during the execution of this work. In particular, WYS wishes to thank Hyun Hee Kim for her unending support and love.

Editors

Dr. William Y. Song, PhD, is the head of Medical Physics, Department of Radiation Oncology, Odette Cancer Centre, Sunnybrook Health Sciences Centre, Toronto, Canada. This is one of the largest medical physics units in the world with 50+ staff. Along with a busy external beam radiotherapy program, the centre sees close to 600 brachytherapy patients a year, making it the busiest program in Canada. Since joining the centre in 2014, he has been an associate professor in the Department of Radiation Oncology, adjunct professor in the Institute of Medical Sciences, Institute of Biomaterials and Biomedical Engineering, and Department of Mechanical and Industrial Engineering, at the University of Toronto. He is also an adjunct professor in the Department of Physics, at the Ryerson University, Toronto, Canada. He earned his PhD in 2006 at the University of Western Ontario, London, Canada, on the topic of image-guided treatment approaches for prostate cancer. Since then, he has pursued research in the field of image guidance systems, 4D motion management technologies, and brachytherapy, resulting in over 50+ peer-reviewed publications and 130+ conference abstracts. Along the way, he became a fully certified medical physicist (American Board of Radiology, 2010), directly supervised(ing) 20+ MSc and PhD graduate students, an ad hoc reviewer for 20+ research journals, and is a member of the Editorial Board for the *Journal of Medical Physics*. In brachytherapy particular, his research focus has been in developing novel applicators and MR image processing techniques that enhances plan quality and plan quality evaluations; one in particular, in cleverly designing MR-compatible metal alloys to create non-isotropic dose distributions that can, in combination with inverse

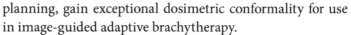

planning, gain exceptional dosimetric conformality for use in image-guided adaptive brachytherapy.

Dr. Kari Tanderup, PhD, is professor at Department of Oncology, Aarhus University Hospital in Aarhus, Denmark. She was educated from Aarhus University with a master's in physics and math in 1997. In 2008 she earned her PhD with a thesis on brachytherapy in cervical cancer. Furthermore, since 2011, she has been appointed as visiting professor at Medical University of Vienna.

At Aarhus University Hospital, Kari Tanderup chairs a research group working with brachytherapy and MR image guidance. Her main research interests are MRI-guided cervix and prostate cancer, clinical studies, and in vivo dosimetry. Within these topics she has authored 90+ papers and has supervised 10+ PhD students. Furthermore, she was committee member for the ICRU (The International Commission on Radiation Units & Measurements) Report 89 on brachytherapy in cervical cancer, and she is associate senior editor for *International Journal of Radiation Oncology, Biology and Physics*.

Kari Tanderup is actively contributing to committee and task group work in ESTRO (European SocieTy for Radiotherapy and Oncology) and is course director in the ESTRO school. She is chairing the GEC ESTRO (Group Européen de Curiethérapie-European Society for Radiotherapy and Oncology) gyn network which represents the most active core European academic centers within gynecological brachytherapy, and embraces members from Central/Eastern Europe, India, Canada, and the United States. Through past and ongoing activities this group is internationally recognized as a leading group in gynecological image-guided brachytherapy. The group has published international guidelines, which are now used worldwide. The GEC ESTRO gyn group has also coordinated the EMBRACE study, which is an international multicenter study on MRI-guided brachytherapy in cervical cancer. The EMBRACE trial has recruited >1400 patients, and this material is currently generating a wealth of clinical evidence on outcome as well as dose and effect relationships.

Dr. Bradley R. Pieters, MD, PhD, is the head of the Brachytherapy department at the Academic Medical Center (AMC) in Amsterdam, The Netherlands. The AMC has a focus on brachytherapy, hyperthermia, and image-guided radiotherapy. He was trained as radiation oncologist at the Radboud University Hospital in Nijmegen, The Netherlands. Because of his interest in brachytherapy he followed at the end of the residency a brachytherapy fellowship at the Daniel den Hoed clinic in Rotterdam and L'Institut Gustav-Roussy in Villejuif, France. Dr. Pieters obtained his MSc in epidemiology in 2006. In 2010 he earned his PhD at the University of Amsterdam after defending his thesis "Pulsed-dose rate brachytherapy in prostate cancer."

Dr. Pieters' main field of interest is general brachytherapy with an emphasis on urologic brachytherapy, gynecologic brachytherapy, and pediatric brachytherapy. His research topics focus on prostate brachytherapy; development of advanced treatment planning optimization algorithms; external beam and brachytherapy dose summation in cervical cancer; and late effects assessment in pediatric brachytherapy. In his role as leader of the brachytherapy research group he supervises PhD students and contributed to more than 40 peer-reviewed papers with the majority concerning brachytherapy topics.

He is one of the coeditors of the *Journal of Contemporary Brachytherapy* and is a member of the Editorial Board of *Brachytherapy*.

For the GEC-ESTRO he contributes as course director for the "Comprehensive and Practical Brachytherapy" course and is member of the GEC-ESTRO Committee.

Contributors

Hamideh Alasti
Radiation Physics Department
Princess Margaret Cancer Centre
and
Department of Radiation Oncology
University of Toronto
Toronto, Canada

Michael Andrassy
R&D Brachytherapy
Eckert & Ziegler BEBIG
Berlin, Germany

Maroie Barkati
Département de Radio-oncologie
Centre Hospitalier de l'Université de
    Montréal
Montréal, Canada

Luc Beaulieu
Département de physiquede génie
    physique et d'optique et Centre
    derecherche sur le cancer
Université Laval
and
Département de radio-oncologie
CRCHU de Québec—Université Laval
Québec, Canada

Sven Beerheide
R&D Brachytherapy
Eckert & Ziegler BEBIG
Berlin, Germany

Alejandro Berlin
Department of Radiation Oncology
Princess Margaret Cancer Centre
University of Toronto
Toronto, Canada

Christoph Bert
Department of Radiation Oncology
Universitätsklinikum Erlangen
Friedrich-Alexander Universität
    Erlangen-Nürnberg
Erlangen, Germany

Peter A.N. Bosman
Centrum Wiskunde & Informatica
Life Sciences Research Group
Amsterdam, The Netherlands

Stephen Breen
Radiation Physics Department
Princess Margaret Cancer Centre
and
Department of Radiation Oncology
University of Toronto
Toronto, Canada

Derek Brown
Department of Radiation Medicine and
    Applied Sciences
University of California San Diego
La Jolla, California

Ashwini Budrukkar
Department of Radiation Oncology
Tata Memorial Hospital
Mumbai, India

Marco Carlone
Radiation Physics Department
Princess Margaret Cancer Centre
and
Department of Radiation Oncology
University of Toronto
Toronto, Canada

Laura Cerviño
Department of Radiation Medicine and
    Applied Sciences
University of California San Diego
La Jolla, California

Kitty Chan
Department of Radiation Oncology
Princess Margaret Cancer Centre
University of Toronto
Toronto, Canada

Robert A. Cormack
Department of Radiation Oncology
Brigham and Women's Hospital
Harvard Medical School
Boston, Massachusetts

Jennifer Croke
Department of Radiation Oncology
Princess Margaret Cancer Centre
University of Toronto
Toronto, Canada

Antonio L. Damato
Department of Medical Physics
Memorial Sloan Kettering Cancer
    Center
New York, New York

Astrid A. de Leeuw
Department of Radiation Oncology
University Medical Center Utrecht
Utrecht, The Netherlands

Talar Derashodian
Department of Radiation Oncology
Hospital Charles Lemoyne
Montreal, Canada

Magatte Diagne
Department of Radiation Oncology
Institut Joliot-Curie Cancer Center
Dakar, Senegal

Colleen Dickie
Department of Radiation Oncology
Princess Margaret Cancer Centre
University of Toronto
Toronto, Canada

Moustafa Dieng
Department of Radiation Oncology
Institut Joliot-Curie Cancer Center
Dakar, Senegal

Lior Dubnitzky
Department of Medical Physics
Sunnybrook Health Sciences Centre
Toronto, Canada

Benjamin Durkee
Department of Radiation Oncology
Swedish American Hospital
A Division of UW Health
Rockford, Illinois

Harry Easton
Department of Medical Physics
Sunnybrook Health Sciences Centre
Toronto, Canada

John Einck
Department of Radiation Medicine and
Applied Sciences
University of California San Diego
La Jolla, California

Aaron Fenster
Department of Radiation Oncology
Case Western Reserve University
Cleveland, Ohio

and

Imaging Research Laboratories
Robarts Research Institute
London, Canada

Irina Fotina
R&D Brachytherapy
Eckert & Ziegler BEBIG
Berlin, Germany

Rick Franich
Département de radio-oncologie et
CRCHU de Québec
CHU de Québec
Québec, Canada

Macoumba Gaye
Department of Radiation Oncology
Institut Joliot-Curie Cancer Center
Dakar, Senegal

Dietmar Georg
Department of Radiation Oncology
Medical University of Vienna
Vienna, Austria

Rachel Glicksman
Department of Radiation Oncology
Princess Margaret Cancer Centre
University of Toronto
Toronto, Canada

Kathy Han
Department of Radiation Oncology
Princess Margaret Cancer Centre
University of Toronto
Toronto, Canada

Taran Paulsen Hellebust
Department of Medical Physics
Oslo University Hospital
and
Department of Physics
University of Oslo
Oslo, Norway

Alana Hudson
Department of Medical Physics
Tom Baker Cancer Center
Calgary, Canada

David Jaffray
Radiation Physics Department
Princess Margaret Cancer Centre
and
Department of Radiation Oncology
University of Toronto
Toronto, Canada

Xun Jia
Division of Medical Physics and
Engineering
Department of Radiation Oncology
University of Texas Southwestern
Medical Center
Dallas, Texas

Marjory Jolicoeur
Department of Radiation Oncology
Hospital Charles Lemoyne
Montreal, Canada

Harald Keller
Radiation Physics Department
Princess Margaret Cancer Center
Toronto, Canada

Martin T. King
Department of Radiation Oncology
Dana-Farber/Brigham & Women's
Cancer Center
Harvard Medical School
Boston, Massachusetts

Christian Kirisits
Department of Radiation Oncology
Medical University of Vienna
Vienna, Austria

Thomas Lanni
Beaumont Health System
Royal Oak, Michigan

Sarbani (Ghosh) Laskar
Department of Radiation Oncology
Tata Memorial Hospital
Mumbai, India

Siddhartha Laskar
Department of Radiation Oncology
Tata Memorial Hospital
Mumbai, India

Yolande Lievens
Department of Radiation Oncology
Ghent University Hospital
Ghent, Belgium

Michael Lotter
Department of Radiation Oncology
University Hospital Erlangen
Erlangen, Germany

Umesh Mahantshetty
Department of Radiation Oncology
Tata Memorial Hospital
Mumbai, India

Eirik Malinen
Department of Medical Physics
Oslo University Hospital
University of Oslo
Oslo, Norway

Alexandr Malusek
Radiation Physics
Department of Medical and Health
Sciences
Linköping University
Linköping, Sweden

Eve-Lyne Marchand
Department of Radiation Oncology
Hôpital Maisonneuve Rosemont
University of Montreal
Montreal, Canada

Cynthia Ménard
Département de Radio-oncologie
Centre Hospitalier de l'Université de
Montréal
Montréal, Canada

Michael Milosevic
Department of Radiation Oncology
Princess Margaret Cancer Centre
University of Toronto
Toronto, Canada

Marinus A. Moerland
Department of Radiation Oncology
University Medical Center Utrecht
Utrecht, The Netherlands

Maryse Mondat
Department of Radiation Oncology
Hospital Charles Lemoyne
Montreal, Canada

Lavanya Naidu
Department of Radiation Oncology
Tata Memorial Hospital
Mumbai, India

Nicole Nesvacil
Department of Radiation Oncology
Medical University of Vienna
Vienna, Austria

Thu Van Nguyen
Department of Radiation Oncology
Hospital Charles Lemoyne
Montreal, Canada

Peter Orio
Department of Radiation Oncology
Dana-Farber/Brigham & Women's Cancer
    Center
Harvard Medical School
Boston, Massachusetts

Thomas Osche
Product Management Brachytherapy
Eckert & Ziegler BEBIG
Berlin, Germany

Amir Owrangi
Department of Radiation Oncology
University of Toronto
and
Department of Medical Physics
Sunnybrook Health Sciences Centre
Toronto, Canada

Geordi Pang
Department of Radiation Oncology
University of Toronto
and
Department of Medical Physics
Sunnybrook Health Sciences Centre
Toronto, Canada

Jose Perez-Calatayud
Physics Section
Department of Radiation Oncology
La Fe University and Polytechnic Hospital
Valencia, Spain

and

Department of Radiotherapy
Clínica Benidorm
Benidorm, Spain

Daniel Petereit
Department of Radiation Oncology
Rapid City Regional Cancer Center
Rapid City, South Dakota

Primoz Petric
Department of Radiation Oncology
National Center for Cancer Care and
    Research
Hamad Medical Corporation
Doha, Qatar

Niclas Pettersson
Department of Radiation Medicine and
    Applied Sciences
University of California San Diego
La Jolla, California

Bradley R. Pieters
Department of Radiation Oncology
Academic Medical Center
University of Amsterdam
Amsterdam, The Netherlands

Tarun K. Podder
Department of Radiation Oncology
Case Western Reserve University
Cleveland, Ohio

Arnoud W. Postema
Department of Urology
Academic Medical Center
University of Amsterdam
The Netherlands

Richard Pötter
Department of Radiation Oncology
Medical University of Vienna
Vienna, Austria

Marie Lynn Racine
Department of Radiation Oncology
Hospital Charles Lemoyne
Montreal, Canada

Ananth Ravi
Department of Radiation Oncology
University of Toronto
and
Department of Medical Physics
Sunnybrook Health Sciences Centre
Toronto, Canada

Susan Richardson
Department of Radiation Oncology
Swedish Cancer Institute
Seattle, Washington

Alexandra Rink
Radiation Physics Department
Princess Margaret Cancer Centre
and
Department of Radiation Oncology
University of Toronto
Toronto, Canada

Mark J. Rivard
Department of Radiation Oncology
Tufts University School of Medicine
Boston, Massachusetts

Manuel Sanchez-Garcia
Department of Nuclear Medicine
Beaujon Hospital
Assistance Publique-Hôpitaux de Paris
Clichy, France

Daniel Scanderbeg
Department of Radiation Medicine and
    Applied Sciences
University of California San Diego
La Jolla, California

Stefan G. Schalk
Department of Electrical and Biomedical
    Engineering
Eindhoven University of Technology
Eindhoven, The Netherlands

Maximilian P. Schmid
Department of Radiation Oncology
Medical University of Vienna
Vienna, Austria

Carmen Schulz
Product Management Brachytherapy
Eckert & Ziegler BEBIG
Berlin, Germany

Ina M. Jürgenliemk-Schulz
Department of Radiation Oncology
University Medical Center Utrecht
Utrecht, The Netherlands

Shyamkishore Shrivastava
Department of Radiation Oncology
Tata Memorial Hospital
Mumbai, India

Adam Shulman
Department of Radiation Oncology
Hamad Medical Corporation
Doha, Qatar

Frank-André Siebert
Department of Medical Physics
Clinic of Radiotherapy
Universitätsklinikum
    Schleswig-Holstein
Kiel, Germany

Anna Simeonov
Radiation Physics Department
Princess Margaret Cancer Centre
and
Department of Radiation Oncology
University of Toronto
Toronto, Canada

Ryan L. Smith
Department of Radiation Oncology
The Alfred Hospital
and
School of Science
RMIT University
Melbourne, Australia

William Y. Song
Department of Radiation Oncology
Sunnybrook Health Sciences Centre
University of Toronto
Toronto, Ontario, Canada

Teo Stanescu
Radiation Physics Department
Princess Margaret Cancer Centre
and
Department of Radiation Oncology
University of Toronto
Toronto, Canada

Vratislav Strnad
Department of Radiation Oncology
University Hospital Erlangen
Erlangen, Germany

Jamema Swamidas
Department of Medical Physics
Tata Memorial Hospital
Mumbai, India

Tony Tadic
Department of Radiation Oncology
University of Toronto
and
Radiation Physics Department
Princess Margaret Cancer Centre
Toronto, Canada

Kari Tanderup
Department of Oncology
Aarhus University Hospital/Aarhus
  University
Aarhus, Denmark

Åsa Carlsson Tedgren
Department of Medical and Health
  Sciences
Linköping University
Linköping, Sweden

and

Department of Medical Physics
Section for Radiotherapy Physics and
  Engineering
The Karolinska University Hospital
Stockholm, Sweden

Dorin A. Todor
Department of Radiation Oncology
Virginia Commonwealth University
Richmond, Virginia

Mehmet Üzümcü
Elekta Brachytherapy Solutions
Veenendaal, The Netherlands

Uulke A. van der Heide
Department of Radiation Oncology
The Netherlands Cancer Institute
Amsterdam, The Netherlands

Rob van der Laarse
Department of Radiation Oncology
Academic Medical Center
University of Amsterdam
Amsterdam, The Netherlands

Jochem R. van der Voort van Zyp
Department of Radiation Oncology
University Medical Center Utrecht
Utrecht, The Netherlands

Té Vuong
Department of Radiation Oncology
Jewish General Hospital
McGill University
Montreal, Canada

Georges Wakil
Department of Radiation Oncology
Hospital Charles Lemoyne
Montreal, Canada

Robert Weersink
Radiation Physics Department
Princess Margaret Cancer Centre
Department of Radiation Oncology
University of Toronto
and
Techna Institute
University Health Network
Toronto, Canada

Hessel Wijkstra
Department of Electrical and Biomedical
    Engineering
Eindhoven University of Technology
Eindhoven, The Netherlands

and

Department of Urology
Academic Medical Center
University of Amsterdam
The Netherlands

Michael J. Zelefsky
Department of Radiation Oncology
Memorial Sloan Kettering Cancer
    Center
New York, New York

Abbreviations

AAPM	American Association of Physicists in Medicine
ABS	American Brachytherapy Society
ACA	Affordable Healthcare Act
ACE	Advanced Collapsed cone Engine
ACR	American College of Radiology
ADC	apparent diffusion coefficient
AESOP	Automated Endoscopic System for Optimal Positioning
AGOI	Association of Gynecologic Oncologists of India
ALA	5-aminolevulinic acid
ALARA	As low as reasonably achievable
AM	additive manufacturing
AMIGO	advances multimodality image guided operating suite
AMPI	Association of Medical Physicists of India
AP	anterior-posterior
APBI	accelerated partial breast irradiation
AROI	Association of Radiation Oncologists of India
ASTRO	American Society for Radiation Oncology
AUA	American Urological Association
BEEUD	biologically effective equivalent uniform dose
BED	biological equivalent dose
BRIT	Board of Radiation and Isotope Technology
CAD	computer-aided design
CBCT	cone beam CT
CC	collapsed cone
CEUS	contrast-enhanced ultrasound
CI	confidence interval
CN	conformation number
CNS	central nervous system
COIN	conformity index
co-RASOR	center-out RAdial Sampling with Off-resonance Reconstruction
CPE	charged particle equilibrium
CPT	current procedural terminology
CPU	central processing unit

CSTAR	Canadian Surgical Technologies & Advanced Robotics
CT	computed tomography
CTV	clinical target volume
CTVHR	high-risk CTV
CUDI	contrast ultrasound dispersion imaging
DA	direct addition
DCE	dynamic contrast enhanced
DECT	dual energy computed tomography
DFS	disease-free survival
DIR	deformable image registration
DMBT	direction (dynamic) modulated brachytherapy
DOF	degree of freedom
DSC	dice similarity coefficient
DVF	deformation vector field
DVH	dose volume histogram
DWI	diffusion weighted imaging
EBRT	external beam radiotherapy
eBT	electronic brachytherapy
EES	extravascular-extracellular space
EGO	enhanced geometric optimization
EGSnrc	National Research Council's (NRC) electron gamma shower software
EM	electromagnetic
EMT	electromagnetic tracking
EPID	electronic portal imaging device
EQD2	equivalent dose in 2 Gy per fraction
ESTRO	European Society for Radiotherapy and Oncology
EUD	equivalent uniform dose
FAA	Federal Aeronautics Administration
FAD	flavin adenine dinucleotide
FIGO	International Federation of Gynecology and Obstetrics
FITC	fluorescein isothiocyanate
FLS	fluorescence-guided surgery
FMEA	failure modes and effects analysis
FSE	fast-spin-echo
GBBS	grid-based Boltzman solver
GEC-ESTRO	Group Européen de Curiethérapie – European Society for Radiotherapy and Oncology
gEUD	generalized equivalent uniform dose
GO	geometric optimization
GPS	global positioning system
GPU	graphics processing unit
GrO	graphical optimization
GTV	gross tumor volume

GYN	gynecology
HCC	hepatocellular carcinoma
HDR	high-dose rate
HDREBT	high-dose rate endorectal brachytherapy
HIPO	hybrid inverse planning optimization
HR	hazard ratio
H&N	head and neck
IAEA	International Atomic Energy Agency
IBS	Indian Brachytherapy Society
IC	intracavitary
IC/IS	intracavitary/Interstitial
ICER	Institute for Clinical and Economic Review
ICG	Indocyanine green
ICRP	International Commission on Radiological Protection
ICRU	International Commission on Radiation Units & Measurements
IGABT	image-guided adaptive brachytherapy
IGBT	image-guided brachytherapy
IGRT	image-guided radiotherapy
IIP	interactive inverse planning
IL	intraprostatic lesions
IMOO	interactive multiobjective optimization
IMPT	intensity modulated proton therapy
IMRT	intensity modulated radiotherapy
IORT	intraoperative radiotherapy
IPIP	inverse planning by integer program
IPSA	inverse planning by simulated annealing
IRB	Institutional Review Board
IS	interstitial
IVD	in vivo dosimetry
keV	kilo electron volt
LAC	linear attenuation coefficient
LDR	low-dose rate
LDV	linear dose-volume
LET	linear energy transfer
MBDCA	model based dose calculation algorithms
MBIR	model-based image reconstruction
MC	Monte Carlo
MCNP	Monte Carlo N-Particle code
MD	medical doctor
MDR	medium-dose rate
MEMs	microelectromechanical systems
MeV	mega electron volt
MIRA	minimally invasive robot assistant

MIRALVA	Mallinckrodt Institute of Radiology Afterloading Vaginal Applicator
MIRD	medical internal radiation dose
mpMRI	multiparametric MRI
mpUS	multiparametric ultrasound
MRC	Medical Research Council
MR	magnetic resonance
MRI	magnetic resonance imaging
MRS	magnetic resonance spectroscopy
MTV	metabolically active tumor volume
MUPIT	Martinez Universal Perineal Interstitial Template
MUV	Medical University of Vienna
NADH	nicotinamide adenine dinucleotide
NASA	National Aeronautics and Space Administration
NCDB	National Cancer Data Base
NDR	natural dose ratio
NDVH	natural dose volume histogram
NEMA	National Electrical Manufacturers Association
NMSC	non-melanoma skin cancer
NPD	natural prescription dose
NRC	Nuclear Regulatory Commission
NSC	Nuclear Safety and Control
NT	normal tissue
NTCP	normal tissue complication probability
OAR	organ at risk
OCC	Odette Cancer Centre
OCT	optical coherence tomography
OIS	oncology information system
OR	operating room
OSLD	optically stimulated luminescence dosimetry
PAI	pubic arch interference
PD	proton-density
PDR	pulsed-dose rate
PEEK	polyether ether ketone
PEG	polyethylene-glycol
PET	positron emission tomography
PET/CT	positron emission tomography/computed tomography
PPL	peripheral pulmonary lesions
PODP	polynomial optimization on dose points
PR	prostatectomy
PSD	plastic scintillator detectors
PSA	prostate-specific antigen
PSMA	prostate-specific membrane antigen
PTV	planning target volume

QI	quality index
RF	radio frequency
RL	radioluminescence
ROI	region of interest
RP	radical prostatectomy
RPN	risk priority number
RSBT	rotating shield brachytherapy
RT	radiation therapist
RTT	radiotherapy technologists
QA	quality assurance
QC	quality control
SAE	Society for Automotive Engineers
SAUR	safety, accuracy, user-friendliness, and reliability
SBRT	stereotactic body radiotherapy
SDD	source-to-dosimeter distance
SEER	surveillance, epidemiology, and end results
SFE	scanning fiber endoscopy
SIRT	selective internal radiotherapy
SNR	signal-to-noise ratio
SPECT	single-photon emission computed tomography
SPECT/CT	single-photon emission computed tomography/computed tomography
STL	standard tessellation language
SUVmax	maximum standardized uptake value
SWE	shear wave elastography
TAUS	transabdominal ultrasound
TCP	tumor control probability
TE	echo time
TICs	time–intensity curves
TG	task group
TLD	thermoluminescent dosimeter
TMH	Tata Memorial Hospital
TPS	treatment planning system
TRAK	total reference air kerma
TRUS	transrectal ultrasound
TSE	turbo spin echo
UCA	ultrasound contrast agents
UCSD	University of California San Diego
UI	Uniformity index
UICC	Union for International Cancer Control
UMCU	University Medical Center Utrecht
US	ultrasound
USP	United States Pharmacopeia
UTE	ultrashort echo time

UV	ultraviolet
UWO	University of Western Ontario
VA	vibro-acoustography
VP	virtual patient
WHO	World Health Organization
WLE	white light endoscopy
2D	2-dimensional
2DUS	2D ultrasound
3D	3-dimensional
3DUS	3D ultrasound
[^{18}F]FAZA	[^{18}F] Fluoroazomycin-arabinoside
[^{18}F]FDG	[^{18}F] Fluorodeoxyglucose
[^{18}F]FMISO	[^{18}F] Fluoromisonidazole
[^{18}F]DCFPyL	[^{18}F] Fluoro-pyridine-3-carbonyl)-amino]-pentyl)-ureido)-pentanedioic acid

CHAPTER 1

Introduction

William Y. Song, Kari Tanderup, and Bradley R. Pieters

CONTENTS

1.1 MOTIVATION OF THE BOOK

The discoveries of radioactivity by Henri Becquerel in 1896 and of radium by Marie Curie in 1898 have had an immense effect on mankind as it turned out that radioactivity has characteristics that are useful in many applications such as in medicine, industry, and nuclear power (http://www.nrc.gov/about-nrc/radiation/around-us/uses-radiation.html). In turn, using radioactivity *properly* has been the impetus of much of the innovative applications seen in medicine ever since and continues to be the main motivation for the emergence of new technologies and practices (Mould, 1993).

Today, modern brachytherapy is a mature subject. Years of accumulated experience have led to a conclusive evidence of its benefit in numerous clinical sites such as gynecological, prostate, breast, ocular, and numerous other cancers. Along the way, major technological advances such as artificial radioactivity (Curie and Joliot-Curie, 1934), afterloading technology (Walstam, 1962), image-guided implants (especially the use of transrectal ultrasound [TRUS]) (Grimm and Sylvester, 2004), computer-assisted treatment planning (Shalek and Stovall, 1961; Williamson, 2006; Thomadsen et al., 2008; Rivard et al., 2009), image-guided adaptive brachytherapy (Haie-Meder et al., 2005; Potter et al., 2006), and quantitative dosimetry (Shalek and Stovall, 1961; Williamson, 2006; Ibbott et al., 2008) have fundamentally shaped the landscape of current state-of-the-art practices. Appropriately and as expected, there exists an abundant supply of excellent review articles (Shalek and Stovall, 1961; Haie-Meder et al., 2005; Potter et al., 2006; Williamson, 2006; Ibbott et al., 2008; Thomadsen et al., 2008; Rivard et al., 2009; Tanderup et al., 2014) and comprehensive books (Thomadsen et al., 2005; Devlin, 2006; Hoskin and Coyle, 2011; Venselaar et al., 2013; Devlin et al., 2016; Montemaggi et al., 2016; European Society for Radiotherapy & Oncology) on the latest established practices on the subject.

A seeming scarcity in the sea of literature, however, is a concise collection of *emerging* technologies and trends in brachytherapy as well as the political/economic landscape in which these innovations must persist, on a global scale. The latter is just as important (and thus interesting) as the technologies themselves, as competing modalities bid for the same attention and patients (Han et al., 2013; Tanderup et al., 2014; Vaidya et al., 2015; Hrbacek et al., 2016; Moon et al., 2016). As the past nearly 120 years of brachytherapy history teaches us, the persistent volatility in the landscape has repeatedly and abundantly allured remarkable innovations and solutions that kept the field successfully competitive. Therefore, in this day and age of ever-changing and fast-paced landscape, a book focused solely on the *emerging* technologies in the field of brachytherapy was thought to be of value to practitioners and students alike on the basis that there seem to be a scarcity of such a concise collection that helps the reader to readily grasp the broader sense of where the field is headed to in the next 5–10 years. It is thus hoped that this book ultimately helps the reader to see the horizon of change and beyond the current state of technology and practices in brachytherapy and understand the direction in which the development of brachytherapy will evolve.

1.2 CONTENTS OF THE BOOK

We have carefully chosen the contents of the book on the basis of our belief that to be of relevance in the future, it must satisfy one or more of the following criteria:

- Improves treatment efficacy

- Reduces radiation exposure and side effects of brachytherapy

- Facilitates treatment of new indications with brachytherapy

- Reduces uncertainties and risks of errors

- Reduces the dependence of the brachytherapists' experiences and skills on treatment quality

- Makes treatment more affordable

In alignment with the spirit of the motivation for this book, we have subdivided the book into five major sections, each containing a group of chapters relevant for the particular topic of interest. (1) Section I—Physics of Brachytherapy: The first section broadly discusses the grassroots innovations in physics. Being that the book is about emerging technologies, this section appropriately contains the largest number of chapters discussing all aspects of innovations and ongoing research from applicators to dosimetry to quality assurance technologies. (2) Section II—Imaging for Brachytherapy Guidance: The second section then follows up with discussions on innovations in the use of imaging for brachytherapy guidance. All usual suspect modalities such as x-ray, ultrasound, computed tomography (CT), magnetic resonance imaging (MRI), and positron emission tomography (PET) are discussed as well as upcoming optical imaging technology that has yet to see regular applications in brachytherapy but holds much potential. (3) Section III—Brachytherapy

Suites: The next section showcases a group of eight clinical sites carefully chosen across Europe, North America, Asia, and Africa, under various resource settings and developmental phases, sharing their vision and experiences on establishing the latest state-of-the-art infrastructure and clinical practices. (4) Section IV—Is Brachytherapy a Competitive Modality? This is followed up by a section that discusses the competitiveness of brachytherapy compared with other radiation modalities such as external beam radiotherapy and particle therapy in terms of treatment efficacy, recent trends, and latest worldwide political/economic landscape. (5) Section V—Vision 20/20: Industry Perspective: The book then finally concludes with the industry perspective on the emerging technologies by two respectable companies that manufacture comprehensive brachytherapy solutions.

1.3 CONCLUSION

As can be seen, we have aimed for a comprehensive yet concise collection of technologies, experiences, and expert opinions on what is *emerging* in brachytherapy today. By the nature of its intent, we do not expect that all ideas and innovations contained in this book will be adopted in the clinic, in the future. Nevertheless, the book provides an opportunity for all practitioners, students, researchers, vendors, and those genuinely interested in a glimpse into the (near) future of brachytherapy practice as the evolution of its impressive technology continues to unfold in history. Ultimately, we sincerely hope that this book will encourage readers to participate in the continued development of brachytherapy that we all came to love.

REFERENCES

Curie I, Joliot-Curie F. Artificial production of a new kind of radio-element. *Nature* 1934;133:201.

Devlin PM. *Brachytherapy: Applications and Techniques*. Lippincott Williams & Wilkins, Wolters Kluwer, Philadelphia, PA, USA, 2006.

Devlin PM, Cormack RA, Holloway CL, Stewart AJ. *Brachytherapy: Applications and Techniques*, Second Edition. Demos Medical Publishing, New York, NY, 2016.

European Society for Radiotherapy & Oncology (ESTRO), *GEC-ESTRO Handbook of Brachytherapy*. Brussels, Belgium. http://www.estro.org/about/governance-organisation/committees-activities/gec-estro-handbook-of-brachytherapy

Grimm P, Sylvester JE. Advances in brachytherapy. *Rev Neurol* 2004;6(Suppl4):37–48.

Haie-Meder C, Potter R, Van Limbergen E et al. Recommendations from Gynaecological (GYN) GEC-ESTRO Working Group (I): Concepts and terms in 3D image based 3D treatment planning in cervix cancer brachytherapy with emphasis on MRI assessment of GTV and CTV. *Radiother Oncol* 2005;74:235–245.

Han K, Milosevic M, Fyles A et al. Trends in the utilization of brachytherapy in cervical cancer in the United States. *Int J Radiat Oncol Biol Phys* 2013;87:111–119.

Hoskin P, Coyle C. *Radiotherapy in Practice—Brachytherapy*, Second Edition. Oxford University Press, USA, 2011.

Hrbacek J, Mishra KK, Kacperek A et al. Practice patterns analysis of ocular proton therapy centers: The International OPTIC Survey. *Int J Radiat Oncol Biol Phys* 2016;95:336–343.

http://www.nrc.gov/about-nrc/radiation/around-us/uses-radiation.html (accessed July 24, 2016).

Ibbott GS, Ma CM, Rogers DW et al. Anniversary paper: Fifty years of AAPM involvement in radiation dosimetry. *Med Phys* 2008;35:1418–1427.

Montemaggi P, Trombetta M, Brady LW. *Brachytherapy: An International Perspective*. Springer International Publishing AG, Switzerland, 2016.

Moon DH, Efstathiou JA, Chen RC. What is the best way to radiate the prostate in 2016? *Urol Oncol* 2016. doi: 10.1016/j.urolonc.2016.06.002.

Mould RF. *A Century of X-Rays and Radioactivity in Medicine: With Emphasis on Photographic Records of the Early Years*. Institute of Physics Publishing, Bristol, 1993.

Potter R, Haie-Meder C, Van Limbergen E et al. Recommendations from gynaecological (GYN) GEC ESTRO working group (II): Concepts and terms in 3D image-based treatment planning in cervix cancer brachytherapy—3D dose volume parameters and aspects of 3D image-based anatomy, radiation physics, radiobiology. *Radiother Oncol* 2006;78:67–77.

Rivard MJ, Venselaar JLM, Beaulieu L. The evolution of brachytherapy treatment planning. *Med Phys* 2009;36:2136–2153.

Shalek RJ, Stovall MA. The calculation of isodose distributions in interstitial implantations by a computer. *Radiology* 1961;76:119–120.

Tanderup K, Eifel PJ, Yashar CM et al. Curative radiation therapy for locally advanced cervical cancer: Brachytherapy is NOT optional. *Int J Radiat Oncol Biol Phys* 2014;88:537–539.

Tanderup K, Viswanathan AN, Kirisits C et al. Magnetic resonance image guided brachytherapy. *Semin Radiat Oncol* 2014;24:181–191.

Thomadsen BR, Rivard MR, Butler W. *Brachytherapy Physics*, Second Edition. Joint AAPM/American Brachytherapy Society Summer School. Medical Physics Monograph No. 31. Medical Physics Publishing, Madison, Wisconsin, USA, 2005.

Thomadsen BR, Williamson JF, Rivard MJ, Meigooni AS. Anniversary paper: Past and current issues, and trends in brachytherapy physics. *Med Phys* 2008;35:4708–4723.

Vaidya JS, Bulsara M, Wenz F et al. Pride, prejudice, or science: Attitudes towards the results of the TARGIT—A trial of targeted intraoperative radiation therapy for breast cancer. *Int J Radiat Oncol Biol Phys* 2015;92:491–497.

Venselaar J, Meigooni AS, Baltas D, Hoskin PJ. *Comprehensive Brachytherapy: Physical and Clinical Aspects*. CRC Press, Taylor & Francis Group, Boca Raton, FL, 2013.

Walstam R. Remotely-controlled afterloading radiotherapy apparatus. *Phys Med Biol* 1962;7:225–228.

Williamson JF. Brachytherapy technology and physics practice since 1950: A half-century of progress. *Phys Med Biol* 2006;51:R303–R325.

I

Physics of Brachytherapy

CHAPTER 2

Sources and Loading Technologies

Mark J. Rivard

CONTENTS

2.1 INTRODUCTION

Brachytherapy is an exciting and beneficial treatment modality that has grown over the past decade with numerous technological developments. This chapter provides a brief review of the following: (1) a description of standard-of-care sources and loading technologies currently available, (2) emerging technologies that are being examined in research settings and by clinical innovators, and (3) developments in brachytherapy source calibrations to bring emerging technologies into the mainstream.

2.2 DESCRIPTION OF STANDARD-OF-CARE SOURCES AND LOADING TECHNOLOGIES

2.2.1 Low-Dose Rate Brachytherapy

2.2.1.1 High-Energy Sources

The first applications of brachytherapy utilized ^{226}Ra, a high-energy photon-emitting radionuclide. Following ^{226}Ra disintegration through the emission of alpha particles, the encapsulation would sometimes fail due to helium production and gaseous ^{222}Rn and other disintegration products. The ^{222}Rn was collected and subsequently encapsulated to provide the second radionuclide used for brachytherapy (Duane, 1917). This era of brachytherapy using naturally occurring radionuclides lasted for over 50 years, until the development of reactor-based (and later accelerator-based) means of generating radionuclides.

The new high-energy radionuclides included ^{60}Co, ^{137}Cs, ^{192}Ir, and ^{198}Au, which were initially all low-dose rate (LDR) sources that delivered less than 2 Gy/h at the prescription point. It was shown through clinical evidence that brachytherapy effectiveness (toward both malignant and healthy tissue) significantly increased as the dose rate surpassed this value. Because of their half-lives, the ^{60}Co and ^{137}Cs sources were used for temporary implants, while ^{192}Ir and ^{198}Au were used for permanent implants. Later, LDR ^{192}Ir wires and ribbons were also used for temporary implants.

Today, high-energy LDR sources are generally restricted to ^{137}Cs. Given the 30-year half-life of ^{137}Cs, there are potentially many sources still available for clinical use. However, in the United States, the availability of LDR ^{137}Cs sources drastically diminished in the past decade due to concerns for a radiological incident following the September 11 terror attacks in 2001 due to the dispersive properties of ^{137}CsCl. Before this event, over a dozen different source models were commercially available. Today, the only readily available source models are from Eckert & Ziegler BEBIG, GmbH (Berlin, Germany). These include the manually delivered model CSM3 source with three radioactive components used as a source train, and the model 67–6520 (previously available from Isotope Products in Valencia, California), which has an isolated source configuration. Images of the model CSM3 and model 67–6520 sources, which may be considered representational of the other sources, are given in Figure 2.1.

2.2.1.2 Low-Energy Sources

Because of concerns for radiation exposure to hospital personnel and the members of the public, lower-energy photon-emitting sources were sought (Henschke, 1957; Friend et al., 1965). The key innovators for this change were Ulrich Henschke and Donald Lawrence

Eckert & Ziegler BEBIG LDR ^{137}Cs model CSM3 source

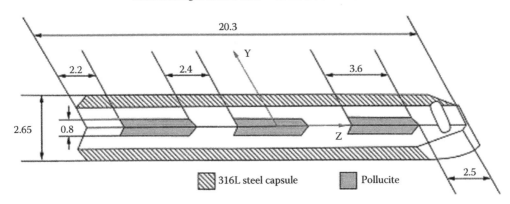

Eckert & Ziegler BEBIG LDR ^{137}Cs model 67-6520 source

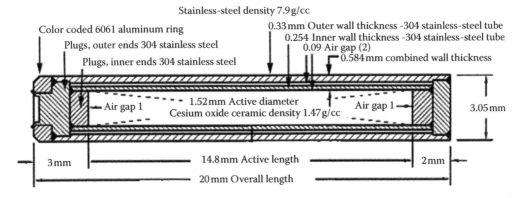

FIGURE 2.1 Dimensional information (in millimeters) and material composition for the model CSM3 (top) and model 67-6520 (bottom) LDR ^{137}Cs sources from Eckert & Ziegler BEBIG. (Images courtesy of joint AAPM/GEC-ESTRO HEBD report and Meigooni et al. *Med Phys* 2009.)

(Aronowitz, 2010). Henschke was an experienced brachytherapist at the Memorial Hospital in New York and Lawrence was a nuclear engineer at the Hanford nuclear reactor in Washington. With guidance from Paul Harper, a surgeon at the University of Chicago and Argonne National Laboratory, Lawrence and Henschke developed low-energy photon-emitting LDR sources containing ^{103}Pd, ^{125}I, and ^{131}Cs (Harper et al., 1958). In addition, other radionuclides without high-energy photon emissions through an entirely different set of emissions, that is, beta particles, were used and included ^{90}Sr/^{90}Y and ^{106}Ru/^{106}Rh.

Low-energy photon emitters have mean photon energies <50 keV, and are used more frequently for interstitial implants, while the beta emitters are used primarily for ophthalmic plaque treatments. ^{125}I usage is decreasing slightly, while ^{103}Pd and ^{131}Cs usage have increased proportionately. Given the skill- and labor-intensive procedure for permanent interstitial implants using these sources, their popularity is decreasing in favor of alternate treatment modalities such as external-beam radiotherapy (EBRT) even though clinical outcomes (such as for early-stage prostate cancer) are better for brachytherapy than with

EBRT. A comparison of the clinical advantages of brachytherapy in comparison to EBRT is discussed in Chapter 27.

2.2.2 High-Dose Rate Brachytherapy

2.2.2.1 High-Energy Radionuclides

For high-dose rate (HDR) brachytherapy, the dose rate is much higher than for LDR, and is typically 10^2 Gy/h. Approximately the same prescription dose is given, but over a much shorter time frame. The key to a miniature HDR brachytherapy source that can be implanted interstitially is having a high specific activity where the number of emitted photons (related to the number of nuclear disintegrations) is large compared to the source volume (related to the mass of the radioactive component). Specific activity (Ci/g) is a measure of this relationship, with values of 9.2 kCi/g for ^{192}Ir and 1.1 kCi/g for ^{60}Co. These values compare to other high-energy radionuclides such as ^{137}Cs (0.086 kCi/g) and ^{198}Au (243 kCi/g). ^{137}Cs is ruled out as an interstitial HDR source due to the large volume that would be necessary (mainly due to its 30-year half-life), which would preclude interstitial use. ^{198}Au is ruled out due to its short half-life (2.7 days), which is impractical for a temporary HDR source. The long half-lives of ^{192}Ir (74 days) and ^{60}Co (5.27 years) help make these radionuclides attractive for high-energy HDR brachytherapy sources.

2.2.2.2 Low-Energy Electronic Brachytherapy

Alternative to using radionuclides, it is possible to fabricate miniature x-ray sources for intracavitary and/or interstitial brachytherapy. These electronic brachytherapy (eBT) sources operate at ~50 kVp and generate photons with energies (and radiological properties in tissue) very similar to ^{103}Pd, ^{125}I, and ^{131}Cs. However, unlike radionuclides, eBT sources can be turned on and off, and they do not present radiological waste problems. Further, they can be shipped without radiological transportation issues.

Another advantage over radionuclides is that their output can be varied by changing the operating voltage or current on the x-ray anode. Output increases typically by the square of the operating voltage and linearly with current. The devices are designed to strike a balance between maximal output while preserving a reasonable operating lifetime. While the eBT sources are low-energy photon emitters, their radiation output can exceed 30 Gy/min in water at 1 cm or 3 mSv/min at 1 m in air. Clearly they are HDR sources! However, owing to their low-energy emissions, medical staff may be present in the room with the patient during treatment as long as high-Z shields are in place (either over the treated region or to protect the staff).

2.2.3 Remote Afterloaders: High-Energy Radionuclides

Owing to the convenience, negligible exposure to hospital personnel, and increased accuracy of computer control of source position in comparison to manual afterloading, remote afterloading with HDR and pulsed-dose rate (PDR) sources has trended upward over the past couple of decades in favor over manual afterloading brachytherapy using LDR sources. Unlike companies providing LDR sources, the vendors of HDR and PDR remote afterloaders provide a dedicated treatment planning system (TPS) to calculate patient-specific dose distributions with multiple imaging modalities. Today, there are five vendors of HDR and two of PDR high-energy remote afterloading systems. The systems generally include standard safety features such as emergency source retract with an emergency backup battery, use of a test run with a dummy source, the ability for height adjustment of the channel indexer, and an on-board radiation detector to confirm the return of the HDR and PDR sources into the internal safe. In addition to the systems by Best Medical International, Inc. (Springfield, Virginia, http://www.teambest.com/) and Oncology Systems, Inc. (Champaign, Illinois, http://www.accusourcehdr.com/index.htm), three of the more developed systems are described below.

2.2.3.1 Elekta Brachytherapy

Elekta Brachytherapy (Veenendaal, The Netherlands) has two remote afterloading systems. The microSelectron® Digital system can contain an HDR ^{192}Ir or PDR ^{192}Ir source, and can be configured with 6, 18, or 30 channels for connecting to brachytherapy applicators or catheters/needles. Up to 90 total channels can be treated independently in one treatment fraction. Distances between source dwell steps can range from 2.5 to 10 mm. The source diameter is 0.9 mm, contained within a stainless-steel capsule attached to a braided stainless-steel wire that extends out from the remote afterloader. The system includes the possibility for remote vendor access. The Flexitron system is similar to the microSelectron Digital, but it can contain an HDR ^{192}Ir source or an HDR ^{60}Co source, and can be configured with 10, 20, or 40 channels for connecting to brachytherapy applicators or catheters/needles. Elekta includes the Oncentra® Brachy TPS for calculating the delivered dose with either system.

2.2.3.2 Varian Medical Systems

Varian Medical Systems (Palo Alto, California) has two remote afterloading systems. The GammaMedplus™ iX system can contain an HDR ^{192}Ir or PDR ^{192}Ir source, and can be configured with either 3 or 24 channels for connecting to brachytherapy applicators or catheters/needles. The source wire is 0.9 mm in diameter. The VariSource™ iX system is similar to the GammaMedplus iX system, but is configured with 20 channels and contains an internal measuring system to verify the extended wire position. The smaller 0.59-mm-diameter source wire is composed of nitinol for durability and flexibility. Varian includes the BrachyVision™ TPS for calculating delivered dose with either system.

2.2.3.3 Eckert & Ziegler BEBIG

The newest company to provide a commercial offering for radionuclide-based HDR remote afterloaders is Eckert & Ziegler BEBIG (Berlin, Germany). They offer the following three systems: GyneSource, MultiSource®, and SagiNova®. All of them can accommodate HDR ^{192}Ir or HDR ^{60}Co sources. The GyneSource system has five channels to connect to simpler applicators such as for gynecological cylinders and intraluminal treatments of the bronchus or esophagus. The MultiSource system has 40 channels for an assortment of clinical applicators. It has a built-in system to allow the user to perform *in vivo* dosimetry for validating treatment delivery. The SagiNova system has 50 channels for an assortment of

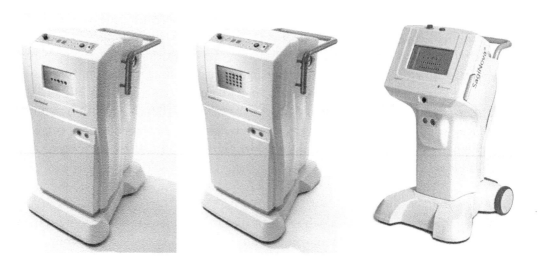

FIGURE 2.2 HDR remote afterloaders from Eckert & Ziegler BEBIG (Berlin, Germany). Left: GyneSource. Middle: MultiSource. Right: SagiNova. All three systems can accommodate either 192Ir or 60Co sources. (Images courtesy of Eckert & Ziegler BEBIG.)

clinical applicators. It has all the features of the prior two systems, and also includes the possibility for remote vendor access and a system of customizable quality assurance tools. The three systems are shown in Figure 2.2. The SagiPlan® TPS or HDRplus™ TPS are available for calculating the delivered dose.

2.2.4 eBT Systems

eBT systems are newer to the brachytherapy community and are not as prevalent as remote afterloaders delivering HDR ^{192}Ir sources. Given their electrically created low-energy photon emissions, a heavy safe is not built into these devices as is necessary for the HDR ^{192}Ir and HDR ^{60}Co remote afterloaders. Owing to the design of the x-ray-generating device, the entry path must generally be straight or shallow, or via an intracavitary applicator. Further, only one channel is used; the treatment is interrupted to connect an additional channel/applicator if necessary. The eBT systems are all similar in that an x-ray source is rigidly positioned on the surface or within a patient while x-rays are administered via a built-in control panel.

While two systems are examined below, there are additional devices that others may consider as eBT systems. These are:

- Esteya® by Elekta Brachytherapy (Veenendaal, The Netherlands)

- Papillon™ by Ariane Medical Systems (Derby, United Kingdom)

- Photoelectric Therapy model by Xstrahl Ltd. (Surrey, United Kingdom)

- Model SRT-100™ by Sensus Healthcare (Boca Raton, Florida)

These devices are quite similar in operation and dosimetrically to superficial x-ray units, can deliver x-rays up to 100 kVp, and (in the author's opinion) do not qualify as eBT systems.

2.2.4.1 INTRABEAM

The INTRABEAM system by Carl Zeiss Meditec, Inc. (Oberkochen, Germany) contains the model XRS4 x-ray source (Figure 2.3) that operates at 40 kVp (40 μA) or 50 kVp (up to 40 μA) with a gold anode. The system is somewhat portable, but is generally fixed in the operating theater. It is mainly used for single-fraction intraoperative radiotherapy (IORT) treatment of early-stage breast cancer, which was popularized under the TARGIT trial (Vaidya et al., 2014). The system positions a spherical applicator (one of several available sizes) into the lumpectomy cavity within the breast with size chosen to match the cavity dimensions. The resulting radiation dose distribution is approximately spherical (the likelihood of forward-directed radiation increases with increasing energy). The other principal clinical application for the INTRABEAM system is for spinal IORT (Wenz et al., 2010). Given the large differences between dose to water and dose to bone for ~30 keV photons, the company has released the radiance® TPS to provide Monte Carlo-based dose calculations for consideration of tissue composition.

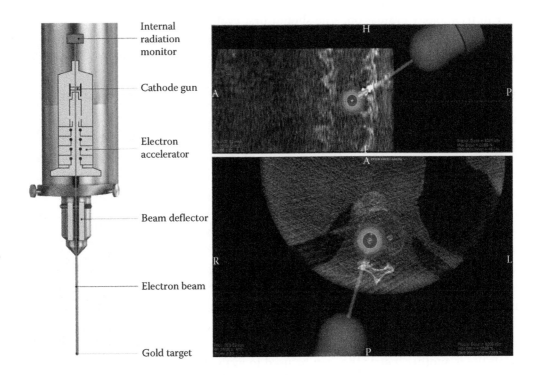

FIGURE 2.3 The model XRS4 x-ray source (left) and a screenshot from the radiance treatment planning system (right) showing the calculated dose distribution for a spinal treatment. (Images courtesy of Carl Zeiss Meditec, Inc.)

2.2.4.2 Axxent

The Axxent® eBT system by Xoft, a subsidiary of iCAD, Inc. (San Jose, California) uses the model S700 x-ray source. This system operates at 50 kVp and 300 µA with a tungsten anode. Given the higher operating current than the INTRABEAM system, the photon emission rate is approximately 10 times higher than the INTRABEAM system. It uses a cooling catheter (0.56 cm outer diameter) to dissipate the 15 W of generated thermal energy, and is to be used for treatments of a single patient, which may entail 10 treatment fractions. Currently available treatment applicators are for breast brachytherapy with the appropriate balloon within the surgical treatment lumpectomy cavity, vaginal cylinders for intracavitary gynecological brachytherapy, and surface cones that may be used to treat the skin or internally as IORT. The dose distributions have been determined through measurements and Monte Carlo simulations (Rivard et al., 2006; Hiatt et al., 2015), and treatment planning is performed by using TPS available to the clinic from other manufacturers.

2.3 EMERGING TECHNOLOGIES

Further brachytherapy advances beyond the LDR and HDR sources, remote afterloaders, and eBT systems are described below. It is helpful to know what developments are in progress toward further advancing the field. Note that several of these brachytherapy sources or devices are under investigational use. As the clinical medical physicist is generally the leader in technology evaluation, s/he should first evaluate the benefits and risks of these brachytherapy sources or devices. Only then can these innovative products be safely integrated into the clinic (Nath et al., 2016).

2.3.1 New Radionuclides

Several radionuclides for brachytherapy have recently been considered beyond those previously mentioned. These include ^{57}Co, ^{75}Se, ^{101}Rh, ^{144}Ce, ^{153}Gd, ^{169}Yb, and ^{170}Tm. While there is desire to use radionuclides with low-energy photon emissions (to minimize radiation exposure to the public and hospital personnel), photoatomic interactions in tissue can be dominated by the photoelectric effect with the dose falloff being quicker than the inverse square of the source distance. An aspect to consider new radionuclides is to have photon energies in the energy region where radiation scatter is higher than attenuation within the first few centimeters in tissue. Based on the study of monoenergetic photons by Luxton and Jozsef (1999), this effect peaks between 60 and 125 keV. In comparison to the inverse-square dose falloff, this can amount to a boost of 35% at 5 cm for a 65-keV photon source. Care must also be taken to ensure that the source capsule can adequately shield any emitted high-energy beta particles so that enormous doses are not given in the first few millimeters from the sources.

Half-life is another metric that can be used to evaluate new radionuclides. For consideration as permanent LDR sources, the half-life should be no longer than that of ^{125}I (59.4 days) since ^{192}Ir (74 days) has proven not to be a viable source in this manner. Based on half-life, only ^{169}Yb (32 days) could serve as an LDR permanent brachytherapy source. As possible HDR sources, the ^{169}Yb half-life (32 days) is considered too short to be practical compared to desires for longer life than available from HDR ^{192}Ir sources. The six other radionuclides have potential as HDR sources as their longer half-lives are not radiobiologically suitable as LDR sources.

In relationship to the half-life and atomic mass, specific activity (kCi/g) is yet another metric for considering new radionuclides for brachytherapy sources. Low values such as for ^{137}Cs (0.1 kCi/g) preclude source miniaturization for HDR applications. Current HDR sources ^{192}Ir (9.2 kCi/g) and ^{60}Co (1.1 kCi/g) set the standard for comparison of potential sources. ^{101}Rh (1.1 kCi/g) is similar to ^{60}Co with a similar half-life (3.3 years). The six other proposed radionuclides have higher specific activities.

^{57}Co has a 272-day half-life, a mean photon energy of 123 keV, and a specific activity of 8.4 kCi/g (Enger et al., 2012). The relatively low photon energy is attractive. ^{75}Se has a 120-day half-life, a mean photon energy of 389 keV, and a specific activity of 15 kCi/g (Weeks and Schulz, 1986). The high specific activity is attractive. ^{101}Rh has a 3.3-year half-life, a mean photon energy of 121 keV, and a specific activity of 1.1 kCi/g (Pakravan et al., 2015). The relatively low photon energy is attractive. ^{144}Ce has a 285-day half-life and has been examined as a potential HDR source by Zilio et al. (2005). ^{144}Ce decays to ^{144}Pr, which has a 17-min half-life and reaches secular equilibrium with ^{144}Ce. The mean photon energies of ^{144}Ce and ^{144}Pr are 89 keV and 1.2 MeV, respectively; however, the emission rates of the two radionuclides are not equal so the mean of the parent/daughter pair is approximately 199 keV. The specific activity of ^{144}Ce is 3.2 kCi/g. The relatively low photon energy is attractive. ^{153}Gd has a 239-day half-life and has been examined as a potential HDR source by Enger et al. (2013). The optimal photon energy (61 keV) is attractive to counter the inverse-square dose falloff. ^{169}Yb has a 32-day half-life, a mean photon energy of 93 keV, and has been examined as a potential HDR source by Lymperopoulou et al. (2005), Granero et al. (2005), and Medich et al. (2006). The high specific activity (24 kCi/g) is attractive. ^{170}Tm has a 129-day half-life and a specific activity of 6 kCi/g (Halmshaw, 1955; Ballester et al., 2010; Enger et al., 2011). The optimal photon energy (66 keV) and the high specific activity are attractive. The radial dose function (dose falloff with the inverse-square effect removed) is depicted in Figure 2.4 for some of these radionuclides.

Of course, all these single-source assessments can be taken further by considering a brachytherapy source containing more than one radionuclide for radiobiologically optimizing tumor response as a function of time (Nuttens and Lucas, 2006, 2008; Wang et al., 2006; Villeneuve et al., 2008).

2.3.2 ^{125}I Seed Directional Radiation

Given their emission of low-energy photons, it is possible to provide internal shielding using high-Z materials within the capsule of an LDR ^{125}I seed (Lin et al., 2008; Chaswal et al., 2012). This approach developed by Thomadsen and colleagues provides directionality for attenuating the primary radiation by >99%, and the effect of shielding diminishes with increasing distance due to the increasing importance of radiation scatter. While this approach is attractive for sparing dose to organs at risk, there remains the challenges of controlling the source orientation during implantation (and keeping it affixed), identifying the source orientation following implantation, and performing brachytherapy treatment

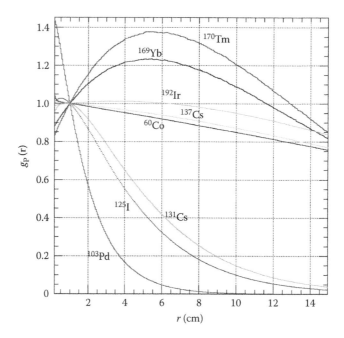

FIGURE 2.4 Radial dose function for a point source with photon energies for several radionuclides. (Image adapted from Figure 2.4 of F. Ballester et al. Study of encapsulated ^{170}Tm sources for their potential use in brachytherapy, *Med. Phys.* 37, 1629–1637, 2010. Courtesy of the authors.)

planning given that all FDA approved and CE Mark TPSs do not accommodate deviations from azimuthal symmetry of source dose.

2.3.3 ^{103}Pd Mesh Directional Radiation

^{103}Pd emissions are attenuated even more than ^{125}I by high-Z shields due to their lower mean photon energy. A brachytherapy mesh (CivaSheet™) has been designed by CivaTech Oncology, Inc. (Durham, North Carolina) with polymer sources (0.5 mm high and 2.5 mm diameter) embedded in a flexible bioabsorbable mesh (Aima et al., 2015). Gold shields in each LDR source provide directional radiation toward one side of the mesh. Unlike the aforementioned directional ^{125}I seeds, individual sources within the CivaSheet are cylindrically symmetric and can be characterized in a conventional brachytherapy TPS. However, the orientation (i.e., front or back) of the mesh is not evident through imaging following implantation.

2.3.4 Directional HDR Delivery Systems

Using high-Z shielding outside the source encapsulation, several designs for a directional HDR source have been examined. ^{192}Ir and ^{60}Co sources have been collimated with applicators in direct contact with the patient for conformal brachytherapy (Perez-Calatayud et al., 2005; Granero et al., 2008; Yang et al., 2011; Webster et al., 2013; Zehtabiana et al., 2015). Even less conventional, eBT and ^{153}Gd sources (Figure 2.5) have been modified for preferentially directing the photon radiation within the patient toward the target of interest while sparing healthy tissues (Adams et al., 2014; Liu et al., 2015). These technologies

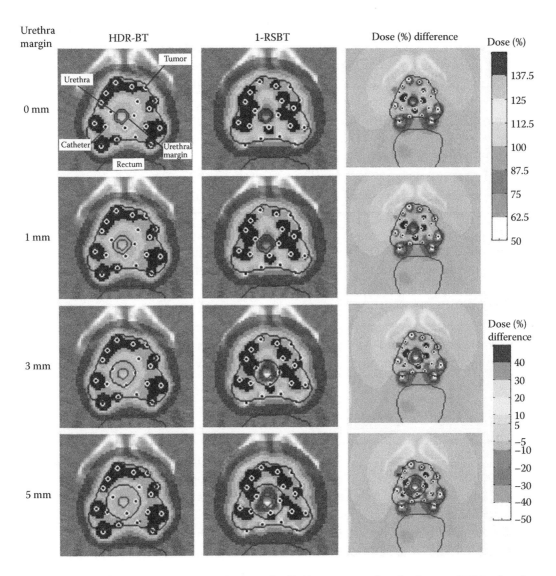

FIGURE 2.5 Axial views of dose distributions for (left) HDR 192Ir brachytherapy (HDR—brachytherapy), (middle) interstitial rotating-shield brachytherapy (I-RSBT) for a shielded 153Gd source, and (right) dose difference relative to the HDR 192Ir treatment plan. (Image adapted from Figure 2.3 of Q. E. Adams et al. *Med. Phys.* 41, 051703, 2014. Courtesy of the authors.)

have many similarities to the collimation as used for EBRT, and are discussed in greater detail in Chapter 3. An example for the treatment of rectal cancer with an endoluminal applicator is given in Section 27.6 of Chapter 27.

2.3.5 Novel Seeds

Instead of focusing on the radiation characteristics of LDR brachytherapy seeds, others have considered entirely different paradigms to improve the brachytherapy experience. Some of these issues are explored more deeply in Chapter 3.

Usually, with permanently implantable medical devices for patients with cancer, there is care to design the product so that it will be compatible with related imaging modalities. For example, brachytherapy seeds are made of non-ferromagnetic materials that are compatible with magnetic resonance imaging. Turning this idea on its head, an LDR ^{125}I seed has been designed with a ferromagnetic core to induce local heating (i.e., hyperthermia) when subject to oscillating magnetic fields (Gautam et al., 2012, 2014).

Chemotherapy is an established treatment modality for many types of cancers, and has been given in combination with radiotherapy (separately or concurrently) for decades. There is recent interest to deliver localized chemotherapy for increasing the toxic effect locally while minimizing systemic toxicity (Löhr et al., 2003). This principle is being used in conjunction with brachytherapy with seed-spacers that elute chemotherapeutic agents (Cormack et al., 2010).

Yet another innovation for brachytherapy seeds would just make them disappear after the radiation has been emitted (i.e., the radioactivity has effectively decayed). This would eliminate seeds causing artifacts and interfering with medical imaging, which is a strong criticism for the permanent seed breast implantation technique for breast cancer (Pignol et al., 2015) where mammography is used to monitor disease status (unlike for prostate cancer where blood tests are utilized). While several patents on dissolvable seeds have been issued, at this point there appear to be no scientific peer-review articles on this interesting topic.

2.3.6 Novel Applicator Sources

Beyond the approach to containing radioactivity in seeds or capsules (as for HDR/PDR sources), liquid brachytherapy sources are available. IsoRay Medical, Inc. (Richland, Washington) manufactures a plastic balloon-shaped applicator where the device is implanted in the tumor bed following brain surgery (Rogers et al., 2006). The dosimetry for a solution containing ^{125}I has been published (Dempsey et al., 1998; Monroe et al., 2001), and the manufacturer has materials to guide treatments when using a solution containing ^{131}Cs. The entire balloon applicator is characterized as a single brachytherapy source. The GliaSite^{TM} device filled with Iotrex (^{125}I) or Cesitrex (^{131}Cs) contains the liquid radioactivity within a double-walled balloon applicator. These applicator sources differ from microsphere brachytherapy where the sources are contained in glass or resin spheres, but are free to move throughout the body due to their liquid carrier (Dezarn et al., 2011).

2.4 NEW BRACHYTHERAPY SOURCE CALIBRATION METHODS

The current international standard for specifying brachytherapy source strength is air-kerma strength or the reference air-kerma rate. This is the only unit with traceability to a primary calibration standard from a national metrology institute (NMI); other antiquated units that are decreasingly being used in the clinic are apparent activity and mg-Ra equivalent. The product of air-kerma strength S_K and the dose rate constant Λ yield the absorbed dose rate to water at 1 cm on the source transverse plane. This latter term is sometimes referred to as the dose rate at the reference position: $S_K \cdot \Lambda = \dot{D}(r_0, \theta_0)$. When using the same

Λ value for any specific source model, the absorbed dose rate may be calculated in a uniform manner worldwide when S_K is obtained. For HDR high-energy sources and LDR low-energy sources, S_K measurements by the clinical medical physicist have associated uncertainties of 1.5% and 1.3% ($k = 1$), respectively. The uncertainties in Λ for these same sources are at least 2% and 3%, which yields total uncertainties of approximately 3% and 4% for $\dot{D}(r_0, \theta_0)$.

Through changing the approach and specifying source strength directly in terms of $\dot{D}(r_0, \theta_0)$, there is the possibility to diminish the associated uncertainty of the dose rate at the reference position (Siebert et al., 2012). Instead of measuring air-kerma strength and transferring this calibration standard to clinic well chambers, NMIs could directly measure $\dot{D}(r_0, \theta_0)$ and transfer this measure to the chambers (Aubineau-Lanièce et al., 2012; Toni et al., 2012), obviating the need for Λ. This new approach would yield total uncertainties of approximately 2% for $\dot{D}(r_0, \theta_0)$, which is a slight improvement over the current approach. However, no clinical TPS permits the direct entry of $\dot{D}(r_0, \theta_0)$ values, and NMIs do not yet provide $\dot{D}(r_0, \theta_0)$ calibrations for all sources. Clearly, this potential advancement would take much coordination to prevent mishaps and mistakes larger than the potential couple-percentage improvements.

An analogous approach for eBT sources has been proposed (DeWerd et al., 2015). For eBT, DeWerd and colleagues suggested replacing Λ with a dose-rate conversion coefficient χ_i, where i represents the applicator type with $i = 0$ referring to the bare eBT source and χ_0 defined as the ratio of $\dot{D}(r_0, \theta_0)$ to the air-kerma rate at 50 cm as eBT sources are not calibrated in terms of S_K. Like the prior method, this proposed method is internally consistent but cannot be used with current brachytherapy TPS as it is not standard practice to enter χ_i values.

2.5 CONCLUSION

From this review of emerging technologies, it is clear that the field of brachytherapy has made substantial advances beyond the ^{226}Ra and ^{222}Rn sources of the twentieth century. Given their lower-energy photon emissions in comparison to EBRT, brachytherapy sources have the possibility to be more conformal. By taking advantage of shielding opportunities that are not handled by the AAPM TG-43 dose calculation formalism (Rivard et al., 2004; Pérez-Calatayud et al., 2012), new brachytherapy sources present possibilities for directional radiation in which the TPS may not be able to calculate. As this issue becomes resolved with advanced brachytherapy TPS (Beaulieu et al., 2012), the possibilities will continue to grow for advanced brachytherapy technologies.

REFERENCES

Q. E. Adams, J. Xu, E. K. Breitbach, X. Li, S. A. Enger, W. R. Rockey, Y. Kim, X. Wu, and R. T. Flynn, Interstitial rotating shield brachytherapy for prostate cancer, *Med. Phys.* 41, 051703, 2014.

M. Aima, J. L. Reed, L. A. DeWerd, and W. S. Culberson, Air-kerma strength determination of a new directional 103Pd source, *Med. Phys.* 42, 7144–7152, 2015.

J. N. Aronowitz, Don Lawrence and the "k-capture" revolution, *Brachytherapy* 9, 373–381, 2010.

I. Aubineau-Lanièce, B. Chauvenet, D. Cutarella, J. Gouriou, J. Plagnard, and P. Aviles Lucas, LNE–LNHB air-kerma and absorbed dose to water primary standards for low dose-rate ^{125}I brachytherapy sources, *Metrologia* 49, S189–S192, 2012.

F. Ballester, D. Granero, J. Perez-Calatayud, J. L. M. Venselaar, and M. J. Rivard, Study of encapsulated [170]Tm sources for their potential use in brachytherapy, *Med. Phys.* 37, 1629–1637, 2010.

L. Beaulieu, Å. Carlsson Tedgren, J.-F. Carrier, S. D. Davis, F. Mourtada, M. J. Rivard, R. M. Thomson, F. Verhaegen, T. A. Wareing, and J. F. Williamson, Report of the Task Group 186 on model-based dose calculation methods in brachytherapy beyond the TG-43 formalism: Current status and recommendations for clinical implementation, *Med. Phys.* 39, 6208–6236, 2012.

V. Chaswal, B. R. Thomadsen, and D. L. Henderson, Development of an adjoint sensitivity field-based treatment-planning technique for the use of newly designed directional LDR sources in brachytherapy, *Phys. Med. Biol.* 57, 963–982, 2012.

R. A. Cormack, S. Sridhar, W. W. Suh, A. V. D'amico, and G. M. Makrigiorgos, Biological in situ dose painting for image-guided radiation therapy using drug-loaded implantable devices, *Int. J. Radiat. Oncol. Biol. Phys.* 76, 615–623, 2010.

J. F. Dempsey et al., Dosimetric properties of a novel brachytherapy balloon applicator for the treatment of malignant brain-tumor resection-cavity margins, *Int. J. Radiat. Oncol. Biol. Phys.* 42, 421–429, 1998.

L. A. DeWerd, W. S. Culberson, J. A. Micka, and S. J. Simiele, A modified dose calculation formalism for electronic brachytherapy sources, *Brachytherapy* 14, 405–408, 2015.

W. A. Dezarn et al., Experimental validation of dose calculation algorithms for the GliaSite™ RTS, a novel ^{125}I liquid-filled balloon brachytherapy applicator, *Med. Phys.* 38, 4824–4845, 2011.

W. Duane, Methods of preparing and using radioactive substances in the treatment of malignant disease, and of estimating suitable dosages, *Boston Med. Surg. J.* 177, 787–799, 1917.

S. A. Enger, M. D'Amours, and L. Beaulieu, Modeling a hypothetical [170]Tm source for brachytherapy applications, *Med. Phys.* 38, 5307–5310, 2011.

S. A. Enger, D. R. Fisher, and R. T. Flynn, Gadolinium-153 as a brachytherapy isotope, *Phys. Med. Biol.* 58, 957–964, 2013.

S. A. Enger, H. Lundqvist, M. D'Amours, and L. Beaulieu, Exploring [57]Co as a new isotope for brachytherapy applications, *Med. Phys.* 39, 2342–2345, 2012.

A. Friend et al., Report on the summary and conclusions of the conference on medical uses of radium and radium substitutes. In: *Medical Uses of Radium and Radium Substitutes.* Chicago, IL: USDHEW, 1965.

B. Gautam, E. I. Parsai, D. Shvydka, J. Feldmeier, and M. Subramanian, Dosimetric and thermal properties of a newly developed thermobrachytherapy seed, *Med. Phys.* 39, 1980–1990, 2012.

B. Gautam, G. Warrell, D. Shvydka, M. Subramanian, and E. I. Parsai, Practical considerations for maximizing heat production in a novel thermobrachytherapy seed prototype, *Med. Phys.* 41, 023301, 2014.

D. Granero, J. Pérez-Calatayud, F. Ballester, A. Bos, and J. Venselaar, Broad-beam transmission data of new brachytherapy sources, Tm-170 and Yb-169, *Radiat. Prot. Dosim.* 118, 11–15, 2005.

D. Granero, J. Perez-Calatayud, J. Gimeno, F. Ballester, E. Casal, V. Crispín, and R. Van der Laarse, Design and evaluation of a HDR skin applicator with flattening filter, *Med. Phys.* 35, 495–503, 2008.

R. Halmshaw, Thulium 170 for industrial radiography, *Br. J. Appl. Phys.* 6, 8–10, 1955.

P. V. Harper et al., Isotopes decaying by electron capture: A new modality in brachytherapy. In: *Second United Nations International Conference on Peaceful Uses of Atomic Energy.* Geneva, Switzerland: United Nations, pp. 417–LP, 1958.

U. K. Henschke, A technic for permanent implantation of radioisotopes, *Radiology* 68, 256, 1957.

J. R. Hiatt, S. D. Davis, and M. J. Rivard, A revised dosimetric characterization of the model S700 electronic brachytherapy source containing an anode-centering plastic insert and other components not included in the 2006 model, *Med. Phys.* 42, 2764–2776, 2015.

L. Lin, R. R. Patel, B. R. Thomadsen, and D. L. Henderson, The use of directional interstitial sources to improve dosimetry in breast brachytherapy, *Med. Phys.* 35, 240–247, 2008.

Y. Liu, R. T. Flynn, Y. Kim, H. Dadkhah, S. K. Bhatia, J. M. Buatti, W. Xu, and X. Wu, Paddle-based rotating-shield brachytherapy, *Med. Phys.* 42, 5992–6003, 2015.

M. Löhr et al., Safety, feasibility and clinical benefit of localized chemotherapy using microencapsulated cells for inoperable pancreatic carcinoma in a phase I/II trial, *Cancer Therapy* 1, 121–131, 2003.

G. Luxton and G. Jozsef, Radial dose distribution, dose to water and dose rate constant for monoenergetic photon point sources from 10 keV to 2 MeV: EGS4 Monte Carlo model calculation, *Med. Phys.* 26, 2531–2538, 1999.

G. Lymperopoulou, P. Papagiannis, L. Sakelliou, N. Milickovic, S. Giannouli, and D. Baltas, A dosimetric comparison of ^{169}Yb versus ^{192}Ir for HDR prostate brachytherapy, *Med. Phys.* 32, 3832–3842, 2005.

D. C. Medich, M. A. Tries, and J. J. Munro III, Monte Carlo characterization of an ytterbium-169 high dose rate brachytherapy source with analysis of statistical uncertainty, *Med. Phys.* 33, 163–172, 2006.

A. S. Meigooni, C. Wright, R. A. Koona, S. B. Awan, D. Granero, J. Perez-Calatayud, and F. Ballester, TG-43 U1 based dosimetric characterization of model 67-6520 Cs-137 brachytherapy source. *Med. Phys.* 36, 4711–4719, 2009.

J. I. Monroe et al., Experimental validation of dose calculation algorithms for the GliaSite™ RTS, a novel ^{125}I liquid-filled balloon brachytherapy applicator, *Med. Phys.* 28, 73–85, 2001.

R. Nath et al., Guidelines by the AAPM and GEC-ESTRO on the use of innovative brachytherapy devices and applications: Report of Task Group 167, *Med. Phys.* 43, 3178–3205, 2016.

V. E. Nuttens and S. Lucas, AAPM TG-43U1 formalism adaptation and Monte Carlo dosimetry simulations of multiple-radionuclide brachytherapy sources, *Med. Phys.* 33, 1101–1107, 2006.

V. E. Nuttens and S. Lucas, Determination of the prescription dose for biradionuclide permanent prostate brachytherapy. *Med. Phys.* 35, 5451–5462, 2008.

D. Pakravan, M. Ghorbani, and A. S. Meigooni, Evaluation of ^{101}Rh as a brachytherapy source, *J. Contemp. Brachytherapy* 7, 171–180, 2015.

J. Pérez-Calatayud, F. Ballester, R. K. Das, L. A. DeWerd, G. S. Ibbott, A. S. Meigooni, Z. Ouhib, M. J. Rivard, R. S. Sloboda, and J. F. Williamson, Dose calculation for photon-emitting brachytherapy sources with average energy higher than 50 keV: Report of the AAPM and ESTRO, *Med. Phys.* 39, 2904–2929, 2012.

J. Perez-Calatayud, D. Granero, F. Ballester, V. Puchades, E. Casal, A. Soriano, and V. Crispín, A dosimetric study of the Leipzig applicators, *Int. J. Rad. Oncol. Biol. Phys.* 62, 579–584, 2005.

J.-P. Pignol, J.-M. Caudrelier, J. Crook, C. McCann, P. Truong, and H. A. Verkooijen, Report on the clinical outcomes of permanent breast seed implant for early-stage breast cancers, *Int. J. Radiat. Oncol. Biol. Phys.* 93, 614–621, 2015.

M. J. Rivard, W. M. Butler, L. A. DeWerd, M. S. Huq, G. S. Ibbott, Z. Li, M. G. Mitch, R. Nath, and J. F. Williamson, Update of AAPM Task Group No. 43 Report: A revised AAPM protocol for brachytherapy dose calculations, *Med. Phys.* 31, 633–674, 2004.

M. J. Rivard, S. D. Davis, L. A. DeWerd, T. W. Rusch, and S. Axelrod, Calculated and measured brachytherapy dosimetry parameters in water for the Xoft Axxent x-ray source: An electronic brachytherapy source, *Med. Phys.* 33, 4020–4032, 2006.

L. R. Rogers et al., Results of a phase II trial of the GliaSite radiation therapy system for the treatment of newly diagnosed, resected single brain metastases, *J. Neurosurg.* 105, 375–384, 2006.

F.-A. Siebert, J. L. M. Venselaar, T. Paulsen Hellebust, P. Papagiannis, A. Rijnders, and M. J. Rivard, Dose-rate to water calibrations for brachytherapy sources from the end-user perspective, *Metrologia* 49, S249–S252, 2012.

M. P. Toni, M. Pimpinella, M. Pinto, M. Quini, G. Cappadozzi, C. Silvestri, and O. Bottauscio, Direct determination of the absorbed dose to water from ^{125}I low dose-rate brachytherapy seeds using the new absorbed dose primary standard developed at ENEA-INMRI, *Metrologia* 49, S193–S197, 2012.

J. S. Vaidya et al., Risk-adapted targeted intraoperative radiotherapy versus whole-breast radiotherapy for breast cancer: 5-year results for local control and overall survival from the TARGIT—A randomised trial, *Lancet* 383, 603–613, 2014.

M. Villeneuve, G. Leclerc, E. Lessard, J. Pouliot, and L. Beaulieu, Relationship between isotope half-life and prostate edema for optimal prostate dose coverage in permanent seed implants, *Med. Phys.* 35, 1970–1977, 2008.

J. Z. Wang, N. A. Mayr, S. Nag, J. Montebello, N. Gupta, N. Samsami, and C. Kanellitsas, Effect of edema, relative biological effectiveness, and dose heterogeneity on prostate brachytherapy, *Med. Phys.* 33, 1025–1032, 2006.

M. J. Webster et al., Dynamic modulated brachytherapy (DMBT) for rectal cancer, *Med. Phys.* 38, 011718, 2013.

K. J. Weeks and R. J. Schulz, Selenium-75: A potential source for use in high-activity brachytherapy irradiators, *Med. Phys.* 13, 728–731, 1986.

F. Wenz, F. Schneider, C. Neumaier, U. Kraus-Tiefenbacher, T. Reis, R. Schmidt, and U. Obertacke, Kypho-IORT—A novel approach of intraoperative radiotherapy during kyphoplasty for vertebral metastases, *Radiat. Oncol.* 5(11), 1–4, 2010.

Y. Yang, C. S. Melhus, S. Sioshansi, and M. J. Rivard, Treatment planning of a skin-sparing conical breast brachytherapy applicator using conventional brachytherapy software, *Med. Phys.* 38, 1519–1525, 2011.

M. Zehtabiana, S. Sinab, M. J. Rivard, and A. S. Meigooni, Evaluation of BEBIG HDR ^{60}Co system for non-invasive image-guided breast brachytherapy, *J. Contemp. Brachytherapy* 7, 469–478, 2015.

V. O. Zilio, O. P. Joneja, Y. Popowski, F. O. Bochud, and R. Chawla, ^{144}Ce as a potential candidate for interstitial and intravascular brachytherapy, *Int. J. Radiat. Oncol. Biol. Phys.* 62, 585–594, 2005.

CHAPTER 3

Applicators

Primoz Petric, Christian Kirisits, Jose Perez-Calatayud, Umesh Mahantshetty, William Y. Song, and Bradley R. Pieters

CONTENTS

3.1 INTRODUCTION

The development of novel brachytherapy applicators is one of the cornerstones of the rapid evolution of image-guided adaptive brachytherapy (IGABT). Optimization of the dose distribution within the brachytherapy target volume is based on the geometric individualization of the brachytherapy implant, accompanied by anatomy-specific tailoring of source dwell positions and dwell times. In contrast to the standard system-based conventional techniques, IGABT provides a platform for anatomy-based personalized adaptation of applicator insertion and treatment planning, according to common concepts, which are in principle independent of the tumor site.

3.2 STANDARD INTRACAVITARY APPLICATORS

Modern intracavitary (IC) applicators for cervical and endometrial cancer are derived from historical systems with long traditions. Notwithstanding some major technical developments, including fixed applicator geometry, thin channels, static shielding, and availability of various accessories, the basic features of modern IC applicators and their dosimetric range have not changed appreciably since they were first described a century ago. Fabrication of magnetic resonance imaging (MRI)-compatible IC applicators enabled the implementation of MRI-based IGABT, which can be considered the single most important element of IC applicator development. Vaginal IC brachytherapy is used extensively in postoperative endometrial cancer treatment and in recurrent tumors of the vagina, extending up to 8 mm in thickness. When compared with historical single-channel applicators, the multichannel systems enable higher flexibility of dose optimization, allowing for individualized treatment adaptation. Further, inflatable multichannel devices, such as the computed tomography (CT)-compatible 13-channel CAPRI applicator (Varian, Palo Alto, Florida) or the MRI-compatible 8-channel anorectal (AR) system (Ancer Medical, Hialeah, Florida), ensure high conformance of the applicator to the tissues, help avoid air pockets, aid immobilization, and increase the source-tissue distance. The concept of multichannel IC brachytherapy is not limited to gynecological tumors and has been applied with specific modifications to other sites. Endorectal IC brachytherapy is an alternative to external beam irradiation, either as a component of neoadjuvant treatment or definitive chemoradiation in selected patients (Appelt et al. 2015; Vuong and Devic 2015). Similar to other IC techniques, the success of rectal IC brachytherapy depends on the geometrical accuracy of the inserted multichannel applicator relative to the target volume and the tumor thickness. Pretreatment MRI, transrectal ultrasound (TRUS), and endoscopy-based preplanning, accompanied by daily IGABT optimization of applicator position can be used to ensure best results (Vuong and Devic 2015). While rigid applicators or inflatable devices with a rigid central part are useful for distal disease, they have limited value for treatment of other rectal tumors due to the relatively tortuous course and compliance of this organ. The dedicated flexible endorectal brachytherapy mold applicator (Elekta AB, Stockholm, Sweden) enables adjustment of its shape to patient anatomy and personalized orientation of the inflatable endocavitary balloon to minimize irradiation of the healthy rectal mucosa. In contrast to the rigid applicators, the flexible device can be navigated to the level of the sigmoid colon and is suitable for treatment of low, mid, and high rectal tumors. CT-based optimization of the source dwell times and positions in the eight equidistant channels, surrounding the central source path, enables conformal dose distributions in the target volume while minimizing the irradiation of adjacent normal tissues. Based on extensive favorable experience from the field of gynecological tumors, it can be expected that systematic implementation of MRI-compatible anorectal applicators will further improve the results in this tumor site. Example of such development is the MRI-compatible AR system (Ancer Medical, Hialeah, Florida). It consists of a compliant external balloon that enables reproducible and conformal applicator placement while ensuring adequate distance from the mucosa to the eight source channels, which are arranged around a noncompliant internal balloon. In contrast to the transition from IC to combined IC and interstitial (IS) strategies in gynecological brachytherapy, a tendency for reversed trend can be observed in breast cancer. IS multi-catheter breast brachytherapy has a long tradition and has been used for boost and accelerated partial breast irradiation with favorable results (Bartelink et al. 2015; Strnad et al. 2015). In the setting of accelerated partial breast irradiation (APBI), IS brachytherapy is being partially replaced by the recently developed IC applicators. Higher training requirements for IS brachytherapy are commonly identified as one of the main reasons for this trend. However, in view of the general complexity of multimodal treatment of breast cancer, these concerns could be regarded as relative. The major limitation of the single-lumen MammoSite® applicator (copyright Hologic, Bedford, Massachusetts), which was introduced over a decade ago, was the inability to avoid high-dose regions in the chest wall and skin in certain situations (Scanderbeg et al. 2009). Multi-lumen MammoSite and Contura applicator (copyright SenoRx Inc., Irvine, California) were developed to overcome these challenges with limited success. Recently, the Strut adjusted volume implant (SAVI®) applicator (copyright Cianna Medical, Aliso Viejo, California) was introduced. When compared with the balloon technique, the multiple channels of SAVI applicator, arranged in an equidistant fashion around a central channel, enable superior optimization of the dose distribution to the skin and chest wall (Scanderbeg et al. 2009; Gurdalli et al. 2011). Longer follow up is needed for adequate comparison between IC and IS techniques.

3.3 LIMITATIONS OF IC APPLICATORS

The limitations of the standard IC and endoluminal applicators can be divided into topographic and dosimetric restrictions, which are intimately interrelated. The shape and size of commercially available devices are rarely ideally suited to individual patient anatomy. This is especially important in the treatment of irregular cavities and cases with narrow anatomy due to tumor or treatment-related fibrosis. Several site-specific solutions have been proposed to overcome these limitations, including individually fabricated molds for gynecological brachytherapy, inflatable balloons for vaginal, rectal, and nasopharyngeal cancer, thermoplastic molds for head and neck cancer, various skin applicators, and so on. Manufacturing (i.e., three-dimensional [3D] printing) of individualized applicators is a more generalized approach that may offer the ideal solution for brachytherapy of various tumor sites and is described in more detail in the following sections and in Chapter 11. Regardless of sophistication and topographic conformity of the IC applicators, clinically achievable and safe expansion of the prescribed dose by IC optimization alone is limited. Different source-channel geometries for uterovaginal brachytherapy (i.e., tandem-ring, tandem-ovoids, Simon–Heyman capsules, etc.) result in different degrees of freedom during insertion and treatment planning. For example, asymmetric placement of two ovoids of different sizes or unequal ovoid spacing can tailor the application to individual patient anatomy and pathological findings to some extent. Further, the ring channel is arranged in a circular manner, while the ovoids offer two lateral source arrays. This translates into different possibilities for activation of "nonstandard" source positions during image-based optimization (i.e., anterior and posterior ring positions) (Kirisits et al. 2005). While these differences are commonly exploited in practice, there is no convincing evidence for their dosimetric and clinical impact. In case of cervical cancer, the prescription isodose can be pushed approximately 4 mm to the lateral direction at the level of point A by means of IC optimization (Kirisits et al. 2005). The range of IC brachytherapy for primary or recurrent vaginal cancer is limited to a maximum of approximately 8 mm from the vaginal mucosa. Similar limitations can be observed in IC/endoluminal brachytherapy of rectal cancer with deep invasion or involvement of the anal complex, deeply invading esophageal or bronchial cancer, and in other primary tumor sites. In general, the target volume that extends beyond the IC brachytherapy range will receive insufficient radiotherapy dose. Unfavorable topography of the target volume in relation to the organs at risk represents a similar challenge even if the target volume is small. While eternal beam radiotherapy (EBRT) with midline blocking has been traditionally used in cervical cancer to boost the under-dosed parametrial regions, the evidence for its effectiveness is limited. IS or combined IC/IS IGABT is an efficient technique that can achieve adequate coverage of extensive and/or topographically unfavorable target volumes. Direction modulated and rotating shield IGABT, used in the setting of IC or combined IC/IS treatments, is a promising future option (Webster et al. 2013; Yang et al. 2013; Adams et al. 2014; Han et al. 2014; Dadkhah et al. 2015; Liu et al. 2015). Current status of these techniques is outlined below.

3.4 IC/IS BRACHYTHERAPY

The concept of image-guided development of novel brachytherapy applicators has been elaborated in the past decade particularly in the field of gynecologic tumors. These developments can be regarded as a paradigm that fits into a bigger trend in brachytherapy and is applicable to other tumor sites.

Freehand or perineal template-based insertion of IS catheters combined with the IC component has a long tradition in pelvic tumors. Recently, applicators that enable the insertion of IS needles through the modified vaginal ring or ovoids were developed. With these devices, the mean distance between the needle insertion points and the target volume is in the order of magnitude of 30 mm, resulting in accurate and reproducible needle placement (Dimopoulos et al. 2006; Kirisits et al. 2006; Jürgenliemk-Schulz et al. 2009; Nomden et al. 2012). Dimopoulos et al. demonstrated that over 90% of inserted needles were placed inside the target volume and mean deviation of the needle tip from its ideal position was only 0.5 (±1) mm (Dimopoulos et al. 2006). The use of these applicators in the context of IGABT enables expansion of the prescription isodose for up to approximately 15 mm lateral to point A (Kirisits et al. 2006). This results in improved target coverage while dose constraints for the organs a risk (OAR) are respected and the total reference air kerma is kept within range, comparable to IC brachytherapy (Dimopoulos et al. 2006; Kirisits et al. 2006; Jürgenliemk-Schulz et al. 2009; Nomden et al. 2012). In a recent comparison of parametrial boost techniques, IC–IS brachytherapy demonstrated dosimetric superiority to EBRT boost (Mohamed et al. 2015).

Best results with IC/IS IGABT of gynecological tumors can be obtained by careful preplanning to achieve optimal implant geometry. In experienced hands, systematic analysis of pre-EBRT clinical and radiological tumor features can be sufficient to predict the location and extent of the remnants at IGABT (Schmid et al. 2013). More sophisticated methods of preplanning allow for accurate preoperative determination of optimal source geometry and resulting dose volume parameters. In one study, a tandem and ring applicator was inserted under paracervical anesthesia, preplanning pelvic MRI obtained, and the applicator removed. Based on the MRI with the applicator in place, virtual optimized IC–IS preplan was created and reproduced at the actual brachytherapy 5 days later (Petric et al. 2014). A similar technique is based on preplanning the insertion of the IC applicator under general anesthesia in week 5 of treatment, followed by two consecutive implants in weeks 6 and 7, reproducing the preplanned geometry (Fokdal et al. 2013). A commonly used hybrid approach consists of delivery of the first brachytherapy fraction with IC technique, followed by analysis of its insufficiencies and determination of optimal IC/IS implant geometry at subsequent fraction(s) (Nomden et al. 2012).

To objectively assess the characteristics of available IC/IS applicators for cervical cancer and to propose *in silico* models of class-solution novel devices, a multicenter study was performed and preliminary results presented (Petric et al. 2010). Clinical target volume $(CTV)_{HR}s$ from 264 cervical cancer patients were co-registered on a reference IC applicator, creating a virtual patient (VP) in digital imaging and communications in medicine (DICOM) format, containing pooled spatial information on CTV_{HR} distribution of the study sample. While the IC applicator enabled adequate dose coverage of around 60% of the VP, the addition of parallel parametrial needles improved the result, reaching up to 95% of the VP. To cover most of the remaining small proportion of the largest and topographically unfavorable VP regions, virtual oblique needles were placed at optimal insertion points, angles, and depths (Figure 3.1) (Petric et al. 2010). These results, combined with extensive clinical experience, enabled the fabrication of class-solution IC/IS applicators with parallel and oblique IS needles (Dimopoulos et al. 2006; Kirisits et al. 2006; Berger et al. 2010), capable of covering almost all tumors seen in clinical practice. However, clinical experience and formal analysis of target volume topography indicate that even the

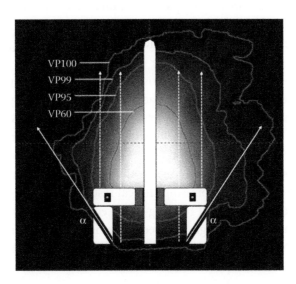

FIGURE 3.1 Frequency distribution map from of high-risk clinical target volumes from 264 cervical cancer patients, co-registered on a tandem-ring applicator in a VP. The representative coronal slice of the VP DICOM object is shown. Each voxel of the VP is assigned a numeric value and gray level (0 = black, 264 = white), according to number of target volumes it occupies. Isosurfaces, connecting voxels with same values, are labeled as percentage of encompassed voxels. VPn represents the VP subvolume, encompassed by the n% isosurface. A schematic example of an IC/IS applicator is projected on the VP, demonstrating the principle of selecting the optimal geometry of a class-solution applicator. Virtual IS needles are presented as arrows.

last generation of applicators fails to achieve optimal dose distributions in approximately 1% of most challenging cervical tumors with extensive paravaginal or parametrial infiltration, in particular along the sacro-uterine ligaments (Petric et al. 2010). Image-guided transperineal implantation still plays an important role in these cases.

Brachytherapy is a standard component of treatment for primary, recurrent, and metastatic tumors of the vagina (Dimopoulos et al. 2012; Mahantshetty et al. 2013). Combined IC/IS application is required to treat lesions with paravaginal extension exceeding 4–8 mm. Intraoperative image guidance using transvaginal/transrectal ultrasonography can be utilized for optimal placement of the implant. The applicators include endovaginal single- and multichannel applicators including molds, Mallinckrodt Institute of Radiology Afterloading Vaginal Applicator (MIRALVA) and MIAMI (TM Varian Medical Systems, Palo Alto CA, USA) devices, and various perineal templates including Syed Neblett, Martinez Universal Perineal Interstitial Template (MUPIT) and customized templates. The IS approach consists of the use of titanium or plastic tubes (5–6 F) or stainless-steel needles (16–18 G). The challenges include optimal implantation of the target for adequate coverage, while avoiding the bladder/bowel perforations and radiation-induced toxicity in the narrow confines of the lower pelvis. Image guidance using transrectal ultrasonography is helpful for optimal implantation. Recent approaches include a laparoscopy-guided technique to avoid perforation of pelvic structures including the bowel (Fokdal et al. 2011) and advanced multimodality image-guided brachytherapy suite (AMIGO) (Kapur et al. 2012). The use of hydrogel spacers to spare the bowel appears promising and is outlined in a separate section (Viswanathan et al. 2013).

Tagliaferri et al. recently reported on the use of multiparametric MRI-guided preplanning for IS IGABT of anal cancer. In their study, MRI/CT-based preplanning with a dummy applicator in place was performed 4–6 weeks after completion of EBRT in patients with residual disease or initial stage T4. Optimal needle positions and insertion depths were determined and the preplanned geometry reproduced at actual implant. Feasibility, safety, and dosimetric outcome were excellent and the authors noted that the use of fully MRI-compatible applicators could simplify the procedure by performing the preplanning and implant during the same session (Tagliaferri et al. 2015). In anal cancer, the use of endosonic rotating probes and adapted IS applicators has been proven feasible and efficient in optimizing the implant and treatment planning, resulting in promising clinical results (Doniec et al. 2006; Christensen et al. 2008). The use of transcervical rotating-probe endosonography in combination with specifically adapted applicator design has been proposed recently as a potential low-cost future alternative to MRI-based cervical cancer IGABT (Petric and Kirisits 2016). Further research and development is required to establish the value of this new approach.

3.5 3D-PRINTED TEMPLATES

An emerging alternative to commercially available IC and IS applicators and molds is the personalized 3D printing of IC/IS templates, tailored to the patient's unique patho-anatomical situation. In gynecological IGABT, this is especially important in cases with a narrow vagina where commercial applicators of standard sizes often do not fit (Lindegaard et al. 2015) and/or in extensive tumors, where customized templates enable highly individualized needle insertion points and angles. This concept ideally involves MRI-based

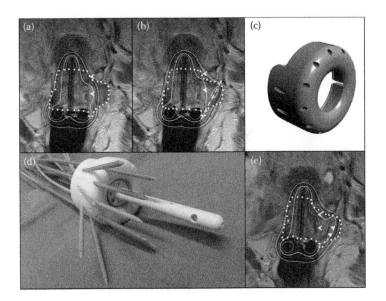

FIGURE 3.2 The concept of applicator 3D printing on an example of locally advanced cervical cancer, treated with MRI-based IGABT. (a) Coronal slice of the MRI with the applicator in place. Tandem and ring were inserted. IS needles were placed parallel to the tandem (open arrows) in the left parametrium and loaded to improve the coverage of an extensive high-risk CTV (CTV_{HR}: white dotted line) by the prescribed dose (outermost isodose line). In spite of the parallel needles, the CTV_{HR} coverage by the prescribed dose remained suboptimal. (b) This implant was used to preplan optimal position of virtual needles, inserted obliquely through the vaginal wall (dashed arrow) resulting in excellent and highly conformal virtual coverage of the CTV_{HR} by the prescribed dose. (c) Based on the preplan, a 3D model of a template ring cap was designed. (d) 3D print of the personalized ring cap was manufactured, enabling insertion of the parallel and oblique IS needles according to the preplan. (e) Optimal dose distribution was achieved by using the 3D-printed template and loading the parallel (open arrows) and oblique needles (solid arrows). Using the described approach, D90 for the CTV_{HR} was improved from 64% (a) to 111% (e) of the planning aim dose. V100 was improved from 76% (a) to 93% (e). Dose to the organs at risk was kept below the dose constraints. (Courtesy of Robert Hudej and Omar Hanuna, Institute of Oncology Ljubljana, Slovenia.)

preplanning, computer-aided design and drafting, and finally 3D printing of brachytherapy templates out of MRI- and biocompatible material (Figure 3.2). The method is demonstrating exciting potential in various tumor sites (Lindegaard et al. 2015; Poulin et al. 2015). Details on 3D printing and rapid prototyping are given in Chapter 11.

3.6 SELECTED SPECIAL CONSIDERATIONS IN IS BRACHYTHERAPY

A concern when performing IS, particularly transperineal, implantations is the possible migration of needles or catheters. The standard way to secure the needles is by fixation to a template or buttons attached to the skin. Regardless of fixation, movement can still occur due to skin displacement, edema formation, or displacement of the organ away from the needle in the longitudinal direction. Displacement of up to 42 mm has been reported for prostate implants (Hoskin et al. 2003). To prevent displacement, a synthetic flexible

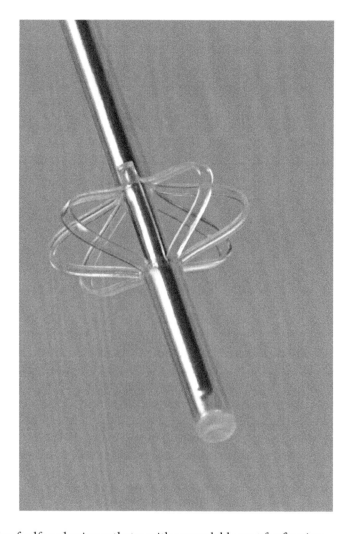

FIGURE 3.3   Tip of self-anchoring catheter with expandable part for fixation.

self-anchoring catheter has been developed (Pieters et al. 2006). Instead of fixating the catheter to the skin, the catheter is fixated into the target organ itself. The fixation is realized by an "umbrella" mechanism at the tip of the catheter (Figure 3.3). The catheter consists of two parts. The inner part with inner lumen is where the radioactive source enters. The outer part slides over the inner part resulting in an umbrella like expansion at the tip, which can be locked by a rotating knob. Before removal, the expansion can be unfolded by turning the knob in the opposite direction. The calculated displacement in prostate implantation was found to be 0 mm ± 1.9 mm in cranio-caudal direction for the first 24 h which had small effect on dosimetric parameters. The V_{100} for prostate target volume coverage decreased for a mean of 2.3% in 48 h. In total, 94% of patients had a D_{90} higher than the prescription dose at incipience and 88% after 48 h (Dinkla et al. 2013). The catheters have been studied mainly in prostate brachytherapy, but can also be used in other areas such as paravaginal implants. Traditionally, bladder brachytherapy is done by the suprapubic approach during which the bladder is opened to insert the flexible 5–6 F catheters (Pieters et al. 2015). An emerging technique is to insert the catheters with laparoscopy or robot assistance (Nap-van Klinken et al. 2014). Specially designed 6 F hollow catheters with a needle that can be manipulated through laparoscopy are used to facilitate insertion in the bladder wall. The catheters are of sufficient length to allow entrance in the abdomen and exit at the opposite site. When the catheter is in place, the needle can be removed, allowing source entry during brachytherapy.

3.7 SHIELDED APPLICATORS

It is evident from the sections above that nowadays, the basic strategy for optimization of the dose distribution within the target volume includes the image-guided individualization of a static IC/IS implant, accompanied by anatomy-specific tailoring of the source dwell positions and dwell times. This approach fails to exploit the third degree of freedom for potential optimization—the direction of the radiation emitted from the source inside the applicators. Modulation of intensity from the directional radiation beam, as opposed to isotropic radiation, can be accomplished by the use of intelligently shaped shields inside the applicators (Figure 3.4) (Han et al. 2014). Currently available shielded ovoids contain a standard pair of fixed or adjustable tungsten-alloy shields, intended to reduce the dose to the rectum and bladder. Utilization of applicators with adjustable and/or rotating shields accompanied by optimal implant geometry and source dwell patterns have been shown to achieve favorable dose conformity in various tumor sites (Webster et al. 2013; Yang et al. 2013; Adams et al. 2014; Han et al. 2014; Dadkhah et al. 2015; Liu et al. 2015). With the recent advent of the model-based dose calculation algorithms, all major commercial treatment planning systems are now capable of integrating high-density metal alloys into the dose calculations

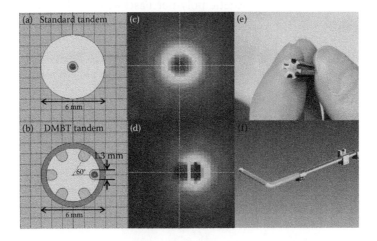

FIGURE 3.4 The proposed direction modulated brachytherapy (DMBT)-concept tandem applicator design. A (a) standard tandem and (b) DMBT tandem cross section with six peripheral holes carved out of a nonmagnetic tungsten alloy rod. The Monte Carlo simulated dose distributions of an Ir-192 source inside a (c) standard tandem and a (d) DMBT tandem. (e) A successfully machined-to-specifications tungsten alloy piece to demonstrate the manufacturability of the applicator. (f) An artistic rendering of the concept applicator in full assembly.

and inverse planning processes, bringing the clinical translation of such technologies closer to reality. While promising, the clinical role and safety of these novel designs remain to be proven before the routine clinical adoption. This is especially important in large tumors with unfavorable topography where improved target coverage can be achieved, but may come at a cost of increased high-dose volumes in the direction of intensified radiation profile and increased total reference air kerma. Although coverage might be comparable, the spatial dose distribution compared to combined IC/IS treatments is substantially different. Nonetheless, shielded applicators are exciting emerging technologies to monitor.

3.8 THERMOPLASTIC MOLDS FOR INTRAOPERATIVE BRACHYTHERAPY

Standard applicators are not ideal for treatment of irregular cavities because the cavity or lumen wall will not be covered adequately. Thermoplastic mold materials can be utilized for these situations. These materials are extensively used for head-and-neck and orbital rhabdomyosarcoma in the so-called Ablative surgery, Mold brachytherapy, REconstruction (AMORE) therapy (Blank et al. 2010). With this technique, similar survival to the standard treatment with external beam irradiation has been reported with less late effects as advantage (Schoot et al. 2015). The characteristic of a thermoplastic material consisting of 5 mm thick sheets is that it will become soft in warm water at 80°C. The material can be cut in desired shape and molded according to the cavity contour. At body temperature, the product will turn hard. Usually, several prepared sheets are placed covering all the walls of a cavity. Catheters are placed in the sheets by making a groove with a soldering iron. Because of the stickiness of the grooves, the catheters can be fixed in the mold. Because of the cavity irregularity the mold will stay in place when filled and pressed against the cavity wall. Traditionally, natural rubber (gutta-percha), was used for thermoplastic molds. This product has been replaced by a synthetic polymer Fastform-Percha® (FastForm Research Ltd.) (Figure 3.5).

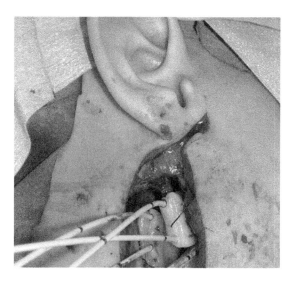

FIGURE 3.5 Several thermoplastic molds with flexible catheters placed in the parapharyngeal space.

3.9 TARGET: ORGAN SPACERS

Biodegradable materials, injected between the target volume and the nearby organs at risk can widen the therapeutic window by increasing the distance between the high-dose region and the normal tissues. While the concept is in principle applicable to many tumor sites, the majority of published experience comes from prostate and gynecological cancer brachytherapy (Prada et al. 2007; Marnitz et al. 2012; Gez et al. 2013; Song et al. 2013; Viswanathan et al. 2013; Mok et al. 2014). Anatomically, the anterior rectal wall is located in the direct vicinity of the prostate in men and cervix/vagina in women. Therefore, brachytherapy to prostate or gynecological tumors will inevitably give some dose to the small volumes of the rectum; in certain cases, a dose equal or higher than the prescribed dose. The spacer, placed outside the Denovilliers' fascia and anterior to the rectum will decrease the dose to the rectal wall. Spacer placement is facilitated by prior hydro dissection with 10–20 mL of injectable saline. There are four products suitable for this purpose (Mok et al. 2014). (1) Hyaluronic acid (HA) is a glycosaminoglycan-based polymer that is present in connective tissue, extracellular matrix, and synovial fluid. The first report on the use of a prostate–rectum spacer was with HA (Prada et al. 2007). The HA hydrogel is formed by cross-linked HA polymers, resulting in a product with high-water affinity and increased elasto-viscosity. Ten milliliter is injected to create a distance of 6–20 mm. The spacer half-life is about 4–8 months. (2) Polyethylene-glycol (PEG) hydrogel consists of PEG oligomers and polymers that retain water to form a gel. Ten milliliter is injected to result in a 7–10 mm distance between the rectum and prostate. After 3 months, the PEG hydrogel begins to resolve. (3) Collagen is a natural protein of the human body and the principal component of connective tissue. By injecting 20 mL, an 8–19 mm separation between rectum and prostate can be achieved. The injected collagen will disappear after 6–12 months. There are no reports of the use of collagen in brachytherapy series. (4) The biodegradable balloon is an inflatable balloon made of a copolymer of poly lactide acid and epsilon caprolactone. Overtime, the polymer will be degraded by hydrolysis. Following a small perineal incision, the balloon is inserted with a tube and sealed after inflation with a variable amount of saline. Up to 2 cm distance between rectum and prostate has been observed (Gez et al. 2013). The balloons are deflated after 3–6 months.

In all the studies employing a prostate–rectum spacer, a marked reduction of rectal dose was seen. In the study of Prada et al., using high dose rate (HDR) brachytherapy, the reduction of the mean rectal maximum dose was from 7.1 to 5.1 Gy (Prada et al. 2007). Strom et al. (2014) found a 22% reduction in D_{2cm3} rectal dose. Similar promising results were obtained in gynecological tumors. The prostate–rectum and vagina–rectum spacers may have a particularly favorable impact in salvage brachytherapy of recurrent disease to protect an already irradiated rectum (Viswanathan et al. 2013; Mahal et al. 2014). Further studies are required to determine the actual clinical significance of these interventions.

3.10 SKIN CANCER APPLICATORS AND ELECTRONIC BRACHYTHERAPY

Skin cancer was one of the first indications for which brachytherapy was applied using Radium-226. In the modern era, typical treatments in skin brachytherapy are performed using molds or flaps. Molds are built with thermoplastic material in which the HDR catheters are embedded, with a usual inter-catheter distance of 1 cm and a minimum distance to the skin of 0.5 cm. Molds fit to the external patient surface. In case of non-excessive irregularity, flaps are used. These consist of regular layers of silicon-based material or linked pellets of 1 cm in diameter in which the catheters are embedded. In both cases, 1 cm inter-catheter distance and a minimum of 0.5 cm distance to the skin are assured. These flaps can be cut to fit the required irradiation area. Typical prescription depth with molds and flaps is 0.5 cm; bigger depths require IS implant (Ouhib et al. 2015). In case of small non-melanoma lesions, previous applications are not adequate and small applicators were developed specifically for this purpose. The most extensively used is the Leipzig-type applicator. It is cup shaped and made of tungsten with the HDR source at its vertex. It tries to mimic the dosimetric properties of a mini-beam, with the benefit of protecting the surrounding healthy tissues due to its shielding. Different diameters, from 1 to 3 or 4.5 cm are available, depending on the manufacturer. The main problem of these applicators is the non-flat dose distribution and large penumbra (Perez-Calatayud et al. 2005; Fulkerson et al. 2014b). To solve this problem and to avoid dependence on HDR source, several solutions have been recently implemented in radionuclide and electronically based applicators.

As far as radionuclide-based applicators are concerned, the development of the Valencia applicator represents the main improvement (Granero et al. 2008). It was developed from the Leipzig applicator by adding a flattening filter to homogenize the dose distribution, resulting in significant improvement of the useful beam and penumbra. Two diameters (2 and 3 cm) are available and the configuration is restricted to cases where the source runs parallel to the skin surface. Because of the filter-induced attenuation, the treatment times are higher. Treatment planning is usually based on atlas or library plans and the typical prescription depth is 0.3 cm with a skin dose of around 135%. The main advantage of this applicator is its efficiency. The treatment is simpler and safer and the required quality assurance is minimal, since it is a simple accessory of the HDR afterloader. Its main disadvantage is the treatment time. When a 3 cm size applicator is used with a typical hyper-fractionated regimen (42 Gy at 0.3 cm in six fractions, two fractions per week), the treatment takes from 7 to 15 min depending on the Ir-192 HDR source air kerma strength (Tormo et al. 2014). In this case, the intra-fraction movement must be well controlled and sometimes the fraction is done in two steps.

Currently, there are three major electronic applicator systems that can be used for skin cancer (Ouhib et al. 2015). The Axxent® (Xoft Inc., USA) has a 50 kVp electronic source and is used mainly for the IC treatments with vaginal cylinders and breast cavity balloons. For skin cancer, the source needs to be mounted on special applicators with a flattening filter (Fulkerson et al. 2014a). The second device is the 50 kVp Intrabeam® (Carl Zeiss, Germany). Primarily used to treat the lumpectomy cavity after breast-conserving surgery, this system allows electronic skin brachytherapy with the specifically developed applicator (Schneider et al. 2014). The third system is Esteya® (Elekta, The Netherlands) (Garcia-Martinez et al. 2014; Pons et al. 2014). The motivation of this system is to mimic the Valencia applicator distribution while an electronical source substitutes the radioactive source. The system is compact and dedicated specifically for skin brachytherapy. It has a 69.5 kVp electronic source and a dose gradient that is slightly shallower (8% per mm) than with Leipzig, Valencia, and Axxent (12% per mm) systems. The dose rate allows for shorter treatment times, lasting approximately one-third of the Valencia applicator times.

3.11 CONCLUSION

Accelerated research and innovations, which we have witnessed in the field of brachytherapy application techniques during the past decade, were catalyzed mainly by implementation of sectional imaging into the brachytherapy process. These developments have improved our ability to deliver safe, efficient, and highly conformal brachytherapy in a cost-effective manner. The potential for future progress in the field is exciting and exceeds the conventional notion of purely technical development. It is expected to gain momentum through various interdisciplinary collaborations and includes development of dedicated applicators for real-time guided insertion using static and rotating endosonography probes, direction-modulated brachytherapy, 3D printing, laparoscopic assistance, drug-eluting technologies, and so on. However, in spite of these major advances, safe delivery of treatment can be assured only if the basic rules of traditional systems and long-standing clinical experience are taken into account.

REFERENCES

Adams QE, Xu J, Breitbach EK et al. Interstitial rotating shield brachytherapy for prostate cancer. *Med Phys* 2014;41(5):051703.

Appelt AL, Pløen J, Harling H et al. High-dose chemoradiotherapy and watchful waiting for distal rectal cancer: A prospective observational study. *Lancet Oncol* 2015;16(8):919–927.

Bartelink H, Maingon P, Poortmans P et al. Whole-breast irradiation with or without a boost for patients treated with breast-conserving surgery for early breast cancer: 20-year follow-up of a randomised phase 3 trial. *Lancet Oncol* 2015;16(1):47–56.

Berger D, Pötter R, Dimopoulos JA, Kirisits C. New Vienna applicator design for distal parametrial disease in cervical cancer. *Brachytherapy* 2010;9:S51–S52.

Blank LECM, Koedooder K, van der Grient HNB et al. Brachytherapy as part of the multidisciplinary treatment of childhood rhabdomyosarcomas of the orbit. *Int J Radiat Oncol Biol Phys* 2010;77:1463–1469.

Christensen AF, Nielsen BM, Engelholm SA. Three-dimensional endoluminal ultrasound-guided interstitial brachytherapy in patients with anal cancer. *Acta Radiol* 2008;49(2):132–137.

Dadkhah H, Flynn RT, Wu X et al. Multi-helix rotating shield brachytherapy for cervical cancer. *Med Phys* 2015;42:3465.

Dimopoulos JCA, Kirisits C, Petric P et al. The Vienna applicator for combined intracavitary and interstitial brachytherapy of cervical cancer: Clinical feasibility and preliminary results. *Int J Radiat Oncol Biol Phys* 2006;66(1):83–90.

Dimopoulos JCA, Schmid MP, Fidarova E et al. Treatment of locally advanced vaginal cancer with radiochemotherapy and magnetic resonance image-guided adaptive brachytherapy: Dose-volume parameters and first clinical results. *Int J Radiat Oncol Biol Phys* 2012;82(5):1880–1888.

Dinkla AM, Pieters BR, Koedooder K et al. Deviations from the planned dose during 48 hours of stepping source prostate brachytherapy caused by anatomical variations. *Radiother Oncol* 2013;107:106–111.

Doniec JM, Schniewind B, Kovács G et al. Multimodal therapy of anal cancer added by new endo-sonographic-guided brachytherapy. *Surg Endosc* 2006;20(4):673–678.

Fokdal L, Tanderup K, Hokland SB et al. Clinical feasibility of combined intracavitary/intersti-
tial brachytherapy in locally advanced cervical cancer employing MRI with a tandem/ring
applicator *in situ* and virtual preplanning of the interstitial component. *Radiother Oncol*
2013;107(1):63–68.

Fokdal L, Tanderup K, Nielsen SK et al. Image and laparoscopic guided interstitial brachytherapy
for locally advanced primary or recurrent gynaecological cancer using the adaptive GEC-
ESTRO target concept. *Radiother Oncol* 2011;100(3):473–479.

Fulkerson R, Micka J, Dewerd L. Dosimetric characterization and output verification for conical brachytherapy surface applicators. Part I. Electronic brachytherapy source. *Med Phys*
2014a;41:022103.

Fulkerson R, Micka J, Dewerd L. Dosimetric characterization and output verification for conical
brachytherapy surface applicators. Part II. High dose rate 192Ir source. *Med Phys* 2014b;41:022104.

Garcia-Martinez T, Chan J, Perez-Calatayud J, Ballester F. Dosimetric characteristics of a new unit
for electronic skin brachytherapy. *J Contemp Brachytherapy* 2014;6:45–53.

Gez E, Cytron S, Ben Yosef R et al. Application of an interstitial and biodegradable balloon system
for prostate–rectum separation during prostate cancer radiotherapy: A prospective multicenter study. *Radiat Oncol* 2013;8:96.

Granero D, Perez-Calatayud J, Jimeno J et al. Design and evaluation of a HDR skin applicator with
flattening filter. *Med Phys* 2008;35:495–503.

Gurdalli S, Kuske RR, Quiet CA, Ozer M. Dosimetric performance of strut-adjusted volume
implant: A new single-entry multicatheter breast brachytherapy applicator. *Brachytherapy*
2011;10(2):128–135.

Han D, Webster MJ, Scanderbeg DJ et al. Direction-modulated brachytherapy for high-dose-rate
treatment of cervical cancer. I: Theoretical design. *Int J Radiat Oncol Biol Phys* 2014;89:666–673.

Hoskin PJ, Bownes PJ, Ostler P, Walker K, Bryant L. High dose rate afterloading brachytherapy for prostate cancer: Catheter and gland movement between factions. *Radiother Oncol*
2003;68:285–288.

Jürgenliemk-Schulz IM, Tersteeg RJHA, Roesink JM et al. MRI-guided treatment planning optimisation in intracavitary or combined intracavitary/interstitial PDR brachytherapy using tandem ovoid applicators in locally advanced cervical cancer. *Radiother Oncol* 2009;93:322–330.

Kapur T, Egger J, Damato A et al. 3-T MR guided brachytherapy for gynecologic malignancies.
Magn Reson Imaging 2012;30(9):1279–1290.

Kirisits C, Lang S, Dimopoulos J et al. The Vienna applicator for combined intracavitary and interstitial brachytherapy of cervical cancer: Design, application, treatment planning, and dosimetric results. *Int J Radiat Oncol Biol Phys* 2006;65(2):624–630.

Kirisits C, Pötter R, Lang S et al. Dose and volume parameters for MRI-based treatment planning in
intracavitary brachytherapy for cervical cancer. *Int J Radiat Oncol Biol Phys* 2005;62:901–911.

Lindegaard JC, Madsen ML, Traberg A, Meisner B, Nielsen SK, Tanderup K, Spejlborg H, Fokdal
LU, Nørrevang O. Individualised 3D printed vaginal template for MRI guided brachytherapy
in locally advanced cervical cancer. *Radiother Oncol* 2016;118(1):173–175.

Liu Y, Flynn RT, Kim Y et al. Paddle-based rotating-shield brachytherapy. *Med Phys*
2015;42:5992–6003.

Mahal BA, Ziehr DR, Hyatt AS et al. Use of a rectal spacer with low-dose-rate brachytherapy for
treatment of prostate cancer in previously irradiated patients: Initial experience and short-
term results. *Brachytherapy* 2014;13:442–449.

Mahantshetty U, Shrivastava S, Kalyani N et al. Template-based high-dose-rate interstitial
brachytherapy in gynecologic cancers: A single institutional experience. *Brachytherapy*
2013;13:352–358.

Marnitz S, Budach V, Weisser F et al. Rectum separation in patients with cervical cancer for treatment planning in primary chemo-radiation. *Radiat Oncol* 2012;7:109.

Mohamed S, Kallehauge J, Fokdal L et al. Parametrial boosting in locally advanced cervical cancer: Combined intracavitary/interstitial brachytherapy vs. intracavitary brachytherapy plus external beam radiotherapy. *Brachytherapy* 2015;14(1):23–28.

Mok G, Benz E, Vallee JP, Miralbell R, Zilli T. Optimization of radiation therapy techniques for prostate cancer with prostate–rectum spacers: A systematic review. *Int J Radiat Oncol Biol Phys* 2014;90:278–288.

Nap-van Klinken A, Bus SJEA, Janssen TG et al. Interstitial brachytherapy for bladder cancer with the aid of laparoscopy. *J Contemp Brachytherapy* 2014;6:313–317.

Nomden CN, De Leeuw AAC, Moerland MA et al. Clinical use of the Utrecht applicator for combined intracavitary/interstitial brachytherapy treatment in locally advanced cervical cancer. *Int J Radiat Oncol Biol Phys* 2012;82(4):1424–1430.

Ouhib Z, Kasper M, Perez-Calatayud J et al. Aspects of dosimetry and clinical practice of skin brachytherapy: The American Brachytherapy Society working group report. *Brachytherapy* 2015;14(6):840–858.

Perez-Calatayud J, Granero D, Ballester F et al. A dosimetric study of the Leipzig applicators. *Int J Radiat Oncol Biol Phys* 2005;62:579–584.

Petric P, Hudej R, Hanuna O et al. MRI-assisted cervix cancer brachytherapy pre-planning, based on application in paracervical anaesthesia: Final report. *Radiol Oncol* 2014;48(3):293–300.

Petric P, Hudej R, Rogelj P et al. Frequency-distribution mapping of HR CTV in cervix cancer: Possibilities and limitations of existent and prototype applicators. *Radiother Oncol* 2010;96(1):70.

Petric P, Kirisits C. Potential role of TRAns Cervical Endosonography (TRACE) in brachytherapy of cervical cancer: Proof of concept. *J Contemp Brachytherapy* 2016;8(3):215–220.

Pieters BR, van der Grient JNB, Blank LECM et al. Minimal displacement of novel self-anchoring catheters suitable for temporary prostate implants. *Radiother Oncol* 2006;80:69–72.

Pieters BR, van der Steen-Banasik E, Van Limbergen E. Urinary bladder cancer. In: van Limbergen E, Pötter R, Hoskin P and Baltas D, eds. *The GEC ESTRO Handbook of Brachytherapy*. 2nd edn. Brussels, Belgium: ESTRO, 2015:3–12.

Pons O, Ballester R, Celada F et al. Clinical implementation of a new electronic brachytherapy system for skin brachytherapy. *J Contemp Brachytherapy* 2014;6:417–423.

Poulin E, Gardi L, Fenster A et al. Towards real-time 3D ultrasound planning and personalized 3D printing for breast HDR brachytherapy treatment. *Radiother Oncol* 2015;114(3):335–338. doi: 10.1016/j.radonc.2015.02.007. Epub February 27, 2015.

Prada PJ, Fernandez J, Martinez AA et al. Transperineal injection of hyaluronic acid in anterior perirectal fat to decrease rectal toxicity from radiation delivered with intensity modulated brachytherapy or EBRT for prostate cancer patients. *Int J Radiat Oncol Biol Phys* 2007;69:95–102.

Scanderbeg D, Yashar C, White G et al. Evaluation of three APBI techniques under NSABP B-39 guidelines. *J Appl Clin Med Phys* 2009;11(1):3021.

Schmid MP, Fidarova E, Pötter R et al. Magnetic resonance imaging for assessment of parametrial tumour spread and regression patterns in adaptive cervix cancer radiotherapy. *Acta Oncol* 2013;52(7):1384–1390.

Schneider F, Clausen S, Tholking J et al. A novel approach for superficial intraoperative radiotherapy (IORT) using a 50 kV X-ray source: A technical and case report. *J Appl Clin Med Phys* 2014;15:167–176.

Schoot RA, Slater O, Ronckers CM et al. Adverse events of local treatment in long-term head and neck rhabdomyosarcoma survivors after external beam radiotherapy or AMORE treatment. *Eur J Cancer* 2015;51(11):1424–1434.

Song DY, Herfarth KK, Uhl M et al. A multi-institutional clinical trial of rectal dose reduction via injected polyethylene-glycol hydrogel during intensity modulated radiation therapy for prostate cancer: Analysis of dosimetric outcomes. *Int J Radiat Oncol Biol Phys* 2013;87(1):81–87.

Strnad V, Ott OJ, Hildebrandt G, Kauer-Dorner D et al. 5-year results of accelerated partial breast irradiation using sole interstitial multicatheter brachytherapy versus whole-breast irradiation with boost after breast-conserving surgery for low-risk invasive and in-situ carcinoma of the female breast: A randomised, phase 3, non-inferiority trial. *Lancet* 2015;387(10015):229–238.

Strom TJ, Wilder RB, Fernandez DC et al. A dosimetric study of polyethylene glycol hydrogel in 200 prostate cancer patients treated with high-dose rate brachytherapy ± intensity modulated radiation therapy. *Radiother Oncol* 2014;111:126–131.

Tagliaferri L, Manfrida S, Barbaro B et al. MITHRA—Multiparametric MR/CT image adapted brachytherapy (MR/CT-IABT) in anal canal cancer: A feasibility study. *J Contemp Brachytherapy* 2015;7(5):336–345.

Tormo A, Celada F, Rodriguez S et al. Non-melanoma skin cancer treated with HDR Valencia applicator: Clinical outcomes. *J Contemp Brachytherapy* 2014;6:167–172.

Viswanathan AN, Damato Al, Nguyen PL et al. Novel use of a hydrogel spacer permits reirradiation in otherwise incurable recurrent gynecologic cancers. *J Clin Oncol* 2013;31(34):446–447.

Vuong T, Devic S. High-dose-rate pre-operative endorectal brachytherapy for patients with rectal cancer. *J Contemp Brachytherapy* 2015;7(2):183–188.

Webster MJ, Devic S, Vuong T et al. Dynamic modulated brachytherapy (DMBT) for rectal cancer. *Med Phys* 2013;40:011718.

Yang W, Kim Y, Wu X et al. Rotating-shield brachytherapy for cervical cancer. *Phys Med Biol* 2013;58:3931–3941.

CHAPTER 4

Applicator Reconstruction

Christoph Bert, Taran Paulsen Hellebust,
and Frank-André Siebert

CONTENTS

4.1 INTRODUCTION

Owing to very highdose gradients of 5%–12%/mm in distances of 1–3 cm from the source channel (Hellebust et al., 2010), precise reconstruction of an applicator is essential (Tanderup et al., 2008). Depending on the treatment site, different imaging options are considered for this part of the treatment workflow. In the following, we first describe the current clinical practice for transrectal ultrasound (TRUS), computed tomography (CT), and magnetic resonance imaging (MRI)-based applicator reconstruction. As emerging technologies, the remaining sections will report on electromagnetic tracking (EMT), special techniques in MRI, and vibro-acoustography (VA).

4.2 CURRENT CLINICAL PRACTICE

4.2.1 TRUS Based

TRUS is the typical imaging modality to supervise needle implantation for permanent low-dose rate (LDR) seed implantation and interstitial prostate high-dose rate (HDR)-brachytherapy with the patient in dorsal lithotomy position. It allows needle (tip) positioning in real time as well as identification of the prostate and urethra as an important organ at risk. Many centers use TRUS only for guidance during the implantation step and perform implant reconstruction or seed localization by CT imaging. Naturally, this requires changes in patient pose, transport of the patient, and thus the risk of relative motion of implanted needles at the advantage of precise and well-accepted implant definition by CT imaging. Moreover, the delineation of the prostate in CT images is challenging due to poor soft-tissue contrast (see also Chapter 16). Some centers, have explored the use of TRUS for implantation guidance in combination with MRI for delineation and implant definition (Dinkla et al., 2013; Rylander et al., 2015).

Several publications show that TRUS-based implant definition is a very accurate, fast, and cost-efficient option (Galalae et al., 2002; Hoskin et al., 2013). Nevertheless, when applying single fraction HDR or focal treatments, the requirements for needle reconstruction rise. Hence, systematic deviations in TRUS, for example, due to shadowing of needles, position offsets due to bright echoes, and offsets in longitudinal position that prohibit the use of TRUS for planning (Siebert et al., 2009; Schmid et al., 2013), are of increasing importance. In recent reports, it was shown that these deviations can be corrected. Schmid et al. (2013) showed that the bright echoes originate from the needles' wall rather than its center. If this systematic offset is corrected, precise spatial identification is possible. In case of a shadowed needle's tip, a correction of cranio-caudal shifts can be based on measuring the residual needle length proximal to the template (Zheng and Todor, 2011). Single session HDR-brachytherapy based on TRUS planning should then be feasible with differences $\lesssim 2\%$ of the prescribed dose compared to a CT-based procedure (Schmid et al., 2013).

4.2.2 CT Based

CT is the main imaging option for implant reconstruction in interstitial brachytherapy for breast, head, and neck, and also for post-planning of seed implants in the prostate. When using plastic catheters, the tracks of catheters can be well identified due to the strong contrast of air or high-Z wires inserted into the catheters for imaging purposes including options for semi-automatic reconstruction (Dise et al., 2015). CT scanners are also widely available in radiation oncology clinics making their use more likely.

Drawbacks of CT imaging are the missing soft-tissue contrast, for example, in gynecologic applications, metal artifacts by applicators or metal implants like hip prosthesis, and the extra dose needed for CT imaging. Solutions include fusion of MRI and CT data sets assuring an overlap of the applicator position rather than, for example, bony anatomy (Berger et al., 2009; Hellebust et al., 2010; De Brabandere et al., 2012).

4.2.3 MRI Based

One of the basic requirements for MRI-based reconstruction is the availability of MRI-compatible equipment. Examples are titanium or plastic needles for interstitial prostate treatments or dedicated applicators for gynecological applications, which are nowadays offered as MRI-compatible (Perez-Calatayud et al., 2009; Petit et al., 2013).

MRI does not provide signals for many applicator materials nor for bony anatomy. Susceptibility artifacts arising due to sudden changes in susceptibility from tissue to implant may influence precise localization of implants such as the needle tip (Muller-Bierl et al., 2004; Wachowicz et al., 2006). Imaging sequences that are dedicated to the clinical task (e.g., seed detection or applicator reconstruction vs. good tissue contrast in prostate seed implants or in gynecologic malignancies) are very important to reduce those artifacts and/or to increase the contrast of the relevant anatomy or applicator (Thomas et al., 2009; Kapur et al., 2012).

To allow reconstruction of the source path in the applicator, several options are currently used. Among them is the use of MRI-visible fluids such as water or glycerin that are injected into the source path to generate MRI contrast (Hellebust et al., 2010). Since the amount of contrast agent is limited, the contrast is often fairly poor. Alternatives include template matching: based on an applicator library containing the three-dimensional (3D) geometry of typical applicators, the nominal shape of the applicator is matched with the 3D MRI dataset. Matching is either based on the void volume visible in the MRI (Hellebust et al., 2010) or based on landmarks in the applicator that provide MRI signal such as holes in the Vienna Applicator that fill *in situ* with body fluids (Berger et al., 2009).

During implant of interstitially placed needles in gynecologic treatments, MRI-guidance is a viable option to reduce the risk of bladder or rectum penetration. Currently, magnetic resonance (MR)-tracking is possible at only 0.5–2 frames per minute (Wang et al., 2015a).

4.3 ELECTROMAGNETIC TRACKING

EMT refers to a technology that allows measuring the position of miniaturized sensors in 3D space often combined with additional degrees of freedom (DOF) for the orientation of the sensor. EMT is used in a variety of medical applications including tracking of surgical instruments (Narsule et al., 2012), guidance of biopsies (Kickuth et al., 2015), and external beam radiation therapy (Cherpak et al., 2009). A recent review is given by Franz et al. (2014). EMT is also an emerging option for interstitial brachytherapy where the sensor can be inserted into the needles/catheters. Precise and quick reconstruction especially of many catheters and/or error detection during the course of treatment planning and delivery are the primary objectives for the introduction of EMT into brachytherapy.

4.3.1 Technological Basis

The basic technology uses a precisely defined inhomogeneous magnetic field emitted by a field generator and a sensor located within that field which measures the magnetic flux traversing it. Due to the known magnetic field properties, the system can establish a correlation between measured flux and the position of the sensor. A typical measurement

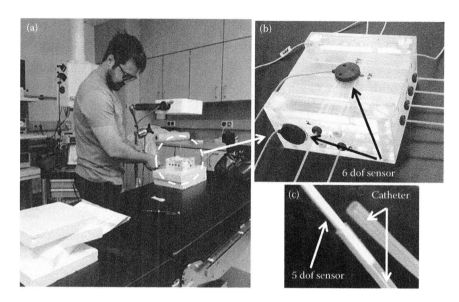

FIGURE 4.1 (a) Typical measurement situation for EMT in a phantom setup. (b) While three six DOF sensors placed on the object to record (breathing) motion and (c) a five DOF sensors inserted into each catheter measures the implant pathway.

geometry is shown in Figure 4.1a with field of views of approximately $50 \times 50 \times 50$ cm^3. For 3D position information, the field generators need to establish at least a sequence of three electromagnetic fields, to allow differentiation of the spatial components. Due to the symmetry of the magnetic dipole of the sensor's inductor, the induced currents also allow determination of two DOF for orientation in a typical standard miniature sensor as shown in Figure 4.1c. Measurements for six DOFs are feasible by combining two five DOF sensors, typically requiring a larger housing (Figure 4.1b).

The systems rely on precisely defined magnetic fields. Therefore, electromagnetic disturbances, for example, caused by additional systems emitting electromagnetic fields, eddy currents induced by the EMT in conductive materials in the field of view, or by ferromagnetic materials decrease the accuracy and precision feasible by EMT. A quantitative derivation for metal distortions was reported by Nixon et al. (1998).

In EMT dedicated environments, the systems can reach a precision of well below 1 mm at accuracies of 0.3–0.9 mm as recently summarized by Zhou et al. (2015) for applications in brachytherapy.

4.3.2 Applications in Brachytherapy

Apart from EMT supported robot-assisted lung brachytherapy (Lin et al., 2008), the application of EMT started recently. Clinical focus of the studies is implant reconstruction and error detection. Error detection is mainly reported by Damato et al. (Wilkinson and Kolar, 2013), who addressed some of the major causes of errors identified by the Nuclear Regulatory Commission (NRC), that is, swapping of channels, shift of catheters, or wrong implant reconstructions from imaging, for example, due to catheters with crossed paths (Damato et al., 2014) (more detail can be found in Chapter 10).

The work reported on implant reconstruction is currently based on phantom studies only. The phantoms were made of gelatin (Poulin et al., 2015), a commercially available ultrasound prostate training phantom (Bharat et al., 2014), or in-house built calibration phantoms that precisely guide the catheters (Zhou et al., 2013; Kellermeier et al., 2015). The majority of the studies investigated the achievable EMT accuracy in comparison to established alternatives such as TRUS or (micro-) CT-based implant reconstruction. The use of tracking technology in combination with optical guidance is discussed in Chapter 13. Alternatively, precision built phantom geometry (Kellermeier et al., 2015) or a robotic arm (Bharat et al., 2014) served as ground truth. The investigations resulted in the already-mentioned accuracy levels of 0.3–0.9 mm, that is, a level comparable to the established technologies (see also Section 4.2). A typical example of EMT versus µCT-based implant reconstruction is shown in Figure 4.2. All studies mentioned above conclude that EMT-based implant reconstruction investigated in phantom geometries works as accurately

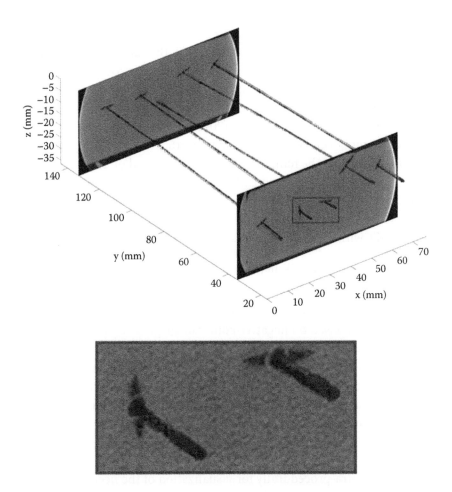

FIGURE 4.2　Implant reconstruction based on EMT (+) and µCT (O). The lower figure is a close-up of the rectangular area indicated above. (Reprinted with permission from Poulin E et al. *Med Phys.* 2015;42:1227.)

as the current clinically used alternatives. EMT is thus an interesting emerging technology for implant reconstruction but requires further clinical investigations before being competitive.

The advantages of EMT lie in the acquisition speed, compatibility with surgery theater use, and most likely even improved accuracy less prone to errors than the currently used options. Acquisition speed is not determined conclusively since most studies used experimental setups. Per channel, acquisition times of approximately 10 s are reported (Bharat et al., 2014; Poulin et al., 2015), resulting in implant reconstruction times of less than 10 min including change time in between catheters and setup (Zhou et al., 2013; Kellermeier et al., 2015).

In the setting of brachytherapy equipment, the achievable accuracy was slightly reduced in some of the studies, for example, due to close vicinity to liquid-crystal displays (Bharat et al., 2014), by metallic materials in the table, or by electronic equipment (Kellermeier et al., 2015). But these influences can most likely be easily avoided by slight adjustments of the workflow or exchange of auxiliary equipment such as a treatment table into EMT compatible alternatives. Of more critical concern is the definition of joint coordinate systems of EMT and other imaging options such as CT that allows for visualization of the anatomy and is thus mandatory for treatment planning. Joint coordinate systems are typically achieved by fusing the information, for example, based on rigid transformations optimized by some similarity measure such as the geometric distance between landmarks visible in both original systems. Bharat et al. (2014) nicely showed that EMT and TRUS can be combined into a common coordinate system by attaching an EMT sensor to the ultrasound probe. The authors of the remaining publications relied on rigid registrations which are, for example, based on CT-defined catheters or at least the tip of catheters. A smooth and quick implementation of EMT into the clinical workflow will require further investigations similar to the achievements of Bharat et al. (2014) for TRUS-imaging, for example, by performing EMT within the CT-room as investigated by Kellermeier et al. (2015).

4.4 MAGNETIC RESONANCE IMAGING

One of the main benefits of MRI over CT is the increased soft-tissue contrast that eases differentiation of tumor and normal tissues. The disadvantage is the reconstruction of applicators or implants based on negative contrasts originating from susceptibility artifacts caused by the metallic stylets, which are influenced by the tissue type and the orientation to the B0-field (Pernelle et al., 2013; Wang et al., 2015a). The motivation of emerging techniques is thus often related to improved MRI-based reconstruction options such that treatment based on fused CT (for reconstruction) and MRI data (for delineation) can be exchanged to MRI-only workflows.

4.4.1 MR-Tracking

MR-tracking is used intra-procedurally for visualization of the needle advance, for example, to supervise bending of the stylet due to resistive forces from the stiff tissues or to avoid penetration of organs at risk. In addition, the same technology was proposed for reconstruction of the catheter geometry. In comparison to conventional stylets, MR-tracking

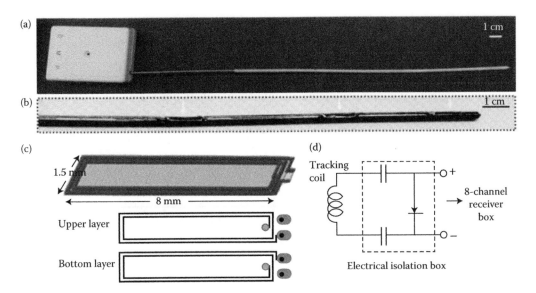

FIGURE 4.3   Details of an MR-tracking enabled conventional stylet. (a) Photograph of the complete system, (b) close-up of the dashed area highlighted in (a) with the position of the flexible printed circuits indicated by arrows. Parts (c) and (d) show the design pattern and diagram of the circuit. (Reproduced with permission from Wang W et al. *Magn Reson Med.* 2015;73:1803–11.)

has one fundamental difference: rather than relying on artifacts generated passively by the metal of the stylet, the stylet is transformed into an active object by means of radio-frequency (RF)-coils. These coils need to produce a local B-field (often referred to as B_1) that has to have very defined properties such that the passive artifacts of the metal are exceeded. Wang et al. (2015a) studied the feasibility of such a technique. They used micro-coils embedded in a printed circuit to convert a standard clinically used metallic stylet into a stylet with active MR-tracking by milling three slots into the stylet that house the RF circuit (see Figure 4.3). By means of micro-coaxial cables guided along the stylet into the white junction box shown in Figure 4.3a the coils are connected to the MRI system considering safety issues such as electrical isolation. Using dedicated MR sequences with echo times of approximately 2 ms, the positions of the micro-coils attached to the stylet can be determined at 40 frames per second with an isotropic resolution of 0.6 mm which is much faster than tracking of stylets using passive methods. In combination with a regular MR image acquired before tracking, software interfaces, and custom-built software solutions, the positions of the three coils and thus the stylet's geometry including its orientation and tip position can be displayed in real time graphically overlaid to the 3D geometry.

The new developed technical option was tested with respect to heating (max. 0.6°C) and with respect to the physics characteristics also in comparison to nonmetallic stylets. The produced B_1 field is not cylindrically symmetric to the stylet due to the position of the milled slot but produced a defined offset of 1 mm that can be corrected for in reconstructed positions. Compared to the negative contrast of metallic artifacts used in passive determination of stylet tip positions, MR-tracking positions were in ±0.5 mm agreement (Wang et al., 2015a).

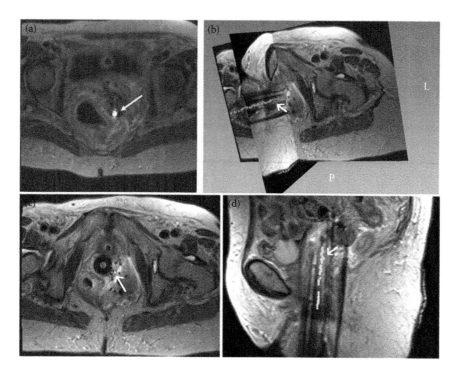

FIGURE 4.4   Reconstructed tip positions (arrows in b–d) of an active stylet (arrow in a) overlaid in 3D to *a priori* scanned MR image. (Reproduced with permission from Wang W et al. *Magn Reson Med*. 2015;73:1803–11.)

The system was also tested in dead chicken carcasses and within an institutional review board approved study in three women with gynecologic cancer. In the clinical setting, the patients underwent the standard treatment workflow based on conventional 12–30 stylets. After placing the catheters, a conventional stylet was replaced against the active one and MR-tracking was performed during its insertion. Figure 4.4 shows the integral result of the real-time MR-tracking procedure. Tracking-based positions overlay with the void signals of the conventionally used susceptibility artifacts.

As discussed in a subsequent report of the same group, the active stylet can also be used offline, that is, after and not during implantation. The stylet is then used for catheter trajectory reconstruction similar to some proposed applications for EMT as described above. In acquisition times of approximately 10 s for each catheter, no clinically significant differences were found comparing MR-tracking-based catheter trajectories with trajectories from CT or MRI-based reconstructions in a feasibility study involving three patients (Wang et al., 2015b).

4.4.2 Virtual Needle Navigation

One of the most integrated brachytherapy suites is the Advances Multimodality Image-Guided Operating (AMIGO) suite at the Brigham and Women's Hospital in Boston, Massachusetts (Kapur et al., 2012). It consists of an operating room (OR) with integrated access to a positron emission tomograph (PET)/CT and a MRI system, the latter is mobile

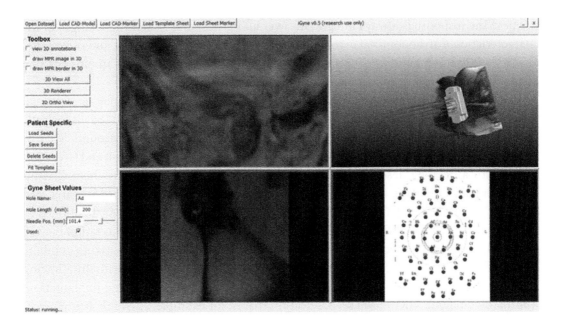

FIGURE 4.5 Screenshot of the *iGyne* software showing the virtual placement of needles in a gynecological case. (Reproduced with permission from Kapur T et al. *Magn Reson Imaging.* 2012;30:1279–90.)

due to ceiling mounts and can thus be brought to the OR for interventional imaging. For gynecologic treatments, the applicator is currently inserted via MRI-guidance using a number of MRI scans of approximately 1 min each to monitor the needle insertion in a stepwise process (see also Section 4.2.3). This procedure reduces the risk of bladder or rectum puncture but does not allow preplanned placement of needles since virtual needle positions within the body cannot be anticipated after the template is sutured to the perineum. This next gap is planned to be bridged with *iGyne* developed as an add-on to the open-source toolkit Slicer (Pernelle et al., 2013).

The idea of *iGyne* is image-guided navigation of applicator/needle insertion. Based on an MR image of the site including the sutured template, the software registers a geometric model of the applicator to the image data (see Figure 4.5). Once this geometric relationship has been established virtual needles can be overlaid with the MRI data, based on the used template coordinates and the penetration depth of the needle. Since these parameters are adjustable, the complete implant can be planned prior to puncturing. Part of this preplanning will be dose calculations such that the optimal placement of needles is not only determined based on geometry but also on dosimetric needs.

4.5 VIBRO-ACOUSTOGRAPHY

Palpation is one of the most important diagnostic procedures in oncology. During palpation, physicians induce a mechanical force on the tissue of interest and their experience allows them to characterize the tissue's mechanical parameters such as elasticity based on sensing resistance. Palpation can only be performed for superficial tissues or tissues in cavities. The response is always integration over all tissues, that is, sensing changes in the characteristics of an abdominal organ always includes response of the skin and subcutaneous fat. Due to the combination of these boundary conditions, palpation is an estimate that often relies heavily on the experience of the physician.

One branch of imaging sciences, in recent years, focuses on elasticity imaging. External forces such as vibrations induce local motion of the tissues, which encode the viscoelastic properties of the medium and are quantified, for example, by Doppler ultrasound, MRI, or hydrophones. One such method is VA, which was initially reported by Fatemi and Greenleaf (1998) with a detailed description of the underlying physics provided by them (Fatemi and Greenleaf, 1999). The motivation for VA in brachytherapy is mainly addressing seed detection during the implantation procedure of prostate cancer patients. As mentioned in Section 4.2.1, currently TRUS is used intra-procedurally to guide the implantation but CT is needed afterward to precisely detect the seed location. VA has the potential to combine both steps, that is, intra-procedural guidance including seed detection potentially allowing better treatments due to real-time and interactive dose calculations during implantation. Alternatives include MRI-based seed detection that is, for example, currently being investigated by Dong et al. (2015).

4.5.1 Technological Basis

VA uses a collimated ultrasound beam to produce a radiation force that causes a vibration of the stimulated tissues. Local stimulation is achieved by superimposing two ultrasound beams at slightly differing wavelengths (see Figure 4.6) that can be created by two confocally arranged transducers. In initial feasibility studies involving gel phantoms and explanted prostates, the focal spot was scanned over the objects by mechanical movement of the probe within the imaging plane in addition to variations in probe–object distance to vary the plane Mitri et al. (2009). The measuring object is typically inserted into water or gel. Data acquisition in that solution is too slow since the image is formed focal point by focal point. Even more limiting is the geometrical size of the transducers, which are too large for transrectal applications (Mehrmohammadi et al., 2014). Developments are thus

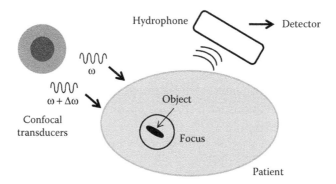

FIGURE 4.6 Generation of a superimposed ultrasound field that induces a radiation force in the focal point. The caused vibrations of the stimulated tissues are transmitted isotropically and can be detected by a hydrophone.

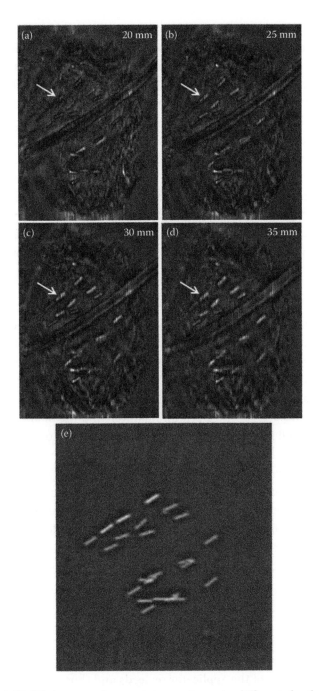

FIGURE 4.7 Quasi-2D VA images of a prostate specimen in different depths of (a) 20 mm, (b) 25 mm, (c) 30 mm, and (d) 35 mm. The seed indicated by an arrow shows the highest contrast at a depth of 30 mm. Part (e) shows the CT scout view of the same specimen. (Reproduced with permission from Mehrmohammadi M et al. *Med Phys.* 2014;41:092902.)

aimed at improving the above-mentioned limitations. Electronic configuration of ultrasound transducer arrays enhances the scanning time since the position of the focal spot can be varied much quicker. Several solutions were developed and tested in the last years (Urban et al., 2011; Kamimura et al., 2012; Mehrmohammadi et al., 2014). Currently, two-dimensional (2D) systems are studied in simulations (Kamimura et al., 2012). Feasibility tests addressing seed detection in excised prostate specimens were reported using quasi-2D arrays attached to a clinical ultrasound system (Mehrmohammadi et al., 2014). The 12×70 piezo elements of these arrays are serially activated in 12×12 regions of which 128 are driven simultaneously to form a concentric, confocal arrangement as with the originally used transducers. The system operates in the focal spot at approximately 51 kHz. The focal spot can be moved one-dimensionally by electronically altering the spatial position of the 12×12 piezo elements. The third dimension is stimulated by mechanically moving the system. As in the initial technical solutions, a hydrophone is used for detecting acoustic emission.

4.5.2 Application for Seed Detection in Brachytherapy

So far, VA has only been tested in phantoms and in *in vitro* prostate specimens. Mitri et al. (2009) tested the method on excised human prostates using the original technical solution based on interfering confocal ultrasound transducers. The prostates were implanted with seeds under TRUS and fluoroscopy guidance before excision from the cadaver mimicking the clinical procedure and thus ensuring representative seed orientations relative to the prostate. The organ was placed into a gelatin phantom and the phantom imaged by VA in depth steps of 1 mm and lateral steps of 0.25 mm covering a volume of $24 \times 50 \times 50$ mm³; 71% of the seeds could be detected. Similarly, experiments were performed using the quasi-2D technique but seeds were implanted into the already excised prostate specimens. Figure 4.7 shows VA images of one of the prostate specimens. The images cover a range of depths and the intensity of the detected seeds vary with depth, that is, the most probable seed location has to be determined by comparing the contrast of several depths. Compared to CT as the clinically accepted ground truth, VA could detect 74%–92% of the implanted fiducials, which is higher than typical TRUS-based seed detection which is in the range of 30%–50% (Su et al., 2011).

4.6 CONCLUSION

EMT, MRI, and VA are promising techniques at the border to clinical introduction that will improve and ease applicator reconstruction in brachytherapy. All techniques allow intra-procedural detection of the applicator or seeds and thus open the window of opportunity for interactive decisions at the time of implantation. In addition, it is likely that the risk of errors such as wrong detection of the needle tip in MR-based gynecological interventions or reconstruction errors due to close proximity of interstitially placed catheters will be reduced. Moreover, it is to be expected that reconstruction uncertainties can be decreased when using these modern techniques. The important next step in the exploration of these technical options will be feasibility studies in a clinical setting to examine the potential on a widespread level.

REFERENCES

Berger D, Dimopoulos J, Potter R, Kirisits C. Direct reconstruction of the Vienna applicator on MR images. *Radiother Oncol.* 2009;93:347–51.

Bharat S, Kung C, Dehghan E, Ravi A, Venugopal N, Bonillas A et al. Electromagnetic tracking for catheter reconstruction in ultrasound-guided high-dose-rate brachytherapy of the prostate. *Brachytherapy* 2014;13:640–50.

Cherpak A, Ding W, Hallil A, Cygler JE. Evaluation of a novel 4D in vivo dosimetry system. *Med Phys.* 2009;36:1672–9.

Damato AL, Viswanathan AN, Don SM, Hansen JL, Cormack RA. A system to use electromagnetic tracking for the quality assurance of brachytherapy catheter digitization. *Med Phys.* 2014;41:101702.

De Brabandere M, Hoskin P, Haustermans K, Van den Heuvel F, Siebert FA. Prostate post-implant dosimetry: Interobserver variability in seed localisation, contouring and fusion. *Radiother Oncol.* 2012;104:192–8.

Dinkla AM, Pieters BR, Koedooder K, van Wieringen N, van der Laarse R, van der Grient JN et al. Improved tumour control probability with MRI-based prostate brachytherapy treatment planning. *Acta Oncol. (Stockholm, Sweden)* 2013;52:658–65.

Dise J, Liang X, Scheuermann J, Anamalayil S, Mesina C, Lin LL et al. Development and evaluation of an automatic interstitial catheter digitization tool for adaptive high-dose-rate brachytherapy. *Brachytherapy* 2015;14:619–25.

Dong Y, Chang Z, Xie G, Whitehead G, Ji JX. Susceptibility-based positive contrast MRI of brachytherapy seeds. *Magn Reson Med.* 2015;74:716–26.

Fatemi M, Greenleaf JF. Ultrasound-stimulated vibro-acoustic spectrography. *Science* 1998;280:82–5.

Fatemi M, Greenleaf JF. Vibro-acoustography: An imaging modality based on ultrasound-stimulated acoustic emission. *Proc Natl Acad Sci USA* 1999;96:6603–8.

Franz AM, Haidegger T, Birkfellner W, Cleary K, Peters TM, Maier-Hein L. Electromagnetic tracking in medicine—A review of technology, validation, and applications. *IEEE Trans Med Imaging.* 2014;33:1702–25.

Galalae RM, Kovacs G, Schultze J, Loch T, Rzehak P, Wilhelm R et al. Long-term outcome after elective irradiation of the pelvic lymphatics and local dose escalation using high-dose-rate brachytherapy for locally advanced prostate cancer. *Int J Radiat Oncol Biol Phys.* 2002;52:81–90.

Hellebust TP, Kirisits C, Berger D, Perez-Calatayud J, De Brabandere M, De Leeuw A et al. Recommendations from Gynaecological (GYN) GEC-ESTRO Working Group: Considerations and pitfalls in commissioning and applicator reconstruction in 3D image-based treatment planning of cervix cancer brachytherapy. *Radiother Oncol.* 2010;96:153–60.

Hoskin PJ, Colombo A, Henry A, Niehoff P, Paulsen Hellebust T, Siebert FA et al. GEC/ESTRO recommendations on high dose rate afterloading brachytherapy for localised prostate cancer: An update. *Radiother Oncol.* 2013;107:325–32.

Kamimura HA, Urban MW, Carneiro AA, Fatemi M, Alizad A. Vibro-acoustography beam formation with reconfigurable arrays. *IEEE Trans Ultrason Ferroelectr Freq Control* 2012;59:1421–31.

Kapur T, Egger J, Damato A, Schmidt EJ, Viswanathan AN. 3-T MR-guided brachytherapy for gynecologic malignancies. *Magn Reson Imaging* 2012;30:1279–90.

Kellermeier M, Herbolzheimer J, Kreppner S, Lotter M, Strnad V, Bert C. SU-F-BRA-02: Electromagnetic tracking in brachytherapy as an advanced modality for treatment quality assurance. *Med Phys.* 2015;42:3533–4.

Kickuth R, Reichling C, Bley T, Hahn D, Ritter C. C-arm cone-beam CT combined with a new electromagnetic navigation system for guidance of percutaneous needle biopsies: Initial clinical experience. *RoFo: Fortschritte auf dem Gebiete der Rontgenstrahlen und der Nuklearmedizin.* 2015;187:569–76.

Lin AW, Trejos AL, Mohan S, Bassan H, Kashigar A, Patel RV et al. Electromagnetic navigation improves minimally invasive robot-assisted lung brachytherapy. *Comput Aided Surg.* 2008;13:114–23.

Mehrmohammadi M, Alizad A, Kinnick RR, Davis BJ, Fatemi M. Feasibility of vibro-acoustography with a quasi-2D ultrasound array transducer for detection and localizing of permanent prostate brachytherapy seeds: A pilot *ex vivo* study. *Med Phys.* 2014;41:092902.

Mitri FG, Davis BJ, Urban MW, Alizad A, Greenleaf JF, Lischer GH et al. Vibro-acoustography imaging of permanent prostate brachytherapy seeds in an excised human prostate—Preliminary results and technical feasibility. *Ultrasonics* 2009;49:389–94.

Muller-Bierl B, Graf H, Lauer U, Steidle G, Schick F. Numerical modeling of needle tip artifacts in MR gradient echo imaging. *Med Phys.* 2004;31:579–87.

Narsule CK, Sales Dos Santos R, Gupta A, Ebright MI, Rivas R Jr., Daly BD et al. The efficacy of electromagnetic navigation to assist with computed tomography-guided percutaneous thermal ablation of lung tumors. *Innovations (Philadelphia, Pa.)* 2012;7:187–90.

Nixon MA, McCallum BC, Fright WR, Price NB. The effects of metals and interfering fields on electromagnetic trackers. *Presence Teleop Virt Environ.* 1998;7:204–18.

Perez-Calatayud J, Kuipers F, Ballester F, Granero D, Richart J, Rodriguez S et al. Exclusive MRI-based tandem and colpostats reconstruction in gynaecological brachytherapy treatment planning. *Radiother Oncol.* 2009;91:181–6.

Pernelle G, Mehrtash A, Barber L, Damato A, Wang W, Seethamraju RT et al. Validation of catheter segmentation for MR-guided gynecologic cancer brachytherapy. *Med Image Comput Comput Assist Interv.* 2013;16:380–7.

Petit S, Wielopolski P, Rijnsdorp R, Mens JW, Kolkman-Deurloo IK. MR guided applicator reconstruction for brachytherapy of cervical cancer using the novel titanium Rotterdam applicator. *Radiother Oncol.* 2013;107:88–92.

Poulin E, Racine E, Binnekamp D, Beaulieu L. Fast, automatic, and accurate catheter reconstruction in HDR brachytherapy using an electromagnetic 3D tracking system. *Med Phys.* 2015;42:1227.

Rylander S, Buus S, Bentzen L, Pedersen EM, Tanderup K. The influence of a rectal ultrasound probe on the separation between prostate and rectum in high-dose-rate brachytherapy. *Brachytherapy* 2015;14:711–7.

Schmid M, Crook JM, Batchelar D, Araujo C, Petrik D, Kim D et al. A phantom study to assess accuracy of needle identification in real-time planning of ultrasound-guided high-dose-rate prostate implants. *Brachytherapy* 2013;12:56–64.

Siebert F-A, Hirt M, Niehoff P, Kovacs G. Imaging of implant needles for real-time HDR-brachytherapy prostate treatment using biplane ultrasound transducers. *Med Phys.* 2009;36:3406–12.

Su JL, Bouchard RR, Karpiouk AB, Hazle JD, Emelianov SY. Photoacoustic imaging of prostate brachytherapy seeds. *Biomed Opt Express.* 2011;2:2243–54.

Tanderup K, Hellebust TP, Lang S, Granfeldt J, Potter R, Lindegaard JC et al. Consequences of random and systematic reconstruction uncertainties in 3D image based brachytherapy in cervical cancer. *Radiother Oncol.* 2008;89:156–63.

Thomas SD, Wachowicz K, Fallone BG. MRI of prostate brachytherapy seeds at high field: A study in phantom. *Med Phys.* 2009;36:5228–34.

Urban MW, Chalek C, Kinnick RR, Kinter TM, Haider B, Greenleaf JF et al. Implementation of vibro-acoustography on a clinical ultrasound system. *IEEE Trans Ultrason Ferroelectr Freq Control* 2011;58:1169–81.

Wachowicz K, Thomas SD, Fallone BG. Characterization of the susceptibility artifact around a prostate brachytherapy seed in MRI. *Med Phys.* 2006;33:4459–67.

Wang W, Dumoulin CL, Viswanathan AN, Tse ZTH, Mehrtash A, Loew W et al. Real-time active MR-tracking of metallic stylets in MR-guided radiation therapy. *Magn Reson Med.* 2015a;73:1803–11.

Wang W, Viswanathan AN, Damato AL, Chen Y, Tse Z, Pan L et al. Evaluation of an active magnetic resonance tracking system for interstitial brachytherapy. *Med Phys.* 2015b;42:7114–21.

Wilkinson DA, Kolar MD. Failure modes and effects analysis applied to high-dose-rate brachytherapy treatment planning. *Brachytherapy* 2013;12:382–6.

Zheng DD, Todor DA. A novel method for accurate needle-tip identification in trans-rectal ultrasound-based high-dose-rate prostate brachytherapy. *Brachytherapy* 2011;10:466–73.

Zhou J, Sebastian E, Mangona V, Yan D. Real-time catheter tracking for high-dose-rate prostate brachytherapy using an electromagnetic 3D-guidance device: A preliminary performance study. *Med Phys.* 2013;40:021716.

Zhou J, Zamdborg L, Sebastian E. Review of advanced catheter technologies in radiation oncology brachytherapy procedures. *Cancer Manag Res.* 2015;7:199–211.

CHAPTER 5

Dose Calculation

Åsa Carlsson Tedgren, Xun Jia, Manuel Sanchez-Garcia, and Alexandr Malusek

CONTENTS

5.1 INTRODUCTION

Model-based dose calculation algorithms (MBDCAs) for brachytherapy are methods that are driven by physics-based models of varying complexity and are capable of providing improved dosimetric accuracy over traditional methods. MBDCA and the imaging methods needed to support them are emerging topics in brachytherapy and are hence covered in detail in this chapter. This chapter will cover both "conventional brachytherapy" with sealed radioactive sources or miniature x-ray tubes of millimeter dimensions and emitting photons with energies from 20 keV to 1.25 MeV and targeted radionuclide therapy with sealed sources of very small (approximately microns) dimensions as, for example, ^{90}Y microsphere brachytherapy, emitting beta particles. Conventional photon brachytherapy has existed since the beginning of radiotherapy and are today still evolving toward the use of new sources and techniques while microsphere brachytherapy is a relatively new technique.

MBDCA for conventional photon and microsphere brachytherapy will be introduced separately. Both applications benefit from new imaging techniques; microsphere brachytherapy from improved accuracy in nuclear medicine techniques to acquire quantitative information on activity distributions and conventional photon brachytherapy from techniques for acquiring information on tissue composition from imaging with higher accuracy than today. Strategies to speed up such calculations, for example, using graphics processing unit (GPU) programming, are also important for their use in practice. Before entering into these areas in more detail, a short background on the status of dose calculations and areas associated therewith for the two areas of brachytherapy are provided.

5.1.1 Conventional Photon Brachytherapy

Dose calculation in treatment planning of photon brachytherapy of today does, with few exceptions, follow the formalism set up by the American Association of Physicists in Medicine Task Group No. 43, AAPM TG-43 (Rivard et al. 2004). This formalism provides information on the distribution of absorbed dose in water around the brachytherapy source and assures that effects of the detailed geometry of the source is accounted for and provides traceability to primary standards (Soares et al. 2009). The AAPM TG-43 formalism was a major leap forward for dosimetric accuracy and safety of brachytherapy; however, as everything outside the source is assumed water, the patient's tissues, source applicators, etc., are not accounted for. MBDCAs can account for non-water heterogeneities based on individual patient information acquired from three-dimensional (3D) images. This has been a topic of great interest to brachytherapy in recent years (see the AAPM TG-186 report for a thorough introduction and review [Beaulieu et al. 2012]).

One reason why the simplistic approach is still in use for brachytherapy is that photon attenuation and scatter build-up in the individual geometry is often a second-order effect compared with the inverse-square fall-off in dose with distance from the source. Another reason is that much empirical knowledge from long and successful use of current methods exists. As is clear from other chapters of this book, brachytherapy of today faces the introduction of new sources, applicators, and treatment techniques. Such developments make the accuracy in dosimetry and thus the use of MBDCAs more important as new applications move away from the dosimetric domain where empirical dose–response knowledge was initially obtained. An early example on the need for MBDCA to account for 3D scattering effects in conjunction with new kinds of treatments was the patient-individual lead shields used with ^{241}Am at 60 keV in gynecological treatments (Nath et al. 1988). Dose distributions in low-atomic-number (Z) materials, like tissue, for photon energies at 50–100 keV are similar to those of higher energies (300–662 keV) over the first centimeter as the higher attenuation at low energies is compensated for by faster build-up of scattered photons. In high-atomic-number materials, like lead and tungsten, however, attenuation is much higher for the lower photon energies due to increased contribution from the photoelectric effect. Nath et al. (1988) found that the individual lead shields also significantly altered the dose distribution in the unshielded direction around the ^{241}Am sources as it was an effect of the 3D nature of dose from scattered photons. The use of MBDCA is further expected to contribute to refined dose-to-outcome correlation in conventional treatments by allowing for treatment planning accounting for individual patient's distribution of tissues; example being how tissue calcifications strongly influences dose distributions at low energies (<50 keV) in prostate and breast implants (see, e.g., Chibani and Williamson 2005).

The introduction of MBDCA to brachytherapy is not trivial. Recent studies in which dose calculation methods that can account for material and geometry boundaries showed that differences using MBDCA to conventional TG-43 vary strongly with photon energy, from being around a few percent in most geometries at the energy of the common isotope ^{192}Ir (emitted photon around 350 keV) and above to up to tenths of percent in soft tissue and more than so in bone and tissue calcifications at low energies (<50 keV) relevant to seeds of ^{125}I and ^{103}Pd and electronic brachytherapy sources.

Furthermore, computed tomography (CT) images used to gather information on the individual patient's tissues in terms of their atomic composition is lacking in accuracy, causing severe problems especially at low energies where it is most needed. It is not either yet resolved if absorbed dose values should be reported in terms of dose to the local medium or to a water cavity in that medium (see, e.g., Carlsson Tedgren and Alm Carlsson 2013). An overview of benefits and current problems with the use of MBDCA as compared to the traditional AAPM TG-43 formalism at the time of treatment planning and later follow-up are provided in Figure 5.1.

5.1.2 Microsphere Beta-Particle Brachytherapy

Most of the blood flow to the healthy liver tissue is provided by the portal vein, while hepatocellular carcinoma (HCC) and liver metastasis feed mainly from the liver artery. Selective internal radiotherapy (SIRT) is a technique that consists of the embolization of a region in the liver containing a tumor and the delivery through the arterial artery of microspheres loaded with ^{90}Y. The ^{90}Y isotope emits electrons with a maximum energy of 2.2 MeV (maximum range of 1.2 cm in soft tissue) delivering a highly localized dose to the target while preserving the liver parenchyma. A comprehensive review of this technique can be found in Kennedy et al. (2007).

Part of the microspheres may find their way to the lung or even to the stomach, limiting the amount of activity that can be safely injected or even precluding treatment. The goal

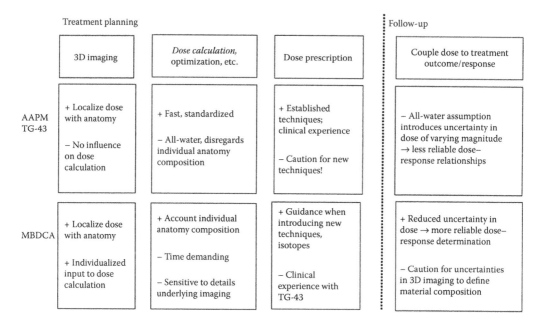

	3D imaging	Dose calculation, optimization, etc.	Dose prescription	Couple dose to treatment outcome/response

Treatment planning / Follow-up

AAPM TG-43	+ Localize dose with anatomy − No influence on dose calculation	+ Fast, standardized − All-water, disregards individual anatomy composition	+ Established techniques; clinical experience − Caution for new techniques!	− All-water assumption introduces uncertainty in dose of varying magnitude → less reliable dose–response relationships
MBDCA	+ Localize dose with anatomy + Individualized input to dose calculation	+ Account individual anatomy composition − Time demanding − Sensitive to details underlying imaging	+ Guidance when introducing new techniques, isotopes − Clinical experience with TG-43	+ Reduced uncertainty in dose → more reliable dose–response determination − Caution for uncertainties in 3D imaging to define material composition

FIGURE 5.1 The benefits of using MBDCA over AAPM TG-43 formalism for the dose calculation part of treatment planning are related to the increased dosimetric accuracy, making possible more individualized treatments and increased reliability in later follow-ups to obtain dose–response relationships. 3D imaging plays an important role as the accuracy in dose calculation output from MBDCA depends on the accuracy with which material composition can be obtained from 3D imaging.

of dosimetry is to provide the spatial dose distribution, $D(\boldsymbol{x})$, stemming from the known activity distribution $A(\boldsymbol{x})$ within the patient.

Classical dosimetry algorithms, like the partition model (Ho et al. 1996) or the medical internal radiation dose (MIRD) scheme (Bolch et al. 1999, 2009), did not have access to the full 3D distribution of activity. Instead, the assumption was made for the activity distribution to be constant within given regions of the patient. As an example, in the partition model, the assumption is made that all activity goes to three compartments: tumoral liver, non-tumoral liver, and lungs. Planar conjugate scintigraphy (IAEA 2014) is often used to estimate the relative proportions of activity in each region. A linear system of three equations with three unknowns is built and solved for the activity in each compartment. Dose is then calculated from its definition by assuming that all energy liberated in a compartment is deposited within the compartment. That is,

$$D = \frac{E\tilde{A}}{m} = \frac{E(A/\lambda)}{m} \approx \frac{49.97\,A}{m} \tag{5.1}$$

where $\tilde{A}$ is the cumulated activity, or total number of disintegrations, A the mean activity in the region, m the region mass, λ the ^{90}Y decay constant, and E the energy liberated per decay. The classical MIRD scheme (Loevinger et al. 1991) extends this formalism to handle an arbitrary number of regions and the cross-irradiation between them. The fact that these methods are based on piecewise constant activity distributions implies that they are only able to provide average doses to organs. When planar imaging is used for quantification (most common case), the uncertainty in the computed dose is dominated by the uncertainty in activity. However, this is not really a limitation of the method as these methods are also used with full 3D SPECT/CT quantification.

5.2 DOSE CALCULATION ALGORITHMS CAPABLE OF ACCOUNTING FOR NON-WATER HETEROGENEITIES IN CONVENTIONAL PHOTON BRACHYTHERAPY

There are several dose calculation algorithms available for brachytherapy that can be utilized to go beyond the TG-43 formalism and account for tissue, seeds, applicators, and shields of high-Z materials with respect to the 3D geometry at hand. Monte Carlo (MC) simulation is the golden standard of dose calculations against which others are to be benchmarked. The major dosimetric effects of introducing MBDCAs are well known; differences to TG-43 being moderate at ^{192}Ir but large (tenths of percentages or more in bone and tissue calcifications) at photon energies <50 keV (see reviews of Rivard et al. 2009, Beaulieu et al. 2012, Carlsson Tedgren et al. 2012, Papagiannis et al. 2014). These reviews can also be consulted for those interested in earlier attempts to increase dose calculation accuracy using one-dimensional (1D) and two-dimensional (2D) correction methods. Implications of characteristics of the brachytherapy photon energy on MBDCAs for the use in treatment planning were treated in some detail by Carlsson Tedgren et al. (2012). The short range of secondary electrons in the photon energy interval 20–400 keV (order of 1–2 mm or less) in relation to the dimensions of voxels of interest to treatment planning applications implies that charged particle equilibrium (CPE) prevails, an assumption utilized by the majority of MBDCA in brachytherapy.

5.2.1 MC Simulations

All common publically available MC codes like Monte Carlo N-Particle (MCNP), EGSnrc, GEANT4, and Penelope have been used for brachytherapy (see, e.g., Landry et al. 2013). The nonpublic MC code Ptran by Williamson and coworkers has had strong influence on brachytherapy computational dosimetry (Williamson 2006). MC utilizes random sampling together with interaction cross sections to solve the Boltzmann transport equation (Landry et al. 2013). To reduce computational time in treatment planning applications, the so-called track length estimators most often replace analog scoring of energy depositions; these derive the photon fluence differential in energy as the track length per volume and calculate dose from there using tabulations of mass–energy absorption coefficients and the CPE assumption (see Williamson 1987, 1988, Landry et al. 2013). Special efforts in treatment planning applications have been made by Chibani and Williamson (2005), Brachydose that is connected to EGSnrc (Taylor et al. 2007) and Algebra that utilizes GEANT4 (Afsharpour et al. 2012). Ptran has been investigated with the correlated sampling technique (Hedtjärn et al. 2002) to further increase calculation efficiency. Accelerated MC codes often include approximations, for instance, incoherent scattering is not fully modeled in the correlated sampling method effort (Hedtjärn et al. 2002), and should hence also be benchmarked against full MC.

5.2.2 Discrete Ordinates Types of Methods or Grid-Based Boltzmann Solvers

Analytical methods to solve the Boltzmann transport equation utilizing "discrete ordinates techniques" were first applied in brachytherapy by Daskalov et al. (2000). They have later been further developed and referred to as grid-based Boltzmann solvers (GBBS) (Beaulieu et al. 2012). The transport equation is discretized in energy, angular, and spatial domains. The angular discretization caused artifacts known as ray-effects, which was a problem for their use in brachytherapy due to the steep dose gradients around such sources, an approach that mitigates this effect is to use a "first collision source term" (Daskalov et al. 2000). Continued developments in this area were made (Gifford et al. 2008, Vassilov et al. 2008) and resulted in Acuros™ that can be obtained with the Brachyvision™ treatment planning software for ^{192}Ir-based brachytherapy from Varian (BV, Varian Medical Systems, Inc., Palo Alto, California). Acuros for brachytherapy calculates the photon fluence, assumes CPE, and derives absorbed dose using mass–energy absorption coefficients. Benchmarking against full MC for clinical applications has been done (Mikell et al. 2012, Zourari et al. 2013). Recent benchmarks of Acuros BV in the single source situation show efficient mitigation of ray-effects (Ballester et al. 2015). Acuros derives the distribution of photon fluence in the local medium. Using mass–energy absorption coefficients either dose to that medium or dose to water in that medium can be reported. Note that dose to water in medium derived in this way yields the dose to a water cavity large enough for establishment of CPE.

5.2.3 Collapsed Cone Point Kernel Superposition

Point kernel superposition methods are based on precalculated distributions of absorbed dose in water around an interaction point of primary photons, named point kernels. The point kernels are derived using MC and by forcing the interaction of primary photons in a point and scoring the resulting distribution of absorbed dose. The point kernels are in subsequent calculations scaled in 3D for non-water heterogeneities and its straightforward use has been successfully applied to brachytherapy (Williamson et al. 1991, Carlsson and Ahnesjö 2000a). These methods were developed to gain accuracy while keeping reasonable calculation times in radiotherapy and are able to calculate dose from primary and once-scattered photons without approximating the physics but involve some approximations for dose from multiple scattered photons (Ahnesjö and Aspradakis 1999, Carlsson and Ahnesjö 2000a). The collapsed cone (CC) version of the method (Ahnesjö 1989) utilizes an angular and spatial discretization approach in combination with fitting the point kernel energy-fluence distribution to exponential functions to gain increased computational efficiency. The CC approach was developed for external beam radiotherapy and later adapted for brachytherapy (Carlsson and Ahnesjö 2000b). The brachytherapy adapted version utilizes the CPE assumption and derives the dose from primary photons analytically by means of ray-tracing and dose from once- and multiple-scatter photons in two subsequent steps. This successive-scattering approach mitigates discretization artifacts of similar type (Carlsson Tedgren and Ahnesjö 2008) as the ray-effects mentioned above for discrete ordinate type of methods represents the use of an extended first- and higher-ordinate scatter source. The brachytherapy version can account for shields of high atomic numbers, including the generation of characteristic x-rays following photoelectric absorption (Carlsson Tedgren and Ahnesjö 2003) and calculate the dose around clinical sources following a primary and scatter separation approach for source characterization (Russell et al. 2005). It is available in Oncentra™ treatment planning system from Elekta (Elekta Ab, Stockholm, Sweden) under the commercial name ACE™ (Advanced Collapsed cone Engine) (Ahnesjö et al. 2017) and validations have shown good results for clinically relevant dose levels (Ma et al. 2015). The ray-effects of ACE are a minor issue in multiple-source implants; however, they constitute a noticeable effect at low (<5%–10%) dose levels around single sources (Ballester et al. 2015).

5.2.4 Dose Reporting Quantity

As for external beam radiotherapy (see, e.g., the review in Chetty et al. 2007), the brachytherapy community has been discussing what is the most suitable dose reporting quantity. With MBDCAs, it is possible to report absorbed dose in terms of dose to the local medium or in terms of dose to a water cavity embedded in that medium. The issue has large implications in brachytherapy, in particular at low energies (<50 keV) as differences between the two quantities can be significant (tenths of percentages or more). In brachytherapy, there is also a discussion on how dose to water should be defined and derived; should the water cavity for reporting be that of a small volume of cellular or DNA dimensions (micro or nanometer) or that of a treatment planning voxel dimensions (mm). The AAPM TG-186 (Beaulieu et al. 2012) found the level of knowledge too premature to make a recommendation and identified this as a matter of interest for future research, recommended that dose to medium is always reported and that the circumstances for defining and reporting dose to water are always reported in detail. A detailed description of the various possibilities, including implications at different brachytherapy energies, has been given by Carlsson Tedgren and Alm Carlsson (2013). Fonseca et al. (2015) compared dose to medium to two different ways (ratios of either mass–energy absorption coefficients or mass–collision stopping powers media to water) of determining dose to water in medium for a clinical ^{192}Ir implant.

5.3 DOSE CALCULATION METHODS FOR MICROSPHERE BRACHYTHERAPY

Modern approaches to dose calculation for microsphere brachytherapy (for classical approaches see Section 5.1) make use of quantitative imaging to directly measure the spatial distribution of activity, $A(x)$. This is feasible and practical thanks to the development of emission tomography devices coupled with x-ray CT (SPECT/CT) (Dewaraja et al. 2012). Two main approaches have been proposed to perform 3D absorbed dose calculations: direct MC methods and dose kernel methods.

Direct MC makes use of general-purpose codes such as MCNP, electron gamma shower (EGS), or GEANT4 to transport the electrons emitted by ^{90}Y within the patient's geometry and store the deposited energy at the voxel level. SPECT/CT images provide the source spatial distribution and attenuation geometry. Examples of such tools are 3D-RD (Prideaux et al. 2007), SIMDOS (Ljungberg et al. 2002), OEDIPE (Chiavassa et al. 2006), and VIDA (Kost et al. 2015). MC methods accurately model the activity and tissue distribution in the patient, being regarded as the golden standard for dosimetry. However, they require significant computational resources that are often not available in the clinic. Recent developments in simplified MC (Hippeläinen et al. 2015) and GPU (see Section 5.5) acceleration may change this in the future.

Dose kernel approaches compute the dose distribution by performing the following integral:

$$D(x) = \frac{\iiint \tilde{A}(s) h_{het}(s,x) ds}{\rho(x)} \tag{5.2}$$

where $\tilde{A}(s)$ is the concentration of cumulated activity, ρ is the mass density, and $h_{het}(s, x)$ is the energy deposition kernel, representing the fraction of energy emitted in point s that is absorbed in point x per unit volume. Equation 5.3 derives from the superposition integral introduced by among others Ahnesjö et al. (1987); see also review by Ahnesjö and Aspradakis (1999) and Sanchez-Garcia et al. (2014) for the derivation details. In a homogeneous medium, m, the kernel is a function only of the distance between source and energy deposition sites

$$h_{het}(s,x) = h_m(|s - x|) \tag{5.3}$$

Under this approximation, the dose kernel is space invariant, so Equation 5.1 becomes a convolution integral, which can be very efficiently computed using fast Fourier transforms. This algorithm is known as dose point Kernel convolution (Sgouros et al. 1990). The method was later extended to explicitly take into account the finite, nonzero volume of voxels and became known as dose voxel-kernel convolution (Bolch et al. 1999), which is widely available in software tools as VoxelDose (Gardin et al. 2003) and RMDP (Guy et al. 2003).

The dose point kernel can be easily calculated using MC simulations (Simpkin and Mackie 1990, Mainegra-Hing et al. 2005, Ferrer et al. 2007, Uusijärvi et al. 2009, Botta et al. 2011). This is conventionally done in water; let h_w be the so-determined kernel.

For a heterogeneous medium, the common approach is to scale h_w by the mean effective density (Cross 1968, Ahnesjö et al. 1987) between s and x, λ_{xs}. This is equivalent to saying that particles transport energy traveling in a straight line from s to x. See Ahnesjö (1989) for the rationale of the approximation. The scaled kernel can be expressed as

$$h_{het}(s,x) = \lambda(x) \lambda_{xs}^2 h_w(\lambda_{xs} |x - s|) \tag{5.4}$$

$$\lambda(x) = \kappa(x) \frac{\rho(x)}{\rho_w} \tag{5.5}$$

$$\lambda_{xs} = \lambda(s, x) = \int_0^1 \lambda(s - l \cdot (s - x)) \, dl \tag{5.6}$$

where $\kappa(x)$ is a factor that takes into account the composition of the medium at point x (Cross 1968). The superposition integral becomes

$$D(x) = \frac{\iiint \tilde{A}(s)\lambda(x)\lambda_{xs}^2 h_w(\lambda_{xs} |x - s|) \, ds}{\rho(x)} \tag{5.7}$$

The CC approximation has been proposed (Sanchez-Garcia et al. 2014) to accelerate the computation of Equation 5.7. Figure 5.2 illustrates the method, consisting of the discretization of the kernel in M spherical cones, and making the approximation that energy deposition takes place along the cone axis only. For each spherical cone, covering a solid angle Ω_i, the CC kernel, $k(r)$, describing the energy transport along the axis, can be written as

$$k(r) = \iint_{\Omega_i} r^2 h(r) \sin\theta \, d\theta \, d\varphi \tag{5.8}$$

Therefore, the dose along each cone axis becomes

$$D(x) = \frac{\int_0^x \tilde{A}(s)\lambda(x)k(\lambda_{xs} |x - s|) \, ds}{\rho(x)} \tag{5.9}$$

The final dose distribution is obtained by summing the dose distributions along each cone axis.

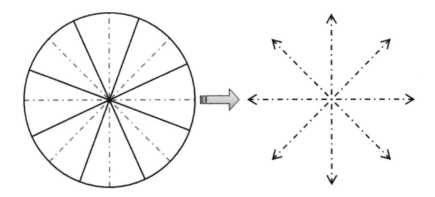

FIGURE 5.2 Illustration of the CC approximation in two dimensions. The circle (i.e., sphere) around a point source is divided into sectors (i.e., spherical cones). In the CC approximation, all energy deposition in the surface (i.e., volume) contained by a sector (i.e., cone) is considered to happen along the sector (i.e., cone) axis.

5.4 DETERMINATION OF TISSUE COMPOSITION FROM MEDICAL IMAGING

Medical imaging modalities like CT, magnetic resonance imaging (MRI), and ultrasound can be used for the estimation of tissue composition. Most of this section is dedicated to CT as it provides information that can be directly used for radiation treatment planning. Direct usage of MRI and ultrasound for the estimation of tissue composition is complicated by the non-quantitative character of these modalities (more information is provided in Section 5.4.6).

The task of a material decomposition is to quantify the amount of individual substances in a material (see Section 5.4.1). In case of CT, material decomposition can be done in the projection space (Section 5.4.2) and image space (Section 5.4.3). Combinations of these methods are also possible.

A spectral CT (Heismann et al. 2012) with N energy channels can provide N independent parameters describing the material of each voxel, since each energy channel provides one equation linking the linear attenuation coefficient (LAC) of the material with the unknown parameters. Typically used parameters are, for instance, the mass density ρ and mass fractions w_i of N base materials. In this case, we have N independent parameters as the mass fractions must fulfil the normalization condition $w_1 + \cdots + w_N = 1$, which follows from the conservation of mass. Several schemes used in practice are described in Section 5.4.3.1.

The energy range of photons used in brachytherapy is from several keV to several hundred keV. Such photons interact with a photoelectric effect, incoherent (Compton) scattering, and coherent (Rayleigh) scattering. The photoelectric effect dominates at low energies, while the Compton scattering dominates at energies larger than tens of keV. This limit, however, depends on the atomic number of the material as the macroscopic cross section for Compton scattering and photoelectric effect depend on Z and Z^n, respectively, where n is between 4 and 5 (Attix 1986). The strong dependence of the photoelectric effect on Z is utilized in material decomposition based on the effective Z of the material (Section 5.4.3.2).

5.4.1 Material Composition

A chemical compound is an entity consisting of two or more different atoms that associate via chemical bonds. For instance, a water molecule consists of two atoms of hydrogen and one atom of oxygen. As the mass of one entity (e.g., a molecule), $m(X)$, is an inconveniently small number, the mass of 1 mol of entities is used in practice. This so-called molar mass, $M(X)$, can be calculated as $M(X) = N_A m(X)$, where $N_A \cong 6.022 \times 10^{23}$ mol^{-1} is the Avogadro's constant (Mohr et al. 2012). In case of an entity consisting of a_i atoms of element X_i, the molar mass is

$$M(X) = \sum_{i=1}^{N} a_i M(X_i) = \sum_{i=1}^{N} a_i A_r(X_i) M_u \qquad (5.10)$$

where $M(X_i)$ and $A_r(X_i)$ are the molar mass and relative atomic mass of element X_i, respectively, $M_u = 10^{-3}$ kg/mol is the molar mass constant and N is the number of elements in the entity. For instance, $M(H_2O) = 2M(H) + 1M(O) = 2 \cdot 1.008 + 1 \cdot 15.995 = 18.011$ g/mol. Mass fraction, w_i, of element X_i in the compound X is then

$$w_i = a_i \frac{M(X_i)}{M(X)} \tag{5.11}$$

For instance, mass fractions of H and O in the water molecule are $2 \cdot 1.008/18.011 = 0.11$ and $1 \cdot 15.995/18.011 = 0.89$, respectively.

A mixture is a material system made up of two or more different substances, which are mixed but not combined chemically. Mixtures are typically characterized by mass fractions of individual components, $w_{\text{mixture}}(\text{component}_i)$. If elemental compositions of these components are known, then the mass fraction $w_{\text{mixture}}(X)$ of an element X in the mixture can be calculated as

$$w_{\text{mixture}}(X) = \sum_i w_{\text{mixture}}(\text{component}_i)w_{\text{component}_i}(X) \tag{5.12}$$

where $w_{\text{component}_i}(X)$ is the mass fraction of element X in the ith component.

In an independent atom approximation, chemical bonds are neglected and the material properties are calculated from properties of individual atoms. For instance, the mass attenuation coefficient of the mixture, μ_m, is estimated as

$$\mu_m = \sum_{i=1}^{n} w_i\mu_{m,i} \tag{5.13}$$

where w_i and $\mu_{m,i}$ are the mass fraction and mass attenuation coefficient of the ith component, respectively. Similar formulas can be written for the mass–energy absorption or mass–energy transfer coefficients. Elemental mass fractions of body tissues are available as tabulated data in ICRU (1992) and Woodard and White (1986).

5.4.2 Material Decomposition in Projection Space

5.4.2.1 Alvarez–Macovski Method

Alvarez and Macovski (1976) presented a method where the material is decomposed into two base materials. The attenuation A of an x-ray beam in a CT scanner can be described as

$$A = \int_E s(E)\exp\left(-\int_L \mu(E,\vec{r})d\vec{r}\right)dE \tag{5.14}$$

where $s(E)$ describes the normalized energy spectrum weighted with the detector efficiency and $\mu(E,\vec{r})$ is the energy-dependent LAC at position $\vec{r}$. The line integral is calculated from the focal spot of the x-ray tube to the detector element.

Consider that the LAC at the position $\vec{r}$ is written as

$$\mu(E,\vec{r}) = c_1(\vec{r})f_1(E) + c_2(\vec{r})f_2(E) \tag{5.15}$$

where $c_i(\vec{r})$ and $f_i(E)$, $1 \le i \le 2$, are functions of position and energy, respectively. For two scans with known weighted spectra $s_1(E)$ and $s_2(E)$, the measured attenuation can be written as

$$\begin{pmatrix} A_1 \\ A_2 \end{pmatrix} = \begin{pmatrix} \int_E s_1(E)[\exp(-f_1(E)]^{C_1}[\exp(-f_2(E)]^{C_2}\,dE \\ \int_E s_2(E)[\exp(-f_1(E)]^{C_1}[\exp(-f_2(E)]^{C_2}\,dE \end{pmatrix} \tag{5.16}$$

where the ray-specific energy-independent coefficients C_1 and C_2 are

$$\begin{pmatrix} C_1 \\ C_2 \end{pmatrix} = \begin{pmatrix} \int_L c_1(\vec{r})\,d\vec{r} \\ \int_L c_2(\vec{r})\,d\vec{r} \end{pmatrix} \tag{5.17}$$

For each ray, Equation 5.16 gives the coefficients C_1 and C_2. Coefficients $c_1(\vec{r})$ and $c_2(\vec{r})$ can then be obtained from Equation 5.17 using a standard reconstruction algorithm, for instance, the filtered backprojection.

A common approach is the decomposition to photoelectric effect and Compton scattering base functions, $\mu(E,\vec{r}) = c_p(\vec{r})f_p(E) + c_c(\vec{r})f_c(E)$. The functions $f_p(E)$ and $f_c(E)$ describe the energy dependence of the normalized cross sections. The main disadvantage of this approach is that reconstructed LACs may be biased as the coherent scattering is neglected in Equation 5.15. An alternative approach is to choose LACs for base materials, $f_i(E) = \mu_i(E)$, for instance, for water and compact bone. The presence of other bases in the imaged object will, however, bias the reconstructed LACs. The main advantage of the Alvarez–Macovski method is that it automatically accounts for the energy spectrum; the reconstructed LACs are free of beam-hardening artifacts.

The Alvarez–Macovski method can be used for the determination of the effective atomic number (Ying et al. 2006)

$$Z_{ac}(\vec{r}) = K\left(\frac{c_p(\vec{r})}{c_c(\vec{r})}\right)^{1/m} \tag{5.18}$$

where K and m are constants determined by calibration and m is between 3 and 4. The usage of the effective atomic number for material decomposition is described in Section 5.4.4.2. Note that it is also possible to apply the two- and three-material decomposition methods (Section 5.4.3.1) on the reconstructed images.

5.4.3 Material Decomposition in Image Space

Material decomposition in image space is performed on reconstructed data. A segmentation of the reconstructed data into individual tissues or organs prior to the material decomposition allows decomposition into organ-specific material bases. For instance, a muscle may be decomposed into lipid, protein, and water triplet, while the prostate may be decomposed into the average prostate tissue and calcium doublet. Such approach is difficult for the material decomposition in projection space.

5.4.3.1 Decomposition into Base Materials

In multienergy CT with N photon energies $E_1, \dots, E_N$, the mixture rule in Equation 5.13 results in N equations, each for a single energy E_j

$$\frac{\mu(E_j)}{\rho} = \sum_{i=1}^{N} w_i \mu_{m,i}(E_j) \tag{5.19}$$

Equation 5.19 together with the normalization condition for mass fractions w_i

$$\sum_{i=1}^{N} w_i = 1 \tag{5.20}$$

form a system of $N + 1$ linear equations for $N + 1$ variables $1/\rho, w_1, \dots, w_N$.

If molar volumes of individual components are conserved in the mixture, then the mass density of the mixture, ρ, can be calculated as

$$\frac{1}{\rho} = \sum_{i=1}^{N+1} \frac{w_i}{\rho_i} \tag{5.21}$$

where ρ_i is the mass fraction of the ith component. The additional Equation 5.21 allows to increase the number of mass fractions w_i in Equations 5.19 and 5.20 from N to $N + 1$. Such equations combined with Equation 5.21 then form a system of $N + 2$ linear equations for $N + 2$ variables $1/\rho, w_1, \dots, w_{N+1}$.

In single-energy CT, the number of photon energies is $N = 1$. Typically, the range of CT numbers is divided into groups corresponding to different materials. A voxel with a CT number gets elemental composition of the corresponding group, the mass density ρ of the mixture in the group can be determined from Equation 5.19 which has the form $\mu = \rho\mu_{m,1}$. Some MC codes allow the definition of materials with fixed values of mass density only. In this case, the mass density can be a constant within the group defined by the CT range. For instance, DeMarco et al. (1998) used six groups representing air, lung, fat, water, muscle, and bone. These six primary groups were further subdivided into 17 material designations with different mass densities. Schneider et al. (2000) used the additional condition for mass density in Equation 5.21; the mass fraction of the first tissue was calculated as

$$w_1 = \frac{\rho_1(H_2 - H)}{(\rho_1 H_2 - \rho_2 H_1) + (\rho_2 - \rho_1)H}$$ (5.22)

where H is the CT number of the voxel and ρ_i and H_i are the mass density and CT number, respectively, of the ith tissue ($1 \leq i \leq 2$). The mass fraction of the second tissue was calculated as $w_2 = 1 - w_1$.

In dual-energy CT (DECT), the number of photon energies is $N = 2$. A decomposition based on Equations 5.19 and 5.20 is often referred to as the two-material decomposition in DECT. It is particularly useful for the identification of contrast agents, for example, iodine. In case of human tissues, it can be used, for instance, for the decomposition of bones to a compact bone and bone marrow doublet, or a prostate tissue can be decomposed to a reference prostate tissue and calcium. Mashouf et al. (2014) describe an application of the two-material decomposition for the determination of the heterogeneity correction factor in brachytherapy. The system of Equations 5.19 was formulated in terms of CT numbers. Water and polyethylene were used as base materials; their CT numbers were determined experimentally. The decomposition based on Equations 5.19 through 5.21 is often referred to as the three-material decomposition in DECT. Malusek et al. (2013) evaluated the possibility of decomposing soft tissues to the lipid, protein, and water triplet. Liu et al. (2009) modified the three-material decomposition method in DECT so that the mass density is derived from the images and not from Equation 5.21, which requires the conservation of molar volumes.

5.4.3.2 Decomposition to Effective Atomic Number and Electron Density in DECT

The LAC, $\mu(E, \vec{r})$, can be written as

$$\mu(E, \vec{r}) = \rho(\vec{r})\mu_m(E, Z(\vec{r}))$$ (5.23)

where $\rho(\vec{r})$ is mass density and $\mu_m(E, Z(\vec{r}))$ is the mass attenuation coefficient that depends on photon energy E and the atomic number Z of the material. Both the mass density and atomic number are functions of position. Heismann et al. (2012) present a solution of Equation 5.23 for the parameterization

$$\mu(E) = \rho\left(\alpha\frac{Z^k}{E^l} + \beta\right)$$ (5.24)

with the constants α, $\beta \approx 0.02$ kg/cm³, $k \approx 3$, and $l \approx 3$ in DECT. For brevity, the dependence of $\mu(E)$, ρ, and Z on $\vec{r}$ is not explicitly mentioned. The solution is given both for monoenergetic E_1 and E_2 and for energy spectra s_1 and s_2. The authors claim that applicability of the parameterization in Equation 5.24 is limited. In practical applications, tables of $\mu_m(E, Z)$ can be used to obtain ρ and Z.

A slightly modified approach is to use electron density $n_e(\vec{r})$ instead of the mass density $\rho(\vec{r})$ in Equation 5.23. In the method by Bazalova et al. (2008), the LAC is approximated by

$$\mu(E) = n_e[Z^4 F(E,Z) + G(E,Z)] \tag{5.25}$$

where $n_e Z^4 F(E,Z)$ and $n_e G(E,Z)$ are the photoelectric and combined Rayleigh and Compton scattering terms, which are obtained by quadratic fits to cross-section data. Again, the dependence of $\mu(E)$, n_e and Z on $\vec{r}$ is not explicitly mentioned. A mixture scanned using a spectrum j ($1 \leq j \leq 2$) with relative bin intensities $\omega_j(E_i)$ has the LAC

$$\mu_j = n_e \sum_{i=1}^{N_j} \omega_j(E_i)[Z^4 F(E_i,Z) + G(E_i,Z)] \tag{5.26}$$

where N_j is the number of energy bins in spectrum j. In case of DECT, Equation 5.26 is solved for Z and n_e where Z is found by an iterative algorithm from the ratio μ_1/μ_2 and n_e is then evaluated from Equation 5.26 for j equal to 1 or 2.

In the method by Bourque et al. (2014), the decomposition $u(E,\vec{r}) = \rho_e(\vec{r}) f(E, Z(\vec{r}))$, where $u \equiv \mu/\mu_w$ is done using a polynomial fit

$$f(Z) \equiv \sum_{m=1}^{M} b_m Z^{m-1} \tag{5.27}$$

the advantage of this approach is that the coefficients b_m can be obtained using an analytic formula. The effective Z and relative electron density are then obtained by assuming that the dual-energy index, $\Gamma \equiv ((u_L - u_H)/(u_L + u_H))$, (Johnson 2011) and the dual-energy ratio, $\Gamma \equiv (u_L/u_H)$, do not depend on electron density.

5.4.4 Estimation of Elemental Composition

Radiation transport codes using the MC method typically require mass density and elemental composition of each voxel.

5.4.4.1 From Base Multiplets

The elemental mass fraction of the ith element in the mixture can be calculated from known mass fractions of a multiplet from Equation 5.12.

5.4.4.2 From Relative Electron Density and Effective Atomic Number in DECT

Electron density, n_e, of a material is defined as the number of electrons per unit volume. It can be calculated as

$$n_e = \rho \sum_{i=1}^{n} w_i Z_i \frac{N_A}{M_i} \tag{5.28}$$

where ρ is the mass density of the material, N_A is the Avogadro's constant, and w_i, Z_i, and M_i are the mass fraction, atomic number, and molar mass of the ith element, respectively. Relative electron density of a material is defined as $\rho_e = n_e/n_{e,w}$, where $n_{e,w} \approx 3.343 \times 10^{23}$ cm^{-3} is the electron density of water.

Effective atomic number, Z_{eff}, of a material for a given physical effect is defined as the atomic number Z that causes the same physical effect. For the attenuation of a photon beam caused by the photoelectric effect, Z_{eff} can be calculated as

$$Z_{eff} = \sqrt[m]{\sum_{i=1}^{n} a_i Z_i^m} \tag{5.29}$$

where a_i is the atomic fraction of elements with atomic number Z_i and m has the value about 3.5 (Johns and Cunningham 1983). To account for the fact that the transport of photons is affected by more interaction processes, the authors use, for instance, $m = 3.3$ (Landry et al. 2016).

Points $X = (Z_{eff}, \rho_e)$ plotted in a scatter plot for each voxel cluster about the center for each material, X_{ref}. The distance from the center can be used for material segmentation. Landry et al. (2016) used the 2D Mahalanobis distance defined as

$$L_M = \left[(X - X_{ref})C^{-1}(X - X_{ref})^T \right]^{1/2} \tag{5.30}$$

where C is the covariance matrix of the noisy measurements.

In the method by Hünemohr et al. (2014), the mass fraction w_i of the ith element is predicted using a linear model with an interaction between Z_{eff} and ρ_e

$$w_i = a_i\rho_e + b_i Z_{eff} + c_i\rho_e Z_{eff} + d_i \tag{5.31}$$

where a_i, b_i, c_i, and d_i are coefficients. These coefficients were obtained by a linear fit to 71 tabulated tissue compositions. The Z_{eff} was calculated from Equation 5.29 for $m = 3.1$, which best-fitted the spectra of the Siemens SOMATOM Definition Flash DECT scanner. The mass density was estimated as

$$\rho = a\rho_e + b \tag{5.32}$$

where coefficients a and b were determined for two ranges of ρ_e: lung tissue range and fat–cortical bone range.

5.4.5 Image Artifact Suppression Methods

Classical filtered backprojection algorithms assume that the signal is formed by an ideal detector that registers monoenergetic primary beam only. Consequently, polyenergetic beams and scattered photons cause distortions (artifact) in the reconstructed image. For instance, a common problem in CT imaging of the prostate is beam hardening caused by pelvic bones. Simple modifications of the classical water beam-hardening algorithm do not fully mitigate the problem. Better results are achieved by algorithms that realistically model the imaging system, including the imaged object. Model-based image reconstruction (MBIR) algorithms use such models to calculate corrections to measured data (Beister et al. 2012). In iterative versions of these algorithms, the accuracy of the reconstructed data is improved in each iteration. Examples of such algorithms are ADMIRE™ (Siemens Healthcare) and VEO™ (GE Healthcare). Recent status and applications of DECT are described, for instance, in Martin et al. (2014) and De Cecco et al. (2015).

Image mixing-based methods combine intensities from several images to suppress artifacts caused by, for instance, metallic radioactive seeds (Côté et al. 2016). They are simpler compared to methods described above, but they may bias resulting intensities. Contrary to the previous methods that have to be implemented by the CT manufacturer or a skilled team of specialists, these methods can be implemented by medical physicists in clinics.

5.4.6 MRI- and Ultrasound-Based Methods

MRI can provide valuable information about the composition of tissues. Current research focuses on providing CT numbers from MRI data using the so-called pseudo- or substitute CT methods. Anatomy-based methods use deformable image registration to register an atlas MR image to a patient's MR image. The transformation is used to transfer CT-related data from the atlas to the anatomy of the patient (Hofman et al. 2008). The main disadvantages of this method are that it uses typical CT data only and thus the method may fail for atypical anatomies. Voxel-based methods assign material properties to each voxel individually, for instance, by tissue classification followed by bulk assignment of electron densities, or direct conversion of MR voxel values to CT numbers or electron density (Catana et al. 2010). The main challenge is the separation of bone and air. The traditional MR sequences typically fail to resolve these two tissues and so ultrashort echo time (UTE) sequences are used to acquire signal from the cortical bone (Johansson et al. 2011). Demol et al. (2015) showed that human tissues can be classified according to the hydrogen content, which correlates with the proton density. Quantitative MRI can, in principle, classify tissues via a threshold segmentation in the T_1, T_2 and proton density space. Soliman et al. (2015) extended this approach by proposing a framework where quantitative MRI is used for the separation of tissue, water, and fat, and susceptibility-weighted imaging with specific postprocessing techniques is used to distinguish calcifications from low-dose rate (LDR) prostate or breast seeds.

Ultrasound may prove useful for the determination of tissue composition (see, for instance, Pazinato et al. 2015). However, there have been no attempts to use it for this purpose in radiation treatment planning as many problems have to be resolved first.

5.5 COMPUTATIONAL SPEED ENHANCEMENT

While MBDCA greatly improve dose calculation accuracy, the associated algorithm complexity adds a burden of low efficiency for use in research and their adoption in clinical settings. Hence, tremendous efforts have been devoted to accelerating computations over the years.

The most straightforward approach to speed up MC-based dose calculation is to parallelize computations. MC is known as an embarrassingly parallelizable task. On a platform with multiple computational units, for example, a CPU cluster, the computation for a user-specified number of particles can be easily parallelized by having each unit be responsible for a subset of particles. The computations on different processing units are independent of each other and only at the end of the calculation will the results be aggregated. The overhead due to data communications, a typical problem encountered in parallel computation context, is therefore minimal. To date, several MC packages typically used for brachytherapy dose calculations have been executed in parallel processing fashion (Zhou and Inanc 2003, Sutherland et al. 2012, Bouzid et al. 2015) and great success has been reported. For instance, computation time for dose calculation of an ^{125}I case using EGSnrc was reduced from about an hour to only a few minutes using a cluster with 30 cores (Sutherland et al. 2012).

One limitation for dose calculation on a CPU cluster comes from the facility requirement. The cost and efforts for high-performance computing facility deployment and maintenance may be a concern for a resource-limited clinic. Recently, a new parallel processing platform called GPU has been employed to accelerate MC dose calculations (Jia et al. 2014). GPU was initially designed as a coprocessor in a desktop computer system to handle graphics-related problems. Nowadays, with its tremendous processing power offered by typically thousands of processors on a single chip, GPU has been recognized as a powerful tool to tackle computationally intensive problems in scientific computing. The needs for advanced GPU in sophisticated graphics problems, for example, computer gaming, have also led to continuous increase of processing power at a reduced cost.

GPU-based MC particle transport simulations in radiation therapy have experienced tremendous advancements over the past several years. bGPUMCD was the first package developed for brachytherapy dose calculations (Hissoiny et al. 2012). In this package, photon transport in the relevant energy range was modeled and electron transport was ignored. A track length estimator was employed to reduce the required number of particles to achieve a satisfactory uncertainty level. bGPUMCD was initially developed for HDR brachytherapy dose calculations in a voxelized patient geometry. A phase-space file-based source model was employed. Later, dose calculations in LDR brachytherapy applications were supported. Particle transport in mixed voxelized and parameterized geometry were enabled to handle dose calculations in the presence of radioactive seeds (Bonenfant et al. 2015). In terms of parallel processing scheme, a number of GPU threads simultaneously transport multiple photons, yielding substantial speed-up compared to conventional CPU-based computations. The computation time to reach 1% uncertainty was only a few seconds for an HDR case and approximately 30 s for LDR. A similar package gBMC was developed for HDR dose calculations (Tian et al. 2015). The computation time was found to be several seconds to reach 1% uncertainty using an analog dose estimator. The high efficiency in these packages has indicated their potential for routine clinical use. In addition, they also greatly facilitated research investigations. For instance, gBMC was used to comprehensively evaluate dose calculation accuracy in an accelerated partial breast irradiation (APBI) brachytherapy application (Szeliski 2010).

Another computationally intensive algorithm is GBBS (Vassilov et al. 2008). At present, parallel computation of this algorithm was enabled in the commercial software Acuros BV in the Varian Brachyvision system. With multiple CPU cores, dose calculation time was a few minutes for a typical case. There is no reported work to implement GBBS on the GPU platform. However, this algorithm in principle could be further accelerated by GPU-based parallel processing. This is because the GBBS algorithm contains a lot of matrix–vector operations during the iterative process to solve the Boltzmann equation. These operations are highly compatible with GPU's single-instruction-multiple-data structure, and hence may greatly benefit from this platform. Regarding CC algorithm, it has been accelerated using the GPU technique in external-beam radiotherapy context (Hissoiny et al. 2009, Jacques et al. 2010, Zhou et al. 2010, Chen et al. 2011). The commercial ACE software in the Elekta Oncentra Brachy system utilises a GPU implementation for ^{192}Ir brachytherapy, see Ahnesjö et al. (2017).

5.6 CONCLUSION

MC and other MBDCAs based on first-principles physical models are available for dose calculations in brachytherapy. Explicit versions are available for clinical use with the common isotope ^{192}Ir. While a transition from today's methods into MBDCAs at the high energy of ^{192}Ir is relatively straightforward and the use of CT information for tissue composition feasible although possible to improve, the situation at low energies is more difficult. The use of MBDCAs at lower photon energies (<50 keV, seeds and electronic miniature x-ray sources) is from the point of view of dosimetric accuracy very important as individual tissue heterogeneities affect values of absorbed dose significantly. Hitherto, however, the use of MBDCA has been prohibited at these energies by problems in associated techniques such as accuracy in obtaining quantitative information on tissue atomic composition from patient imaging and lack of insight into most suitable dose reporting quantity. Even after having agreed upon a dose reporting quantity and with an interim solution in the form of tissue segmentation based on organ contouring, the differences in calculated values of absorbed dose values at low energies are so significant that a recalibration in prescribed doses would be needed and studies using large groups of retrospective data necessary in this low energy domain.

Advanced dose calculation methods create a need for increased computational efficiency to make clinical treatment planning applications possible. This is today best accomplished by the use of GPU techniques. Imaging is a key component for improved dosimetry in both conventional and microsphere brachytherapy. Increased accuracy in dosimetry is of utmost importance in introducing new techniques, applicators, and new sources (as covered in detail in Chapter 2) as these applications often go outside the domain where today's dose–response experience were gained.

REFERENCES

Afsharpour H, Landry G, D'Amours A, Enger S A, Reniers B, Poon E, Carrier J-F, Verhaegen F and Beaulieu L. 2012. ALGEBRA: Algorithm for the heterogeneous dosimetry based on GEANT4 for brachytherapy. *Phys. Med. Biol.* 57, 3273–80.

Ahnesjö A. 1989. Collapsed cone convolution of radiant energy for photon dose calculation in heterogeneous media. *Med. Phys.* 16, 577–92.

Ahnesjö A, Andreo P and Brahme A. 1987. Calculation and application of point spread functions for treatment planning with high energy photon beams. *Acta Oncol.* 26, 49–55.

Ahnesjö A and Aspradakis M M. 1999. Review: Dose calculations for external photon beams in radiotherapy. *Phys. Med. Biol.* 44, R99–155.

Ahnesjö A, van Veelen B and Carlsson Tedgren Å. 2017. Collapsed cone dose calculations for heterogeneous tissues in brachytherapy using primary and scatter separation source data. *Comput. Methods Prog. Biomed.* 139, 17–29.

Alvarez R E and Macovski A. 1976. Energy-selective reconstructions in x-ray computerised tomography. *Phys. Med. Biol.* 21, 733–44.

Attix F H. 1986. *Introduction to Radiological Physics and Radiation Dosimetry*, New York: Wiley & Sons.

Ballester F, Vijande J, Carlsson Tedgren Å, Granero D, Haworth A, Smith R, Mourtada F et al. 2015. A generic high-dose-rate Ir-192 brachytherapy source for evaluation of model-based dose calculations beyond the TG-43 formalism. *Med. Phys.* 42, 3048–62.

Bazalova M, Carrier J-F, Beaulieu L and Verhaegen F. 2008. Dual-energy CT-based material extraction for tissue segmentation in Monte Carlo dose calculations. *Phys. Med. Biol.* 53, 2439–56.

Beaulieu L, Carlsson Tedgren Å, Carrier J-F, Davis S D, Mourtada F, Rivard M J, Thomson R M, Verhaegen F, Waring T and Williamson J F. 2012. Report of the Task Group 186 on model-based dose calculation methods in brachytherapy beyond the TG-43 formalism: Current status and recommendations for clinical implementation. *Med. Phys.* 39, 6208–36.

Beister M, Kolditz D and Kalender W A. 2012. Iterative reconstruction methods in X-ray CT. *Phys. Med.* 28, 94–108.

Bolch W E, Bouchet L G, Robertson J S, Wessels B W, Siegel J A, Howell R W, Erdi A K et al. 1999. MIRD pamphlet No. 17: The dosimetry of nonuniform activity distributions radionuclide S values at the voxel level. Medical Internal Radiation Dose Committee. *Eur. J. Nucl. Med.* 40, 11S–36S, Online: http://www.ncbi.nlm.nih.gov/pubmed/99350833.

Bolch W E, Eckerman K F, Sgouros G and Thomas S R. 2009. MIRD pamphlet No. 21: A generalized schema for radiopharmaceutical dosimetry-standardization of nomenclature. *J. Nucl. Med.* 50, 477–84, Online: http://www.ncbi.nlm.nih.gov/pubmed/19258258.

Bonenfant É, Magnoux V, Hissoiny S, Ozell B, Beaulieu L and Despres P. 2015. Fast GPU-based Monte Carlo simulations for LDR prostate brachytherapy. *Phys. Med. Biol.* 60, 4973–86.

Botta F, Mairani A, Battistoni G, Cremonesi M, Di Dia A, Fásso A, Ferrari A et al. 2011. Calculation of electron and isotope dose point kernels with Fluka Monte Carlo code for dosimetry in nuclear medicine therapy. *Med. Phys.* 38, 3944.

Bourque A E, Carrier J-F and Bouchard H. 2014. A stoichiometric calibration method for dual energy computed tomography. *Phys. Med. Biol.* 59, 2059.

Bouzid D, Bert J, Dupre P F, Benhalouche S, Pradier O, Boussin N and Visvikis D. 2015. Monte-Carlo dosimetry for intraoperative radiotherapy using a low energy x-ray source. *Acta Oncol.* 54, 1788–95.

Carlsson Å K and Ahnesjö A. 2000a. Point kernels and superposition methods for scatter dose calculations in brachytherapy. *Phys. Med. Biol.* 45, 357–82.

Carlsson Å K and Ahnesjö A. 2000b. The collapsed cone superposition algorithm applied to scatter dose calculations in brachytherapy. *Med. Phys.* 27, 2320–32.

Carlsson Tedgren Å and Ahnesjö A. 2008. Optimization of the computational efficiency of a 3D collapsed cone dose calculation algorithm for brachytherapy. *Med. Phys.* 35, 1611–8.

Carlsson Tedgren Å and Alm Carlsson G. 2013. Specification of absorbed dose to water using model based dose calculation algorithms for treatment planning in brachytherapy. *Phys. Med. Biol.* 58, 2561–79.

Carlsson Tedgren Å, Verhaegen F and Beaulieu L. 2012. On the introduction of model-based algorithms performing nonwater heterogeneity corrections into brachytherapy treatment planning. In: *Comprehensive Brachytherapy: Physical and Clinical Aspects*, eds. Venselaar J,

Baltas D, Meigooni A S, Hoskin P J, Chapter 11, Boca Raton, London, New York: CRC Press, Taylor & Francis Group, pp. 145–60.

Carlsson Tedgren Å K and Ahnesjö A. 2003. Accounting for high Z shields in brachytherapy using collapsed cone superposition for scatter dose calculation. *Med. Phys.* 8, 2206–17.

Catana C, van der Kouwe A, Benner T, Michel C J, Hamm M, Fenchel M, Fischl B, Rosen B, Schmand M and Sorensen A G. 2010. Toward implementing an MRI-based PET attenuation-correction method for neurological studies on the MR-PET brain prototype. *J. Nucl. Med.* 51, 1431–8.

Chen Q, Chen M and Lu W. 2011. Ultrafast convolution/superposition using tabulated and exponential kernels on GPU. *Med. Phys.* 38, 1150–61.

Chetty I J, Curran B, Cygler J E, DeMarco J J, Ezzell G, Faddegon B, Kawrakow I et al. 2007. Report of the AAPM Task Group No. 105: Issues associated with clinical implementation of Monte Carlo-based photon and electron external beam treatment planning. *Med. Phys.* 34, 4818–53.

Chiavassa S, Aubineu-Lanièce I, Bitar A, Lisbona A, Barbet J, Franck D, Jourdain J R and Bardiés M. 2006. Validation of a perzonalized dosimetric evaluation tool (Oidepe) for targeted radiotherapy based on the Monte Carlo MCNPX code. *Phys. Med. Biol.* 51, 601–16.

Chibani O and Williamson J F. 2005. MCPI: A sub-minute Monte Carlo dose calculation engine for prostate implants. *Med. Phys.* 32, 3688–98.

Côté N, Bedwani S and Carrier J-F. 2016. Improved tissue assignment using dual-energy computed tomography in low-dose rate prostate brachytherapy for Monte Carlo dose calculation. *Med. Phys.*, 43, 2611–2618.

Cross W G. 1968. Variation of beta dose attenuation coefficient in different media. *Phys. Med. Biol.* 13, 611–8.

Daskalov G M, Baker R S, Little R C, Rogers D W O and Williamson J F. 2000. Two-dimensional discrete ordinates photon transport calculations for brachytherapy dosimetry applications. *Nucl. Sci. Eng.* 134, 121–34.

De Cecco C N, Laghi A, Schoepf U J and Meinel F. 2015. *Dual Energy CT in Oncology*, Cham, Switzerland: Springer International Publishing.

DeMarco J J, Solberg T D and Smathers J B. 1998. A CT-based Monte Carlo simulation tool for dosimetry planning and analysis. *Med. Phys.* 25, 1–11.

Demol B, Viard R and Reynaert N. 2015. Monte Carlo calculation based on hydrogen composition of the tissue for MV photon radiotherapy. *J. Appl. Clin. Med. Phys.* 16, 117–30.

Dewaraja Y K, Frey E C, Sgouros G, Brill A B, Robertson P, Zanzonico P B and Ljungberg M. 2012. MIRD pamphlet No. 23: Quantitative SPECT for patient-specific 3-dimensional dosimetry in internal radionuclide therapy. *J. Nucl. Med.* 53, 1310–25.

Ferrer L, Chouin N, Bitar A, Lisbona A and Bardiés M. 2007. Implementing dosimetry in GATE: Dose-point kernel validation with GEANT4 4.8.1. *Cancer Biother. Radiopharm.* 22, 125–9.

Fonseca G P, Carlsson Tedgren Å, Reniers B, Nilsson J, Persson M, Yoriyaz H and Verhaegen F. 2015. Dose specification for 192Ir high dose rate brachytherapy in terms of dose-to-water-in-medium and dose-to-medium-in-medium. *Phys. Med. Biol.* 60, 4565–79.

Gardin I, Bouchet L G, Assié K, Caron J, Lisbona A, Ferrer L, Bolch W E and Vera P. 2003. Voxeldose: A computer program for 3-D dose calculation in therapeutic nuclear medicine. *Cancer Biother. Radiopharm.* 18, 109–15.

Gifford K, Price M J, Horton J L, Wareing T A and Mourtada F. 2008. Optimization of deterministic transport parameters for the calculation of dose distributions around a high dose-rate 192Ir brachytherapy source. *Med. Phys.* 35, 2279–85.

Guy M, Flux G D, Papavasileiou P, Flower M A and Ott R J. 2003. RMDP: A dedicated package for ^{131}I SPECT quantification, registration and patient-specific dosimetry. *Cancer Biother. Radiopharm.* 18, 61–9.

Hedtjärn H, Alm Carlsson G and Williamson J F. 2002. Accelerated Monte Carlo based dose calculations for brachytherapy planning using correlated sampling. *Phys. Med. Biol.* 47, 351–76.

Heismann B J, Schmidt B T and Flohr T. 2012. *Spectral Computed Tomography*, Bellingham, Washington: SPIE (Society).

Hippeläinen E, Tenhunen M and Solhlberg A. 2015. Fast voxel-level dosimetry ^{177}Lu labelled peptide treatments. *Phys. Med. Biol.* 60, 6685–700.

Hissoiny S, D'Amours A, Ozell B, Despres P and Beaulieu L. 2012. Sub-second high dose rate brachytherapy Monte Carlo dose calculations with bGPUMCD. *Med. Phys.* 39, 4559–67.

Hissoiny S, Ozell B and Despres P. 2009. Fast convolution-superposition dose calculation on graphics hardware. *Med. Phys.* 36, 1998–2005.

Ho S, Lau Y W, Leung T W, Chan M, Ngar Y K, Johnson P J and Li A K. 1996. Partition model for estimating radiation doses from yttrium-90 microspheres in treating hepatic tumours. *Eur. J. Nucl. Med.* 39, 947–52, Online: http://www.ncbi.nlm.nih.gov/pubmed/8753684.

Hofman M, Steinke F, Scheel V, Charpiat G, Furquhar J, Aschoff P, Brady M, Scholkopf B and Pichler B J. 2008. MRI-based attenuation correction for PET/MRI: A novel approach combining pattern recognition and atlas registration. *J. Nucl. Med.* 49, 1875–83.

Hünemohr N, Paganetti H, Greilich S, Jäkel O and Seco J. 2014. Tissue decomposition from dual energy CT data for MC based dose calculation in particle therapy. *Med. Phy.* 41, 61714, doi:10.1118/1.4875976.

IAEA. 2014. *Quantitative Nuclear Medicine Imaging: Concepts, Requirements and Methods Ed. International Atomic Energy Agency (IAEA)*. Vienna, Austria: IAEA, Online: http://www-pub.iaea.org/books/IAEABooks/10380/Quantitative-Nuclear-Medicine-Imaging-Concepts-Requirements-and-Methods.

ICRU. 1992. Photon, Electron, Proton and Neutron Interaction Data for Body Tissues. ICRU Report 46. International Commission on Radiation Units and Measurements, Bethesda, Maryland.

Jacques R, Taylor R, Wong J and McNutt T. 2010. Towards real-time radiation therapy: GPU accelerated superposition/convolution. *Comput. Methods Programs Biomed.* 98, 285–92.

Jia X, Ziegenhein P and Jiang S B. 2014. GPU-based high-performance computing for radiation therapy. *Phys. Med. Biol.* 59, R151.

Johansson A, Karlsson M and Nyholm T. 2011. CT substitute derived from MRI sequences with ultrashort echo time. *Med. Phys.* 38, 2708–14.

Johns H E and Cunningham J R 1983. *The Physics of Radiology*, Springfield, IL, USA: Charles C Thomas Publisher. ISBN-13: 978-0398046699.

Johnson T. 2011. *Dual Energy CT in Clinical Practice*, Berlin: Springer.

Kennedy A, Nag S, Salem R, Murthy R, McEwan A J, Nutting C, Benson A et al. 2007. Recommendations for radioembolization of hepatic malignancies using yttrium-90 microsphere brachytherapy: A consensus panel report from the radioembolization brachytherapy oncology consortium. *Int. J. Radiat. Oncol. Biol. Phys.* 68, 13–23.

Kost S D, Dewaraja Y K, Abramson R G and Stabin M. 2015. VIDA: A voxel-based dosimetry method for targeted radionuclide therapy using Geant4. *Cancer Biother. Radiopharm.* 30, 16–26.

Landry G, Gaudreault M, van Elmpt W, Wildberger J E and Verhaegen F. 2016. Improved dose calculation accuracy for low energy brachytherapy by optimizing dual energy CT imaging protocols for noise reduction using sinogram affirmed iterative reconstruction. *Z. Med. Phys.* 26(1), 75–87.

Landry G, Rivard M, Williamson J F and Verhaegen F. 2013. Monte carlo methods and applications for brachytherapy dosimetry and treatment planning. In: *Monte Carlo Techniques in Radiation Therapy*, ed. Hendee W, Boca Raton, Florida: CRC Press, Taylor & Francis.

Liu X, Yu L, Primak A N and McCollough C H. 2009. Quantitative imaging of element composition and mass fraction using dual-energy CT: Three material decomposition. *Med. Phys.* 36, 1602.

Ljungberg M, Sjögreen K, Liu X, Frey E C, Dewaraja Y K and Strand S-E. 2002. A 3-dimensional absorbed dose calculation method based on quantitative SPECT for radionuclide therapy: Evaluation for ^{131}I using Monte Carlo simulation. *J. Nucl. Med.* 43, 1101–9.

Loevinger R, Budinger T and Watson E. 1991. *MIRD Primer for Absorbed Dose Calculations*, Revised Edition, New York: Society of Nuclear Medicine.

Ma Y, Lacroix F, Lavallée M-C and Beaulieu L. 2015. Validation of the Oncentra brachy advanced collapsed cone engine for a commercial 192Ir source using heterogeneous geometries. *Brachytherapy* 14, 939–52.

Mainegra-Hing E, Kawrakow I and Rogers D W O. 2005. Calculation of photon energy deposition kernels and electron point kernels in water. *Med. Phys.* 32, 685–99.

Malusek A, Karlsson M, Magnusson M and Alm Carlsson G. 2013. The potential of dual-energy computed tomography for quantitative decomposition of soft tissue to water, protein and lipid in brachytherapy. *Phys. Med. Biol.* 58, 771.

Martin D, Boll D T, Mileto A and Nelson R C. 2014. State of the art: Dual-energy CT of the abdomen. *Radiology* 271, 327–42.

Mashouf S, Lechtman E, Lai P, Keller B M, Karotki A, Beachey D J and Pignol J P. 2014. Dose heterogeneity correction for low-energy brachytherapy sources using dual-energy CT images. *Phys. Med. Biol.* 59, 5305–16.

Mikell J K, Klopp A H, Gonzales G M, Kisling K D, Price M J, Berner P A, Eifel P J and Mourtada F. 2012. Impact of heterogeneity-based dose calculation using a deterministic grid-based Boltzmann equation solver for intracavitary brachytherapy. *Int. J. Radiat. Oncol. Biol. Phys.* 83, 417–22.

Mohr P J, Taylor B N, Newell D B. 2012. CODATA recommended values of the fundamental physical constants: 2010. *Journal of Physical and Chemical Reference Data* 41, 043109. doi:10.1063/1.4724320.

Nath R, Peschel R E, Park C H and Fischer J J. 1988. Development of an [241]Am applicator for intracavitary irradiation of gynecologic cancers. *Int. J. Radiat. Oncol. Biol. Phys.* 14, 969–78.

Papagiannis P, Pantelis E and Karaiskos P. 2014. Current state of the art brachytherapy treatment planning dosimetry algorithms. *Br. J. Radiol.* 87, 20140163.

Pazinato D, Stein B, Almeida W, Werneck R, Mendes Júnior P, Penatti O, Torres R, Menezes F H and Rocha A. 2016. Pixel-level tissue classification for ultrasound images. *IEEE J. Biomed. Health Inform.* 20(1), 256–67.

Prideaux A R, Song H, Hobbs R F, He B, Frey E C, Ladenson P W, Wahl R L and Sgouros G. 2007. Three-dimensional radiobiologic dosimetry: Application of radiobiological modeling to patient-specific 3-dimensional imaging-based internal dosimetry. *J. Nucl. Med.* 48, 1008–16.

Rivard M J, Coursey B M, DeWerd L A, Hanson W F, Huq M S, Ibbott G S, Mitch M G, Nath R and Williamson J F. 2004. Update of AAPM Task Group No. 43 Report: A revised AAPM protocol for brachytherapy dose calculations. *Med. Phys.* 31, 633–74.

Rivard M J, Venselaar J and Beaulieu L. 2009. The evolution of brachytherapy treatment planning. *Med. Phys.* 36, 2136–53.

Russell K R, Carlsson Tedgren Å K and Ahnesjö A. 2005. Brachytherapy source characterization for improved dose calculations using primary and scatter dose separation. *Med. Phys.* 32, 2739–52.

Sanchez-Garcia M, Gardin I, Lebtahi R and Dieudonné A. 2014. A new approach for dose calculation in targeted radionuclide therapy (TRT) based on collapsed cone superposition: Validation with [90]Y. *Phys. Med. Biol.* 59, 4769–84.

Schneider U, Bortfeld T and Schlegel W. 2000. Correlation between CT numbers and tissue parameters needed for Monte Carlo simulations of clinical dose distributions. *Phys. Med. Biol.* 45, 459–78.

Sgouros G, Barest G, Thekkumthala J, Chui C, Mohan R, Bigler R E and Zanzonico P B. 1990. Treatment planning for internal radionuclide therapy: Three-dimensional dosimetry for nonuniformly distributed radionuclides. *J. Nucl. Med.* 31, 1884–91.

Simpkin D J and Mackie T R. 1990. EGS4 Monte Carlo determination of the beta dose kernel in water. *Med. Phys.* 17, 179–86.

Soares C G, Douysset G and Mitch M G. 2009. Primary standards and dosimetry protocols for brachytherapy sources. *Metrologia* 46, S80–98.

Soliman A, Elzibak A, Fatemi A, Safigholi H, Han D, Ravi A, Morton G and Song W. 2015. TU-AB-201-11: A novel theoretical framework for MRI-only image guided LDR prostate and breast brachytherapy implant dosimetry. *Med. Phys.* 42, 3596.

Sutherland J G, Furutani K M, Garces Y I and Thomson R M. 2012. Model-based dose calculations for (125)I lung brachytherapy. *Med. Phys.* 39, 4365–77.

Szeliski R. 2010. *Computer Vision: Algorithms and Applications,* London: Springer.

Taylor R E P, Yegin G and Rogers D W O. 2007. Benchmarking BrachyDose: Voxel based EGSnrc Monte Carlo calculations of TG-43 dosimetry parameters. *Med. Phys.* 34, 445–57.

Tian Z, Scanderbeg D, Zhang M, Yashar C M and Jia X. 2015. GPU-based Monte Carlo dose calculations for high-dose-rate brachytherapy and evaluations of TG-43 accuracy in APBI treatments with SAVI devices. *Brachytherapy* 14(Suppl. 1), S23–4.

Uusijärvi H, Chouin N, Bernhardt P, Ferrer L, Bardiés M and Forssell-Aronsson E. 2009. Comparison of electron dose-point kernels in water generated by Monte Carlo codes, PENELOPE, GEANT4, MCNPX and ETRAN. *Cancer Biother. Radiopharm.* 24, 461–7.

Vassilov O N, Wareing T A, Davis I M, McGhee J, Barnett D, Horton J L, Gifford K and Failla G. 2008. Feasibility of a multigroup deterministic solution method for 3D radiotherapy dose calculations. *Int. J. Radiat. Oncol. Biol. Phys.* 72, 220–7.

Williamson J F. 1987. Monte Carlo evaluation of kerma at a point for photon transport problems. *Med. Phys.* 14, 567–76.

Williamson J F. 1988. Monte Carlo simulation of photon transport phenomena: Sampling techniques. In: *Monte Carlo Simulations in the Radiological Sciences,* ed. Morin R L, Boca Raton, Florida: CRC Press, pp. 53–102.

Williamson J F. 2006. Brachytherapy technology and physics practice since 1950: A half-century of progress. *Phys. Med. Biol.* 51, R303–25.

Williamson J F, Baker R S and Li Z. 1991. A convolution algorithm for brachytherapy dose computations in heterogeneous geometries. *Med. Phys.* 18, 1256–65.

Woodard H Q and White D R. 1986. The composition of body tissues. *Br. J. Radiol.* 59, 1209–18.

Ying Z, Naidu R, Crawford C R. 2006. Dual energy computed tomography for explosive detection. *Journal of X-ray Science and Technology* 14, 235–256.

Zhou B, Cedric X, Chen D Z and Sharon X. 2010. GPU-accelerated Monte Carlo convolution/superposition implementation for dose calculation. *Med. Phys.* 37, 5593–603.

Zhou F and Inanc F. 2003. Integral-transport-based deterministic brachytherapy dose calculations. *Phys. Med. Biol.* 48, 73–93.

Zourari K, Pantelis E, Moutsatsos A, Sakelliou L, Georgiou E, Karaiskos P and Papagiannis P. 2013. Dosimetric accuracy of a deterministic radiation transport based ^{192}Ir brachytherapy treatment planning system. Part III. Comparison to Monte Carlo simulation in voxelized anatomical computational models. *Med. Phys.* 40, 011712-1-9.

CHAPTER 6

Dose Optimization

Rob van der Laarse and Peter A.N. Bosman

CONTENTS

6.1 INTRODUCTION

Generally speaking, the goal of dose optimization in brachytherapy is to generate treatment plans that result in delivering a sufficiently high dose to the planned target volume (PTV) while sparing the surrounding organs at risk (OARs) and normal tissues (NTs) as much as possible. Thus, a treatment plan in the context of dose optimization represents a dose distribution with which the patient can be treated. In this chapter, the following terminology is used when discussing optimization. Variables are the quantities that can be varied by an optimization algorithm, parameters are the settings of the algorithm, objectives are the quantities that are optimized, and constraints are the quantities that describe which combinations of variable settings are not allowed.

The variables that can be used in order to optimize the dose distribution associated with a treatment plan differ between the different modes of brachytherapy, see Table 6.1. This table lists brachytherapy modes that have been in use for many years and also includes emerging ones. In the field of low-dose rate (LDR) seed implants of the prostate, notable novel developments are (i) chemobrachytherapy, which is the use of spacers loaded with nanoparticles containing a slow-release anticancer drug such as docetaxel (Kumar et al. 2015) and (ii) thermobrachytherapy, which is the use of seeds containing a ferromagnetic core with a sharp Curie transition near 50°C, coated with a layer containing the ^{125}I (Gautam et al. 2014). Moreover, two emerging brachytherapy treatment modes are listed in Table 6.1: electronic brachytherapy and radioembolization. Electronic brachytherapy uses a miniature x-ray source instead of a radioactive one for high-dose rate (HDR) brachytherapy (Eaton 2015). Dose optimization aspects of electronic brachytherapy are dependent on

TABLE 6.1   Degrees of Freedom in Optimizing a Treatment Plan

Mode of Brachytherapy	Variables
LDR (e.g., seed implant and iridium wire implant)	• Source positions • Strength of individual sources Additionally for seed implants: • Spacers that locally deliver an anticancer drug: chemobrachytherapy • Seeds that contain a ferromagnetic core: thermobrachytherapy
Manual afterloading (e.g., cesium tubes in gynecological cervix applications)	• Source positions • Strength of individual sources • Placement of shields to reduce dose to OAR (e.g., rectum and bladder)
Stepping source HDR/PDR (e.g., prostate, cervix, and breast)	• Dwell positions • Dwell times • Placement of shields to reduce dose to OAR (e.g., rectum and bladder)
Electronic brachytherapy	• x-ray energy • Dose rate • Focus-skin distance (if applicable) If the miniature x-ray source is positioned inside the target volume: • Dwell positions • Dwell times
Radioembolization	• Selection of blood vessel(s) to target • Number of spheres inserted (amount of radioactivity) in each blood vessel

TABLE 6.2   Differences between Distance and Volume Implants/Applications

Type of Implant/Application	Shape of NDVH Curve	Treatment Site
Distance implant	No peak	Intraluminal bronchus implant and bladder implant
Distance application	No peak	Vaginal-, rectum applicator, mold, Freiburg flap, and skin application with electronic brachytherapy
Volume implant	Peak	Prostate implant with seeds or HDR/PDR, biplane breast implant, head and neck implant
Mixed volume and distance application	Small peak	Gynecological cervix

the type of equipment. With equipment for skin treatment (Garcia-Martinez et al. 2014), no dwell time optimization takes place, while for volume implants (Rana et al. 2015), the x-ray source is stepping through one or more catheters and planning optimization techniques remain largely similar compared to radioactive sources.

Radioembolization uses microspheres loaded with a radioactive isotope (Yttrium-90, Holmium-166) for the treatment of hepatic malignancies (Smits et al. 2015). These spheres are injected into the hepatic artery and lodge in and around the tumors. Methods for the calculation of the dose around these microspheres are given in Chapter 5.

In the case of stepping source brachytherapy, the goal of optimization depends on whether the implant/application is of the distance or volume type (van der Laarse and Luthmann 2001), see Table 6.2. Optimization of a distance implant/application is aimed at obtaining the prescription dose (PD) at a given distance from the dwell positions. Optimization of a volume implant/application is aimed at shaping the prescription isodose surface to match the PTV surface while preventing large high-dose volumes around the dwell positions and minimizing dose delivered to the OARs and NTs. This difference in optimization goals translates into the natural dose volume histogram (NDVH) of a volume implant/application having a more or less pronounced peak, which is absent in the NDVH of a distance implant. A detailed introduction in the field of dose optimization in stepping source HDR brachytherapy is given in Baltas and Kolkman-Deurloo (2013). An exhaustive literature review on the same subject up to 2010 is presented by De Boeck et al. (2014).

Implants and applications are also characterized by the type of source placement in or near the PTV, see Table 6.3. Details on these implants and applications are given by Devlin (2007).

6.2 EVALUATING A TREATMENT PLAN

Optimization of a treatment plan involves finding the *best possible* values for the variables that are relevant to the mode of brachytherapy at hand (see Table 6.1). In theory, this optimization problem is well-posed and can be tackled with optimization algorithms if a treatment plan can be evaluated and scored so that plans can be compared and ranked. In practice, however, the notion of quality of a treatment plan is typically found to be a subjective measure as its definition can differ between institutions and physicians. Treatment

TABLE 6.3   Different Types of Source Placement

Type of Source Placement	Source Placement	Type of Implant/ Application	Example of Treatment Site
Interstitial	Surgically in the tumor tissue	Volume	Prostate, breast, head, and neck
Intracavitary	In a body cavity	Mixed or distance	GYN cervix and rectum
Intraluminal	In a lumen	Distance	Esophagus and bronchus
Surface	Adjacent to tumor tissue	Distance	Skin mold, Freiburg flap, and eye plaques
Intraoperative	In target during surgery,	Volume or distance	Breast and Freiburg flap
Electronic brachytherapy	X-ray source adjacent to tumor tissue, e.g., skin	Distance	Skin
	X-ray source stepping through catheter	Volume or distance	Breast
Radioembolization	Via blood stream in tumor tissue	Volume	Liver

plans could be objectively evaluated by predicting the tumor control probability (TCP) and the normal tissue control probability (NTCP) (Carlson et al. 2013). However, the field of radiobiological modeling on which the TCP and NTCP are based, has not yet advanced to a state where such predictions can be made sufficiently accurately and reliably (O'Rourke et al. 2009). Consequently, the use of TCP and NTCP for dose optimization is not yet clinically applicable on a routine basis.

Current figures of merit of a treatment plan that *are* in clinical use for a volume implant, are the quality index (QI), the uniformity index (UI), the conformation number (CN), and the conformity index (COIN), for a description of these figures of merit see Baltas and Kolkman-Deurloo (2013). However, not one of these figures of merit represents *all* aspects of a treatment plan that are of importance. So using only one of them to evaluate treatment plans is of limited value. Moreover, even *together* these figures of merit still do not allow treatment plans to be evaluated such that all aspects of importance are covered. To combat this issue, a set of DVH indices can be constructed so that each index in this set scores a specific dose distribution property of a treatment plan. Such a set is provided in Table 6.4. The indices of Table 6.4 are explained in detail in Baltas and Kolkman-Deurloo (2013). The natural prescription dose (NPD) and natural dose ratio (NDR) are introduced in Moerland et al. (2000). However, these indices still do not provide a universal solution because physicians do not agree universally on their relative importance. To improve the agreement between physicians in the future, acceptable ranges for each one of the above indices for a given tumor site should be determined first. The next step then is to determine the relative importance of the DVH indices.

Besides these well-known DVH-based indices, another property of a dose distribution that can be derived from the DVHs is important in practice: the so-called non-recoverable overdose volumes. The risk of overdose for a volume depends strongly on its shape, for example, it is higher for a spherical volume than for an elongated volume. High doses along an elongated OAR, such as the urethra in a prostate implant, are better expressed in their length along the organ than in their volume. This risk of overdose for a volume should also be quantified and taken into account.

TABLE 6.4   DVH Indices, Scoring a Single Property of a Treatment Plan of a Volume Implant

DVH Index	Expression	Dose Distribution Property
Quality index (QI)	Calculated from NDVH, QI ≥ 1	Dose homogeneity over implanted volume. Value 1 is worst case
Coverage inside index (CI)	PTV(D > PD)/PTV, CI ≤ 1	Coverage with dose > PD of dose distribution over PTV. Value 1 is optimal
Coverage outside index (CO)	PTV(D > PD)/V(D > PD), CO ≤ 1	Extension of PD outside the PTV. Value 1 is optimal
Natural prescription dose (NPD)	Located at border of dose distribution inside which the implanted volume lies and outside which the inverse square law reigns	Normalization based on the dose distribution
Natural dose ratio (NDR)	NDR = NPD/PD	Difference between prescription based on the dose distribution, NPD and prescription based on the PTV coverage, PD NDR > 1: PTV is overdosed NDR < 1: PTV is underdosed
D0.1cc, D1cc, and D2cc for each OAR	Dose enveloping the 0.1, 1, and 2 cm³ volume with the highest dose of the OAR	Volumes with highest doses to OAR

It should be noted that the issues raised above severely complicate the design of clinically useful automated dose optimization algorithms (and software tools) because there is a good chance that plans that result from automated dose optimization using alternative figures of merit do not perfectly reflect the wishes of the physician. The need to ultimately still be able to cater to these wishes makes it equally important to have (software) tools available that can be used to efficiently and effectively fine-tune plans interactively. Moreover, as long as an ultimate objective quality evaluation measure is not available for treatment plans, catering to these wishes is the *true* optimization problem being solved in daily clinical practice.

6.3 METHODS OF OPTIMIZATION

Two methods for dose optimization in brachytherapy are generally distinguished, forward and inverse (Pouliot et al. 2013). Forward methods manipulate the plan variables directly to change the dose distribution. They can, therefore, arguably also be called manual or direct methods. Inverse methods manipulate plan variables by using optimization algorithms to automatically change the variables so as to align the dose distribution as closely as possible with an *a priori* specified dose distribution. They can, therefore, arguably be called semi-automatic, or indirect, optimization methods. To change the values of the variables with these methods, the optimization problem itself must be changed, that is, the specification of the objectives and constraints of the desired dose distribution must be altered.

Forward optimization methods are mostly used when the implanted volume defines the target volume, for example, in the case of a distance implant or when the catheters of a volume implant have been reconstructed from radiographs and regions of interest (ROIs) are not indicated. Inverse optimization methods can be applied when contours of the PTV and OARs of an implant are available from three-dimensional (3D) imaging such as Computed Tomography (CT), Magnetic Resonance (MR), or Ultrasound (US).

6.3.1 Forward Optimization

The forward methods such as polynomial optimization on dose points (PODP), geometric optimization (GO), and graphical optimization (GrO) *a.k.a.* drag-and-drop of isodose lines, have been in use for many years. In POPD (van der Laarse and de Boer 1989), dose points are placed around the target (distance implant) or inside the target midway between the catheters (volume implant). Optimization is aimed at establishing the prescribed dose in these points. In GO (Edmundson 1989), the dwell times are set inversely proportional to the dose contribution from the other dwell positions. Two versions of GO are used, one for distance implants and one for volume implants (Baltas and Kolkman-Deurloo 2013). Both versions are used clinically, because of their simplicity and because in the case of very irregular implants other optimization methods fail. POPD optimizes the dose distribution over the implanted volume to match the user-defined doses in dose points around the catheters or inside the implanted volume. GrO is an interactive method where the user drags a point on an isodose line to a new position. Next, the dose distribution is locally adapted such that the isodose line runs through the new position. Its implementation is similar to that described in the appendix "Increasing the maximum dose or decreasing the minimum dose on a ROI" of Dinkla et al. (2015b). These well-known methods will not be discussed further.

The emerging forward methods are enhanced geometric optimization (EGO) (Dinkla et al. 2015b) and its automated version AutoEGO (Dinkla et al. 2015b). Table 6.5 presents an overview of optimization methods.

6.3.1.1 Enhanced Geometric Optimization

EGO is an extension of GO. It was developed because GO usually delivers less desirable dose distributions compared to PODP and GrO. Two parameters were introduced, volume strength s_{EGO} and range r_{EGO} (Dinkla et al. 2015b), which determine how the dose distribution that is obtained with GO is modified. First, the concept of dwell weights is introduced as relative dwell times, normalized to 1 for the longest dwell time. For each activated dwell position i, EGO modifies the dwell weight w_i obtained by GO for distance implants, by increasing its difference from weight 1. The magnitude of modification is a user-defined factor, called the volume strength s_{EGO}, that ranges between 0 and 1. The larger the value of s_{EGO}, the more dose is moved from the center of the implant to its borders. The adaptation of GO for volume implants determines the dwell weight in each dwell position by excluding the dose contribution from the other dwell positions in the same catheter. EGO introduces the parameter r_{EGO}, which now sets a range of exclusion. When calculating the dwell weight in position i from the doses delivered by the surrounding dwell positions, the dwell positions in the same catheter within range r_{EGO} of dwell position i are excluded. As a result, the dose contributions from these nearby dwell positions in the same catheter are excluded. This enhances the influence of the dose contributions from the other catheters, thus increasing the dose between the catheters. When r_{EGO} is zero, all dwell positions are included and the implant is optimized as a distance implant. In this way, r_{EGO} allows a transition between a distance implant ($r_{EGO} = 0$) and a volume implant ($r_{EGO} = $ large, e.g., 6 cm). Suitable values of s_{EGO} and r_{EGO} depend on the visual inspection of the resulting dose distribution. The current implementation of EGO uses sliders for s_{EGO} and r_{EGO} and practically real-time displays the corresponding dose distribution after a slider movement (Figure 6.1).

6.3.1.2 AutoEGO

When optimizing a volume implant, finding suitable values for s_{EGO} and r_{EGO} by interaction with the EGO sliders is not straightforward because a change in r_{EGO} usually requires re-evaluating s_{EGO}. AutoEGO uses an optimization algorithm to determine the values for s_{EGO} and r_{EGO} that give the maximum value of QI, which scores the homogeneity of the dose distribution (Table 6.5).

6.3.2 Inverse Optimization

Inverse optimization methods optimize objectives that are defined for the dose distribution over the PTV and for the high doses in the OARs and NTs. Software packages based on inverse optimization methods for volume implants such as inverse planning by simulated annealing (IPSA) and hybrid inverse planning optimization (HIPO) are in widespread use and are well documented (De Boeck et al. 2014, Dinkla et al. 2015a). Here, they will be discussed only in a general sense.

6.3.2.1 IPSA and HIPO

The optimization algorithms underlying IPSA (Lessard and Pouliot 2001, Alterovitz et al. 2006) and HIPO (Karabis et al. 2005, 2009) optimize a so-called total objective function that is the sum of several objective functions. Each of these objective functions penalizes linearly the violation of dose limits (upper and lower) in PTVs and OARs (Holm et al. 2012). Furthermore, both packages restrict the free modulation of dwell times to allow smoother distributions of dwell time over dwell positions in order to reduce the size of high-dose volumes (Baltas et al. 2009, Cunha et al. 2015).

IPSA and HIPO are considered *a priori* methods, meaning that they are well-suited to provide so-called class solutions: solutions that are found by using standard values for dose limits. However, each patient is different, so the standard values are almost never exactly right, warranting patient-specific adjustments. This is especially true in brachytherapy treatment planning because of the mismatch between optimization objectives formulated and used by optimization algorithms and the final acceptance criteria used by physicians. The upside of an inverse method providing a class solution is that the outcome is typically obtained much faster than with the repetitive use of a forward method such as GrO. The downside, however, is that the trial-and-error fine-tuning process of an inverse method is less intuitive. That is, to change the outcome, the combination of objectives and dose limits needs to be manipulated and the inverse optimization to be run again, and since the objective functions are mutually dependent, it is difficult to predict the resulting changes

TABLE 6.5 Optimization Methods

Optimization Program	Method	Algorithm	Advantages/Disadvantages
HDR/PDR. **PODP** Polynomial optimization on dose points	Deterministic, analytical method, forward planning on activated dwell positions	Optimization on prescribed doses in dose points using singular value decomposition	• Straightforward for distance implants • Requires incorporation of GO for volume implants • Global method (no local control over dose distribution)
HDR/PDR. **GrO** Graphical optimization	Heuristic, forward planning on activated dwell positions	Drag and drop of isodose lines	• Generally applicable for distance and volume implants • Local method, local control over dose distribution • Strongly dependent on skill of planner
HDR/PDR. **GO** Geometrical optimization	Heuristic, forward planning on activated dwell positions	First set all weights to 1.0. Then determine the dose at each dwell position by the other ones. Use reverse of that dose as dwell weight.	• Generally applicable for distance and volume implants • Optimization often suboptimal • Global method
HDR/PDR. **EGO*** Enhanced geometric optimization	Heuristic, forward planning on activated dwell positions	Interactively enhancing the amount of GO optimization	• Generally applicable for distance (including molds) and volume implants • Global method • Equivalent results in cases where PODP is used, but without the need of placement of dose points
HDR/PDR. **AutoEGO***	Heuristic, forward planning on activated dwell positions	Optimizes EGO to the maximum value of the QI	• Applicable for volume implants • Global method
LDR/HDR/PDR **HIPO needle or catheter placement** Hybrid inverse planning optimization	Stochastic, inverse planning on optimum position of seeds or catheters in conjunction with optimization of dose distribution	Simulated Annealing SA	• Applicable for needle or catheter positions in volume implants • Global method • Trial and error fitting of objectives
LDR/HDR/PDR **HIPO dose distribution**	Deterministic, inverse optimization of dose distribution by seed position in needles, or dwell position and time in catheters	Linear programming solver LBFGS, linear penalization for any dose deviation	• Applicable for volume implants • Global method • Trial and error fitting of objectives

(Continued)

TABLE 6.5 (Continued) Optimization Methods

Optimization Program	Method	Algorithm	Advantages/Disadvantages
LDR/HDR/PDR IPSA Inverse planning by simulated annealing	Stochastic, inverse optimization of dose distribution by seed position in needles, or dwell position and time in catheters	Simulated annealing SA, linear penalization for any dose deviation	• Applicable for volume implants • Global method • Trial and error fitting of objectives
HDR/PDR IPIP* Inverse planning by integer program **HDR/PDR LDV** Linear dose-volume-based integer program	Inverse optimization of CTV coverage by specifying the dosimetric criteria V(x%PD) for target and OARs	Mixed integer program (with IPIP using a heuristic component)	• Applicable for volume implants • Global method • Uses clinical constraints V(x%PD) for target and OARs
HDR/PDR IIP* Interactive inverse planning	Heuristic, inverse planning on lower and upper limit doses for PTV and OAR contours	Interactively adapting the dose distribution near dose points with maximum or minimum dose on contours of PTV and OARs	• Applicable for volume implants, in combination with auto EGO • Local method • Allows use of clinical constraints such as V(x%PD, D(2cc) • Allows use of radiobiological variables such as EQD2 as optimization parameter
HDR/PDR IMOO* Interactive multiobjective inverse optimization	Deterministic inverse optimization of dose distribution	Interactive multiobjective optimization algorithm	• Global method, but used iteratively • Uses objectives not directly linked to clinical requirements
Radioembolization using pretreatment dosimetry	Forward planning of the amount of activity to deliver in one or more arteries	Scout dosimetry to determine the amount of activity to deliver in one or more arteries	• Local method • Vascular anatomy complicates delivery of amount of activity via arteries

Methods indicated by * are published but not yet part of a commercial treatment planning system.

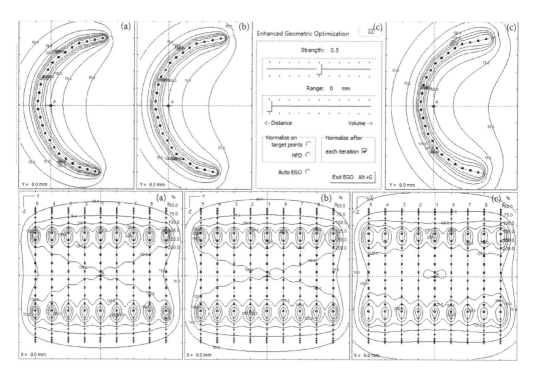

FIGURE 6.1 Comparison of optimization methods for a curved mold with thickness 5 mm. The prescription point A is set at 3 mm depth. (a) All dwell weights are 1.0. (b) GO on distance. (c) EGO with strength parameter set to 0.5. The range parameter is 0 mm, so all dwell positions are taken into account.

in the optimized dose distribution. Also, the correlation between the dosimetric indices that are used to evaluate a plan and the configuration of the total objective function is not always evident. For these reasons, many treatment centers use a combination of forward and inverse planning, for example, after running IPSA or HIPO, GrO is used for the final touch to satisfy any remaining wishes of the physician.

6.3.2.2 Inverse Planning by Integer and Linear Dose-Volume Programming

Inverse planning by integer program (IPIP) is an inverse treatment planning method that directly uses clinical constraints on volume instead of dose (Siauw et al. 2011). For the PTV, optimization is constrained to (i) at least a given volume, for example, $V_{PTV}(100\%PD)$ > 90% of the PTV or (ii) at most a given volume, for example, $V_{PTV}(150\%PD)$ < 45% of the PTV. For an OAR, the optimization is constrained to at most a given dose or volume, for example, $V_{OAR}(75\%PD)$ < 1 cm^3. A disadvantage of IPIP is its objective function which maximizes coverage within limits to OAR dose, thus without an upper bound for $V_{PTV}(100\%PD)$. This may result in large increases to the OAR dose, just staying within its limits, in order to optimize PTV coverage. Therefore, iterating over alternative OAR limits is often required. The actual solving of IPIP is done using linear programming with heuristic relaxation of the IPIP constraints.

The linear dose-volume (LDV)-based method (Gorissen et al. 2013) maximizes the fraction of the PTV receiving the prescribed dose, while constraining the dose in a given fraction of each OAR. This model provides also an upper bound for $V_{PTV}(100\%PD)$.

6.3.2.3 Interactive Inverse Planning

Originally, the interactive inverse planning (IIP) method was developed as an inverse version of GrO to be used after IPSA or HIPO (Dinkla et al. 2015b). Instead of tailoring a dose distribution to a specific patient anatomy by a series of drag and drop actions on isodose lines, IIP adapts the dose distribution to a specified upper or lower dose limit for a given ROI by iteratively changing the dwell times in the dwell positions near the highest and lowest dose on the contours of the ROI, see Figure 6.2. An ROI in this context is a PTV or an OAR. Currently there is much interest to optimize to the volume of an ROI that should receive a given dose (Hsu et al. 2010) or to the dose that should envelop a given volume of the ROI (Dinkla et al. 2015a). Using the effect of the specification of upper or lower dose limits of the ROIs on the cumulative DVHs, IIP was extended with these two options, see Figure 6.3.

If the ROI belongs to the PTV, optimization is constrained to ensure a given coverage of the PTV by either dose or volume, for example, $V_{PTV}(95\%PD)$ or $D(90\%PTV)$, respectively, and, if the ROI is an OAR, to at most a given dose or volume, for example, $V_{OAR}(80\%PD)$ or $D(2 \text{ cm}^3OAR)$, respectively. The same approach is applied to optimize on the radiobiological dose unit Equivalent Dose for 2 Gy, EQD2.

6.3.2.4 Interactive Multiobjective Optimization

The interactive multiobjective optimization (IMOO) method (Ruotsalainen et al. 2010) has clinically the same approach as IIP, as the planner interactively changes the values of objectives based on doses in surface points on the PTV and OARs and evaluates the resulting dose distributions. Their objective functions, however, have no simple relation to the required dose distribution, so the treatment planner must iteratively evaluate Pareto optimal solutions. Although a graphical interface is presented which assists the treatment planner, still more training of the treatment planner is required with this approach than with IIP.

6.3.2.5 Optimization of Radioembolization Treatment Plans

The metastatic tumor load of the liver is often multifocal. With SPECT studies, the tumor regions can be distinguished from the healthy tissues. Optimization of radioembolization treatment plans is done by using the outcome of a pretreatment scout microsphere delivery in each artery to a tumor region, to calculate the amount of activity to be delivered by that artery. Posttreatment dosimetry also indicates whether certain regions may benefit from an additional treatment, either by a second radioembolization or by another treatment modality.

The established method of radioembolization is with microspheres containing Yttrium-90 (beta emitter, average energy 0.93 MeV). The scouting is done with a different, gamma-emitting, microsphere of about the same dimension, usually labeled with

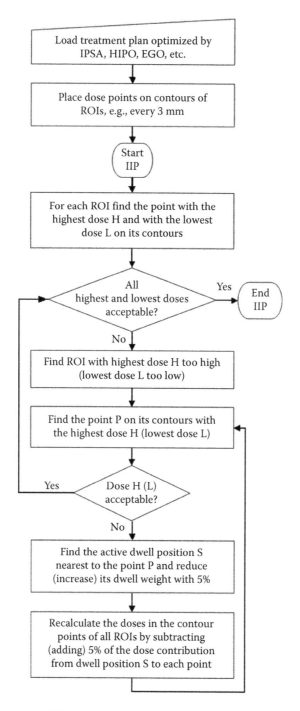

FIGURE 6.2   Flow diagram of IIP to fine-tune a treatment plan by reducing the highest dose on the surfaces of PTV and OARs, that is exceeding a given upper bound. The steps to increase the lowest dose that exceeds a lower bound are given between brackets.

(a) Dose limits of target and organs

All planes are selected

Angle specification (gantry angle [0-360])

Start angle: 0   Stop angle: 360

Margin in selected planes and angle segment.
First set margin value then push button

Disable dwell pos. outside margin ○ Margin: 3   mm
Enable dwell pos. inside margin ○

Maximum and minimum doses in target and organs

	cGy Lower limit	Upper limit	%PD	D(90)	D(98)	D(100)	D(5)	D(2)	D(1)	D(0.1)
PTV	469	1518	☑	111.1	102.7	87.2	-1	-1	-1	-1
Bladder	167	1301	☐	-1	-1	-1	365.4	467.4	540.5	859.9
Rectum	134	525	☐	-1	-1	-1	294.2	352.1	386.4	443.1
Urethra	542	861	☐	-1	-1	-1	772.3	1279.5	621.9	707.4

(xx) = %vol.ROI | cm^3, [xx.x] = cGy | %PD

Settings — Low/Up Limits — % PD ○ cGy ⊙

Prescription — Prescr

Display — Dose(volume) ⊙ EQD2 ○ Volume(dose) ○

Undo / Redo IIP — Undo Redo Undo all

(b) Dose limits of target and organs

All planes are selected

Angle specification (gantry angle [0-360])

Start angle: 0   Stop angle: 360

Margin in selected planes and angle segment.
First set margin value then push button

Disable dwell pos. outside margin ○ Margin: 3   mm
Enable dwell pos. inside margin ○

Maximum and minimum doses in target and organs

	cGy brachy Lower limit	Upper limit	D(90)	D(98)	D(100)	D(5)	D(2)	D(1)	D(0.1)
PTV	469	1518	84.8	80.4	72.8	-1	-1	-1	-1
Bladder	167	1301	-1	-1	-1	65.5	74.7	82.3	125.8
Rectum	134	525	-1	-1	-1	60	64.4	67.2	72.3
Urethra	542	861	-1	-1	-1	112.2	207.7	91.9	103

(xx) = %vol.ROI | cm^3, [xx.x] = Gy (EQD2)

Settings — Low/Up Limits — % PD ○ cGy ⊙

Prescription — Prescr

Display — Dose(volume) ○ EQD2 ⊙ Volume(dose) ○

Undo / Redo IIP — Undo Redo Undo all

FIGURE 6.3 (a) IIP interface for an HDR treatment plan with dose (volume) as optimization variable. PD 550 cGy. The upper and lower limit section is in cGy of the brachytherapy treatment. D(90), D(98), and D(100) are percent of PD around the volumes of 90%, 98%, and 100% of the PTV. D(5), ... D(0.1) are the doses on the surface of the volumes of 5 cm³ ... 0.1 cm³ with the highest doses in the OARs. The dark background of D(90) in the PTV indicates a value lower than a preset lower bound and the dark background of D(0.1) in the bladder a value higher than a preset upper bound. The planner can reduce the 859.9 Gy of the bladder D(0.1) by selecting the box with the 859.9 Gy value and overwriting it with the new value. Next, IIP will iteratively lower the upper limit of the bladder (1301 cGy) until the new value is reached. (b) Same plan as 6.3a, but with EQD2 as optimization variable. The HDR brachytherapy PD of 550 cGy per fraction, delivered four times, gives an EQD2 of 79 Gy for the PTV, including the contribution from the external beam treatment already delivered. The upper and lower limit section is in physical dose cGy of the brachytherapy treatment. All other doses are expressed in EQD2 (Gy). The planner can reduce the 125.8 Gy of the bladder EQD2 D(0.1) by selecting the box with the 125.8 Gy value and overwriting it with a new value. Next, IIP will iteratively lower the cGy brachy upper limit of the bladder (1301 cGy) until the new value is reached.

technetium-99m, using SPECT combined with CT imaging. Posttreatment dosimetry is based on ^{90}Y-PET and ^{90}Y-Bremsstrahlung SPECT imaging. An accuracy of 5% of the true ^{90}Y activity in the liver is possible but not easily reached (Rong et al. 2012).

An emerging radioembolization method uses holmium-166 microspheres. Optimization of the radioembolization with these microspheres is more straightforward because ^{166}Ho emits high-energy beta particles (max. energy 1.8 MeV) for tumor irradiation and gamma photons (81 keV) for nuclear imaging. Furthermore, because it is a highly paramagnetic metal, it can be visualized on Magnetic Resonance Imaging (MRI). Pretreatment dosimetry requires a small scout dose of ^{166}Ho-microspheres in each artery leading to a tumor region to determine the therapeutic dose to be delivered by that artery. Posttreatment dosimetry using MRI imaging gives accurate absorbed dose distributions to tumor and normal liver tissues (Smits et al. 2013).

6.3.3 Emerging Objectives
6.3.3.1 Physical Objectives
Nowadays, instead of evaluating upper and lower dose limits of an ROI, a dose distribution in brachytherapy is increasingly assessed by evaluating the volume of the ROI enveloped by a given dose and the lowest or highest dose enveloping a given volume of the ROI (Hsu et al. 2010). While these objectives are considered to be clinically important, most inverse optimization methods do not accept these objectives. The volume enveloped by at least a given dose is an objective applied to the PTV, for example, $V_{PTV}(90\%PD)$ and the volume enveloped by at most a given dose is an objective applied to an OAR, for example, $V_{OAR}(80\%PD)$. These objectives are available to directly steer the optimization in IPIP and in IIP. In other programs, these volumes as function of dose must be obtained from the cumulative DVHs, after the optimization is finished. The lowest dose enveloping a given volume is applied to the PTV, for example, $D_{PTV}(90\%PTV)$ and the highest dose around a given volume is applied to an OAR, for example, $D_{OAR}(2$ cm$^3)$ the 2 cm^3 of the OAR with the highest dose. This is implemented as an objective in IIP (Figure 6.3). In other methods, these doses as function of volume must again be obtained from the cumulative DVHs.

It is to be expected that in the near future directly optimizing to the volume of a ROI enveloped by a given dose, or to the lowest or highest dose enveloping a given volume of a ROI will become widely available.

6.3.3.2 Radiobiological Variables and Objectives
The inclusion of radiobiological variables in the optimization process is very important, because the radiobiological effect of the dose on the tumor determines the outcome of the treatment. The variable physical dose can be replaced by biological equivalent dose (BED) or equivalent dose in 2 Gy per fraction (EQD2). Also, a number of radiobiological objectives have been developed, such as the biologically effective equivalent uniform dose (BEEUD), generalized equivalent uniform dose (gEUD), and TCP and NTCP.

Currently, only IIP allows the use of EQD2 as a direct replacement of physical dose in a point or voxel. In the same way, BED could also be used as a direct replacement for physical dose. Objectives such as BEEUD (Kehwar and Akber 2008) and gEUD (Giantsoudi et al. 2013) have been investigated for clinical use but until now did not make it as objectives into routine HDR/pulsed-dose rate (PDR) optimization methods. Radiobiological modeling to evaluate a treatment plan by predicting the TCP and NTCP is not yet clinically applicable on a routine basis, as the required radiobiological knowledge on which they are based is still limited (Ebert and Zavgorodni 2000).

In the near future, the replacement of physical dose by EQD2 is to be expected for at least some clinical sites. In the current approaches to calculate EQD2, the contribution of the external beam part is considered to be constant over each ROI. To add the EQD2 contribution from the external beam treatment voxel by voxel to the brachytherapy EQD2 dose distribution requires voxel by voxel dose accumulation in External Beam Radiation Therapy (EBRT) through daily imaging and 3D-deformable image registration to match the brachytherapy dose distribution. For more details on this emerging technique, see Chapter 7.

6.4 DEVELOPMENTS IN TREATMENT PLANNING

6.4.1 Influence of Rapid Developments in Other Fields

The increasing use of 3D imaging in brachytherapy will decrease the use of forward optimization, as the outlines of the ROIs allow the use of class solutions via inverse methods to arrive at initial plans quickly. In the cases where forward optimization will still be used, EGO will be a good candidate for replacing GO, as with EGO the planner can adjust the dose distribution in real time.

The computing power of treatment planning systems increases continuously. Planning software will increasingly apply parallel programming in order to use a cluster of CPUs (Sutherland et al. 2012, Bouzid et al. 2015) as well as the potential additional computing power of the processing units on video cards (GPUs) (Jia et al. 2014). This will result in Monte Carlo-based TPSs being suitable for 3D dose calculations in a clinical setting (see Chapter 5 for more details). At least for the near future, however, optimization algorithms will continue using less accurate analytical solutions for the dose calculation, due to speed requirements. Only after optimization, the final result could then be calculated with a Monte Carlo (MC) dose calculation.

6.4.2 Protocol-Based Treatment Planning

Brachytherapy is in many aspects not protocol based, that is, there is usually no straight path to arrive at a brachytherapy plan that is suitable for treatment. The underlying reason for this is the mismatch between the criteria used by physicians to rate plans and the criteria that are actually implemented and used in optimization. Manually optimizing a dose distribution by GrO will only lead to a good plan if the planner is experienced in use of the GrO itself.

More consistent results will be obtained with IIP, because of the use of user-defined upper and lower dose limits for a ROI instead of just drag-and-drop of points on isodose lines. But even inverse optimization methods such as IPSA and HIPO still frequently require subsequent manual adaptations to reach a plan that will be approved by a physician. Only if protocols can be developed that describe how to arrive at an optimal treatment plan, does brachytherapy treatment planning become less of an art and more of a science. A recent protocol-based optimization method of a tandem and ovoid treatment of locally advanced cervical cancer comes close (Sharma et al. 2015). It consists of three distinct steps. In step 1, the dose distribution is manually optimized according to the 2000 ABS guidelines (Nag et al. 2000) while keeping the classical pear-shaped prescription isodose surface. In step 2, this pear-shaped prescription isodose surface is converted to a ROI. Finally, in step 3, an IIP method with dose-volume constraints is applied to all ROIs, including the pear-shaped one. In this way, the pear-shaping of the prescription isodose volume is kept as much as is allowed by the other dose-volume constraints. Steps 2 and 3 can also be done by IIP, because IIP can optimize by locally adapting a given, for example, pear-shaped, dose distribution to the dose-volume constraints.

6.4.3 Knowledge-Based Optimization

Without an internationally agreed upon objective measure of a plan's quality, brachytherapy treatment planning will always require multiple stages, the last one of which is an interactive fine-tuning process to ensure that the wishes of the physician are met. How much tuning is needed, depends on the magnitude of the discrepancy between the measures used in inverse optimization and what the physician has in mind. Research into radiobiological modeling on which the TCP and NTCP are based may provide an outcome in the future, but there are also emerging alternative approaches. Of particular interest are knowledge-based approaches that analyze previously approved plans and incorporate extracted knowledge in the optimization process. To this end, machine learning algorithms can be used, which are becoming increasingly powerful and popular (Bishop 2006, Schmidhuber 2015). For brachytherapy treatment planning, machine learning can be used to learn what makes a good brachytherapy plan from previously approved plans (by individual physicians or by groups of physicians). This is something that is typically considered to be difficult to formulate explicitly beforehand by a physician. This knowledge can then be used to improve inverse optimization and generate plans that are closer to what a physician desires.

6.4.4 Robust Optimization

Currently, optimization methods for brachytherapy plan design use formulations of objectives that assume that the real-world realization of a plan perfectly matches this plan. However, there are actually various uncertainties at play in clinical practice including uncertainties in dose calculations, locations of dwell positions, and catheter displacements between the moment of imaging and the moment of actual dose delivery. The latter becomes even more prevalent if plans are made even before needles or applicators are placed (e.g., because a patient-specific 3D-printed applicator can be constructed based on imaging data). Even if a plan is optimal according to a certain set of objectives, it may still be an undesirable plan because it is not *robust*, meaning that the execution of that plan may very likely give a result different from that what was planned. To overcome this important issue, the so-called robust optimization is required (Bertsimas and Sim 2004, Chu et al.

2005). This introduces additional dimensions to the evaluation of the quality of a plan, which further complicates the design of automated optimization methods. However, current and near-future advances in computing power and algorithmic design make robust optimization an important emerging possibility.

6.5 CONCLUSION

Consensus by physicians on the quality of a brachytherapy treatment plan is a major issue. In combination with the lack of accordance between the formulation of objectives and the resulting dose distributions in current inverse optimization methods, a gap often still exists between a dose distribution resulting from inverse optimization and the final, clinically approved one.

To narrow this gap, a number of options are emerging (i) knowledge-based optimization by learning from previously approved plans to enhance the consensus by physicians, (ii) protocol-based optimization to enhance the consistency of plans, and (iii) interactive inverse optimization such as IIP to interactively adapt an already optimized plan. Related to this issue is the robustness of actual dose delivery, that is, the difference between the planned, clinically approved, dose distribution and the factually delivered one. In the near future, robust optimization will narrow this difference.

In several optimization methods, options will emerge to replace the formulation of objectives in terms of absorbed dose by objectives defined in another physical or radiobiological basis. Also, the adding of the dose distribution expressed in EQD2 to the external beam plan with a voxel by voxel accumulation approach will become possible due largely to 3D image-guide treatment regimes.

Electronic brachytherapy and radioembolization are developing rapidly. However, much work is still to be done especially in the field of optimization of radioembolization treatment plans.

Finally, chemobrachytherapy and thermobrachytherapy are promising developments, although it is currently still too early to predict whether or not they will become routine.

REFERENCES

Alterovitz, R., E. Lessard, J. Pouliot, I.-C.J. Hsu, J.F. O'Brien, and K. Goldberg. Optimization of HDR brachytherapy dose distributions using linear programming with penalty costs. *Med. Phys.* 33, 2006: 4012–4019.

Baltas, D., and I.-K.K. Kolkman-Deurloo. Optimization and evaluation. In *Comprehensive Brachytherapy: Physical and Clinical Aspects*. Eds. Venselaar J.L.M., D. Baltas, A.S. Meigooni, and P.J. Hoskin, Chapter 12, 161–175. Boca Baton, Florida: CRC Press, 2013.

Baltas, D. et al. Influence of modulation restriction in inverse optimization with HIPO of prostate implants on plan quality: Analysis using dosimetric and radiobiological indices. In *IMFBE Proceedings*. Munich, Germany: Springer, 2009. pp. 283–286.

Bertsimas, D., and M. Sim. The price of robustness. *Oper. Res.* 52(1), 2004: 35–53.

Bishop, C.M. *Pattern Recognition and Machine Learning*. Berlin, Germany: Springer, 2006.

Bouzid, D., P.F. Dupre, S. Benhalouche, O. Pradier, N. Boussin, and D. Visvikis. Monte-Carlo dosimetry for intraoperative radiotherapy using a low energy x-ray source. *Acta Oncol.* 54, 2015: 1788–1795.

Carlson, D.J., Z.J. Chen, P. Hoskin, Z. Ouhib, and M. Zaider. Radiobiology for brachytherapy. In *Comprehensive Brachytherapy: Physical and Clinical Aspects*. Eds. Venselaar J.L.M., D. Baltas, A.S. Meigoon and P.J. Hoskin, 253–270. Boca Baton, Florida: CRC Press, 2013.

Chu, M., Y. Zinchenko, S.G. Henderson, and M.B. Sharpe. Robust optimization for intensity modulated radiation therapy treatment planning under uncertainty. *Phys. Med. Biol.* 50(23), 2005: 5463–5477.

Cunha, A., T. Siauw, I.-C. Hsu, and J. Pouliot. A method for restricting intracatheter dwell time variance in high-dose-rate brachytherapy plan optimization. *Brachytherapy*, 15(2), 2015: 246–251. doi: 10.1016/j.brachy.2015.10.009.

De Boeck, L., J. Beliën, and W. Eyged. Dose optimization in high-dose-rate brachytherapy: A literature review of quantitative models from 1990 to 2010. *Oper. Res. Health Care* 3, 2014: 80–90.

Devlin, P. *Brachytherapy Applications and Techniques*. Philadelphia, Pennsylvania: Lippincott Williams and Wilkins, 2007.

Dinkla, A.M. et al. A comparison of inverse optimization algorithms for HDR/PDR prostate brachytherapy treatment planning. *Brachytherapy* 14, 2015a: 279–288.

Dinkla, A.M. et al. Novel tools for stepping source brachytherapy treatment planning: Enhanced geometrical optimization and interactive inverse planning. *Med. Phys.* 42(1), 2015b: 348–353.

Eaton, D.J. Electronic brachytherapy—Current status and future directions. *Br. J. Radiol.* 88, 2015: 1049.

Ebert, M.A., and S.F. Zavgorodni. Modeling dose response in the presence of spatial variations in dose rate. *Med. Phys.* 27, 2000: 393–400.

Edmundson, G.K. Geometry based optimization for stepping source implants. In *Proceedings Brachytherapy Meeting Remote Afterloading: State of the Art*. Dearborn, Michigan: Nucletron Corporation, 1989. pp. 184–192.

Garcia-Martinez, T., J.-P. Chan, J. Perez-Calatayud, and F. Ballester. Dosimetric characteristics of a new unit for electronic skin brachytherapy. *J. Contemp. Brachytherapy* 6, 2014: 45–53.

Gautam, B., G. Warrell, D. Shvydka, M. Subramanian, and E.I. Parsai. Practical considerations for maximizing heat production in a novel thermobrachytherapy seed prototype. *Med. Phys.* 41, 2014: 023301-1–023301-10 doi: 10.1118/1.4860661.

Giantsoudi, D. et al. A gEUD-based inverse planning technique for HDR prostate brachytherapy: Feasibility study. *Med. Phys.* 40, 2013: 041704-1–041704-12.

Gorissen, B.L., D. den Hertog, and A.L. Hoffmann. Mixed integer programming improves comprehensibility and plan quality in inverse optimization of prostate HDR brachytherapy. *Phys. Med. Biol.* 58, 2013: 1041–1057.

Holm, A., T. Larsson, and A. Carsson Tedgren. Impact of using linear optimization models in dose planning for HDR brachytherapy. *Med. Phys.* 39, 2012: 1021–1028.

Hsu, I.-C. et al. Phase II trial of combined high-dose-rate brachytherapy and external beam radiotherapy for adenocarcinoma of the prostate: Preliminary results of RTOG 0321. *Int. J. Radiat Oncol. Biol. Phys.* 78, 2010: 751–758.

Jia, X., P. Ziegenhein, and S.B. Jiang. GPU-based high-performance computing for radiation therapy. *Phys. Med. Biol.* 59, 2014: R151–R182.

Karabis, A., P. Belotti, and D. Baltas. Optimization of catheter position and dwell time in prostate HDR brachytherapy using HIPO and linear programming. In *IFMBE Proceedings 25/I*. Munich, Germany: Springer, 2009. pp. 612–615.

Karabis, A., S. Giannouli, and D. Baltas. HIPO: A hybrid inverse treatment planning optimization algorithm in HDR brachytherapy. *Radiother. Oncol.* 76(Suppl 2), 2005: 29.

Kehwar, T.S., and S.F. Akber. Assessment of tumor control probability for high-dose-rate interstitial brachytherapy implants. *Rep. Pract. Oncol. Radiother.* 13, 2008: 74–77.

Kumar, R. et al. Nanoparticle-based brachytherapy spacers for delivery of localized combined chemoradiation therapy. *Int. J. Radiat. Oncol. Biol. Phys.* 91, 2015: 393–400.

Lessard, E., and J. Pouliot. Inverse planning anatomy-based dose optimization for HDR-brachytherapy of the prostate using fast simulated annealing algorithm and dedicated objective function. *Med. Phys.* 28, 2001: 773–779.

Moerland, M.A., R. van der Laarse, R.W. Luthmann, H.K. Wijrdeman, and J.J. Battermann. The combined use of the natural and the cumulative dose volume histograms in planning and evaluation of permanent prostatic seed implants. *Radiother. Oncol.* 57, 2000: 279–84.

Nag, S. et al. The American Brachytherapy Society recommendations for high-dose-rate brachytherapy for carcinoma of the cervix. *Int. J. Radiat. Oncol. Biol. Phys.* 48, 2000: 201–211.

O'Rourke, S.F.C., H. McAneney, and T. Hillen. Linear quadratic and tumour control probability modelling in external beam radiotherapy. *J. Math. Biol.* 58, 2009: 799–817.

Pouliot, J., R. Sloboda, and B. Reniers. Two-, three- and four dimensional brachytherapy. In *Comprehensive Brachytherapy: Physical and Clinical Aspects*. Eds. Venselaar J.L.M., D. Baltas, A.S. Meigooni and P.J. Hoskin, Chapter 9, 117–136. Boca Raton, Florida: CRC Press, 2013.

Rana, Z. et al. Comparative dosimetric findings using accelerated partial breast irradiation across five catheter subtypes. *Radiat Oncol.* 10, 2015: 160. doi: 10.1186/s13014-015-0468-7.

Rong, X. et al. Development and evaluation of an improved quantitative (90)Y bremsstrahlung SPECT method. *Med. Phys.* 39, 2012: 2346–2358.

Ruotsalainen, H., K. Miettinen, J.-E. Palmgren, and T. Lahtinen. Interactive multiobjective optimization for anatomy-based three-dimensional HDR brachytherapy. *Phys. Med. Biol.* 55, 2010: 4703–4719.

Schmidhuber, J. Deep learning in neural networks: An overview. *Neur. Netw.* 61, 2015: 85–117.

Sharma, M., E.C. Fields, and D.A. Todor. A novel two-step optimization method for tandem and ovoid high dose rate brachytherapy treatment for locally advanced cervical cancer. *Brachytherapy* 14, 2015: 571–577.

Siauw, T., A. Cunha, A. Atamtürk, I.-C. Hsu, J. Pouliot, and K. Goldberg. IPIP: A new approach to inverse planning for HDR brachytherapy by directly optimizing dosimetric indices. *Med. Phys.* 38, 2011: 4045–4051.

Smits, M.L. et al. In vivo dosimetry based on SPECT and MR imaging of 166Ho-microspheres for treatment of liver malignancies. *J. Nucl. Med.* 54, 2013: 2093–2100.

Smits, M.L.J. et al. Radioembolization dosimetry: The road ahead. *Cardiovasc. Intervent. Radiol.* 38, 2015: 261–269.

Sutherland, J.G., K.M. Furutani, Y.I. Garces, and R.M. Thomson. Model-based dose calculations for 125I lung brachytherapy. *Med. Phys.* 39, 2012: 4365–4377.

van der Laarse, R., and R.W. de Boer. Computerized high dose rate brachytherapy treatment planning. In *Proceedings Brachytherapy Meeting Remote Afterloading: State of the Art*. Dearborn: Nucletron Corporation, 1989. pp. 169–183.

van der Laarse, R., and R.W. Luthmann. Computers in brachytherapy dosimetry. In *Principles and Practice of Brachytherapy Using Afterloading Systems*. Eds. Joslin C.A.F., A. Flynn, and E.J. Hall, 49–79. London: Arnold, 2001.

CHAPTER 7

Image Processing for Brachytherapy

Niclas Pettersson and Laura Cerviño

CONTENTS

7.1 INTRODUCTION

Patient imaging in brachytherapy has undergone substantial development in the last decades with the introduction of volumetric imaging modalities such as ultrasound (US), computed tomography (CT), and magnetic resonance imaging (MRI). These developments have enabled a transition from a situation where the geometry of the applicators was used both to prescribe the treatment and to display the dose distribution, to a situation where images of each patient's anatomy are used to optimize and display the dose distribution.

Imaging provides information in two important steps during the treatment planning process: it is the basis for target and organs at risk (OARs) definitions, and it assists in the determination and optimization of radiation source positions. The optimal imaging procedure would take place with the patient in the treatment position after the applicators or seeds have been implanted, and the imaging modality (or modalities) would provide soft-tissue contrast for all relevant targets and OARs, and simultaneously allow reconstruction of the implant geometry. In the clinical situation, however, this is not necessarily achieved.

The high spatial resolution of x-ray images makes this modality well suited for applicator reconstruction. For arbitrary applicator shapes or multi-applicator implants with unknown configuration, reconstruction methods utilizing images from two or more angles are widely used. However, x-ray imaging in general provides little anatomical information about target volumes and OARs. US, CT, and MRI are commonly used three-dimensional (3D) imaging modalities in brachytherapy. They provide volumetric patient anatomy information and, if they are acquired after implantation, they may also allow for applicator reconstruction.

Given the image complexity and multimodality approach of image-guided brachytherapy, image processing is essential to make optimal use of all available image information.

7.2 IMAGE REGISTRATION

In image-guided fractionated radiotherapy, patient images are acquired several times throughout the treatment course. If two image sets are registered to each other, that is, if the spatial relationships between all voxels in the two image sets are determined, image information from the different images can be set in relation to each other. In image registration, one image set (moving image) is spatially transformed to match another image set (static or fixed image). Image registration can be done for different purposes, but here we will mainly be focused on image registration for dose distribution accumulation.

A particular type of image registration is rigid image registration, which consists of allowing the moving image to be rotated and translated to match the fixed image (Maintz and Viergever, 1998; Oliveira and Tavares, 2014). This can be expected to work well for registration of non-deformable tissues such as bony structures. However, substantial independent movement of other structures can still be present (Figure 7.1) and need a more advanced image registration.

In deformable image registration, the voxels in the moving image are allowed to move according to individual deformation vectors to match the fixed image. Different deformation models exist. One method to find the best alignment between the fixed and moving images is to use the images' grayscale information. This is called intensity-based registration and is primarily applicable for registration between images from the same modality. Another method is to define corresponding pairs of anatomical landmarks (points, surfaces/contours, or volumes) in both images and use those to align the images. This is called feature-based registration and can also be used for registration of images from different modalities. Combinations of models can be used; for example, one could start with one model for a fast but less accurate registration, which is then improved upon by more advanced methods (Maintz and Viergever, 1998; Crum et al., 2014; Kessler, 2014; Oliveira

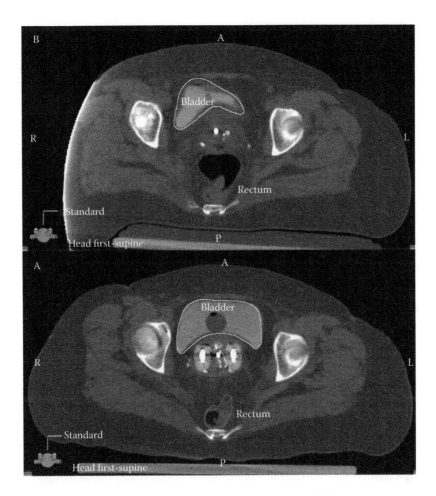

FIGURE 7.1 Transversal CT slices after rigid image registration of bony structures in a cervical cancer patient at two different brachytherapy fractions. There are significant changes in position of both the urinary bladder and the rectum relative to the bony structures.

and Tavares, 2014). After image registration, each voxel in the moving image will have information on where it is located in the fixed image. This information is described by the deformation vector field (DVF).

7.2.1 Image Processing in Brachytherapy of Cervical Cancer

By taking advantage of the excellent soft-tissue contrast of MRI, cervical cancer brachytherapy has undergone substantial development in the last decade. Treatment is usually delivered in several fractions, which means that, even if the same treatment plan was delivered at each fraction, the OAR dose distributions could be different due to internal organ motion and deformation, as well as different placement of the applicator(s). Consequently, much effort has been dedicated to accumulate dose distributions from different fractions (Christensen et al., 2001; Bondar et al., 2010; Andersen et al., 2012, 2013; Jamema et al., 2013; Lang et al., 2013; Nesvacil et al., 2013; Simha et al., 2014; Jamema et al., 2015; Osorio et al., 2015; Teo et al., 2015; van Dyk et al., 2015; Zhen et al., 2015).

7.2.1.1 Dose Distribution Accumulation

The simplest evaluation of the total OAR dose distribution is the direct addition (DA) of dose-volume histogram (DVH) parameters from each fraction. However, since high doses may occur in different parts of the OAR for different fractions, this approach can potentially lead to overestimation of the total dose. Since the probability of a hotspot occurring in the precise same location for several fractions decreases with the number of fractions, DA overestimation will typically increase with increasing number of fractions.

Recently, deformable image registration of MR images was used to estimate the difference in D_{2cm3} between voxel-wise accumulated dose distributions and DA for OARs (Andersen et al., 2013; Jamema et al., 2015). Andersen et al. (2013) used an in-house biomechanical-based registration method for the bladder, while Jamema et al. (2015) used both a commercially available intensity-based method and an in-house contour-based algorithm for the bladder and the rectum. They found that although DA systematically overestimated the OAR D_{2cm3}, it provided a "good estimate" of the cumulative dose.

There are additional complicating factors when trying to estimate the total cumulative dose distribution from combined external beam radiotherapy (EBRT) and brachytherapy. Particularly, the dose accumulation involves registering brachytherapy images containing applicators to applicator-free EBRT images. For the special case of homogeneous EBRT dose distribution in target volumes and parts of OARs close to or inside the planning target volume (PTV), GEC-ESTRO has suggested that the total cumulative EBRT and brachytherapy dose can be calculated by adding the prescribed EBRT dose to the accumulated brachytherapy dose (Pötter et al., 2006). However, effort is being spent onto more accurately calculating the cumulative dose via deformable image registration. Conventional deformable registration methods will fail to accurately register images containing applicator(s) to applicator-free images. In an effort to overcome this challenge, Teo et al. (2015) have used intensity-based deformable image registration of CT images where they pre-processed the OARs and applicators to improve algorithm performance. They found that DA was a good approximation for EBRT and brachytherapy addition of D_{2cm3}, but that this estimate could be improved in cases with less homogeneous EBRT dose distribution.

7.2.2 Image Processing in Brachytherapy of Prostate Cancer

Brachytherapy is a widely used treatment modality for prostate cancer where treatments are delivered either in a mono-modality setting or in combination with EBRT. Prostate brachytherapy has predominantly been delivered as a real-time, image-guided procedure using transrectal US with the patient in lithotomy position.

7.2.2.1 Prostate Delineation

While prostate US imaging in general provides adequate contrast between the prostate and the surrounding tissue, the base and apex can be difficult to delineate. Also, the presence of applicator needles will degrade US image quality. Several authors have tried to address these problems through image registration between pretreatment MR images, where the prostate is more easily identifiable, and US images acquired during the implant procedure (Reynier et al., 2004; Mitra et al., 2012; Sparks et al., 2013).

7.2.2.2 Dose Distribution Accumulation

In US-based brachytherapy, the image field-of-view is limited, and consequently, only parts of the bladder and the rectum can be contoured for DVH calculation. This limits the brachytherapy OAR dose distribution information and may be one of the reasons for the limited number of investigations on organ motion and cumulative dose distributions for high-dose-rate (HDR) prostate brachytherapy. In an investigation with a similar purpose for low-dose-rate (LDR) brachytherapy, Zhang et al. (2010) used deformable image registration to register post-implant CT images to a CT acquired for intensity-modulated radiotherapy treatment planning. They then recalculated both dose distributions taking radiobiological effects into account before adding the dose distributions voxel-wise using the DVF obtained from the deformable image registration.

7.2.2.3 Ultrasound-Probe-Induced Prostate Deformation

The presence of a transrectal US probe influences not only the shape of the rectum wall, but also the prostate and surrounding anatomy. In a study of the anatomical variation and rectal dose distribution difference for HDR treatments with and without US probe, Rylander et al. (2015) did rigid image registration between US and MR images, using the prostate gland and the most posterior needles as landmarks. They evaluated the change in distance between the prostate and the anterior rectal wall and found that the median distance increased by 10 mm at the base of the prostate and decreased by 2 mm at the apex. In a similar study of LDR brachytherapy, probe-pressure-induced prostate deformation was quantified by x-ray-based seed reconstruction with and without the US probe present (Liu et al., 2015). They found that by removing the probe, the prostate had been decompressed in the anterior–posterior direction by 2 mm on average and contracted in the superior–inferior direction by 1 mm on average.

7.2.3 Image Processing in Brachytherapy of Head-and-Neck Cancer and Breast Cancer

Image registration for brachytherapy has also been used in other sites. Hilts et al. (2015) used a combination of rigid and deformable intensity-based registration for CT image sets before and after LDR seed implantation to aid in breast cancer seroma visualization. For permanent LDR treatment of breast cancer, Watt et al. (2015) used intensity-based deformable registration to investigate the effect of ipsilateral arm position on post-implant dosimetry. Osorio et al. (2011) used deformable registration to evaluate the cumulative EBRT and interstitial brachytherapy dose distribution for oropharyngeal cancer.

7.2.4 Current Challenges in Brachytherapy Image Registration

7.2.4.1 General Image Registration Challenges

One of the challenges in image registration is that the accuracy is difficult to assess because of the lack, in the general case, of a ground truth to compare against. The most intuitive evaluation is to look at the spatial agreement using easily identifiable features or landmarks in both images. Metrics such as the Hausdorff distance and the Dice similarity coefficient (DSC) have been suggested for more quantitative analyses (Dice, 1945; Ghose et al., 2015) when comparing target images and registered deformed images. It should, however, be noted that, even when there are perfect correspondences between volumes or surfaces, there is no guarantee that individual voxels have been accurately tracked (Rohlfing, 2012). Therefore, much effort has been dedicated to the evaluation of deformation algorithm performance (Serban et al., 2008; Castillo et al., 2009; Gu et al., 2010; Nie et al., 2013; Wognum et al., 2014; Graves et al., 2015; Kirby et al., 2016).

One evaluation approach is to use physical phantoms where algorithm behavior can be investigated on the voxel level. Phantoms capable of producing two-dimensional (2D) or 3D deformations have been designed (Serban et al., 2008; Graves et al., 2015). In one study, Wognum et al. (2014) used a porcine bladder model to investigate nine registration algorithms. After attaching 30–40 fiducials to the outer wall of excised bladders, they varied the filling between 100 and 400 cm^3 and acquired CT images. Using the attached fiducials as the ground truth, they showed that although the DSC was high (>0.95 for eight out of nine algorithms), the spatial correspondences of the markers were low. Median deviations were typically between 5 and 15 mm, where the registration between the bladders containing 100 and 400 cm^3 was the most challenging for all investigated algorithms. There are, however, not yet deformable physical phantoms specific to brachytherapy (see Chapter 10 for further detail). When designing such a phantom, one should ideally allow for the use of different applicators as well as the capability to place the applicators at different locations.

Another evaluation approach is to deform a fixed image with a virtual DVF and then do image registration between the fixed image and the created deformed moving image (Nie et al., 2013; Zhen et al., 2015). Both volume-based and voxel-to-voxel evaluations can now be done by comparing the DVF from the image registration to the ground truth represented by the virtual DVF.

7.2.4.2 Challenges in Pelvic Image Registration

There are challenges that limit the accuracy of image registration of OARs, especially in the pelvic region. The rectum, the sigmoid, and the urinary bladder are highly flexible organs that may exhibit large volume changes, substantial deformations, and large displacements making it difficult to accurately register them between images (Godley et al., 2009; Bondar et al., 2010; Osorio et al., 2015; Teo et al., 2015). Several approaches have been suggested to overcome one or several of these problems. Teo et al. (2015) registered CT images using intensity-based methods. They contoured the bladder and rectum in the images, assigned them a CT number of 1000 HU, and then used the artificially increased soft-tissue contrast to guide the registration. Furthermore, they also pre-processed air pockets in the rectum and bladder contrast as well as the volume occupied by the applicator. Osorio et al. (2015) used a combination of delineation of organs and autosegmentation of anatomical features for feature-based deformable registration of MR images. They created one local DVF for each of the defined OARs and for the autosegmented features, and one additional DVF for the remaining regions. The separate DVFs were then added into one final DVF using a weighted sum.

7.2.4.3 Applicator-Induced Challenges

The use of applicators will lead to additional challenges for deformation algorithms since they may create substantial local anatomy deformations that are difficult to accurately model. Furthermore, most deformable image registration algorithms rely on the assumption that there is a one-to-one correspondence between image voxels. This creates complications when different applicators are used for different fractions, or when registration is performed between images with and without applicator(s). One approach to overcome these difficulties has been proposed by Zhen et al. (2015). They developed a semi-automatic algorithm to segment the applicators and remove them from each CT image. Image registration was then performed on the applicator-free images (Figure 7.2). Evaluation of the algorithm consisted of both volume-based metrics such as the DSC as well as by applying virtual DVFs to the fixed image and comparing this to the DVFs from the image registration.

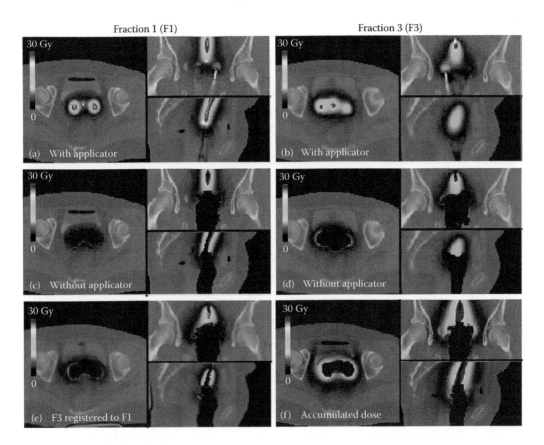

FIGURE 7.2 Registration and dose accumulation of CT images for cervical cancer brachytherapy using ovoid and tandem applicators. Images are shown in three orthogonal slices. (a) Dose distribution for fraction 1 (F1) with applicator; (b) dose distribution for fraction 3 (F3) with applicator; (c) dose distribution for F1 without applicator; (d) dose distribution for F3 without applicator; (e) CT image for F3 after image registration to F1; and (f) the cumulative dose distribution for F1 and F3.

In a similar study, Berendsen et al. (2014) registered MR images from brachytherapy and EBRT by manually contouring the applicator in the brachytherapy image. They then introduced a volume-based penalty term during the registration to minimize the influence of the volume of the applicator void.

7.2.4.4 Consequences for Dose Accumulation

The challenges outlined above will negatively impact the accuracy of how well individual voxels are tracked during image registration and will therefore also impact the accuracy of the corresponding dose distribution accumulation. This issue is illustrated in the study by Jamema et al. (2015), who recently investigated dose accumulation for cervical cancer brachytherapy. They observed that intensity-based registration of MR images "resulted frequently in implausible deformations because it was not regulated by mechanical properties of organ walls" and quantified the effect on cumulative dose by doing intensity-based image registration twice, alternating the static and moving image sets. The resulting absolute differences in fractionation-corrected D_{2cm3} for the bladder and the rectum are shown in Figure 7.3.

The resulting image registration errors will lead to uncertainty in cumulative dose. This uncertainty will be particularly large in high dose/high dose gradient regions.

Another issue that has been recognized is that when registering structures with large volume changes, the choice of fixed/moving image will affect the result of dose accumulation (Andersen et al., 2013; Jamema et al., 2015). This will be the case even if image registration is accurately done for each voxel.

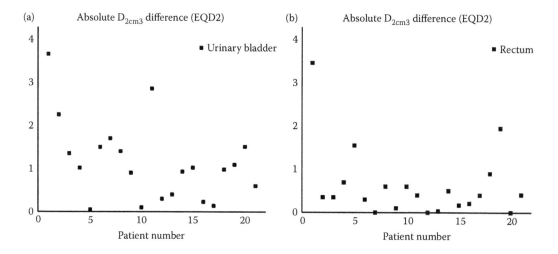

FIGURE 7.3   Absolute difference in the fractionation-corrected D_{2cm3} for the urinary bladder (a) and the rectum (b) after image registration alternating the roles of fixed and moving images. The prescribed dose in each fraction was 7.0 Gy. (Data kindly provided by Jamema, S.V., U. Mahantshetty, K. Tanderup et al. *Radiotherapy and Oncology*, 2013; 107(1): 58–62.)

7.3 AUTOMATIC SEED LOCALIZATION AND AUTOMATIC APPLICATOR RECONSTRUCTION

Accurate reconstruction of source positions, either as dwell positions inside an applicator or as radioactive seed locations, is a fundamental part of brachytherapy treatment planning. Frequently used applicators with known geometry, such as ring applicators, are typically stored in the treatment planning system as library applicators, which can be imported into patient images. Therefore, their sizes and shapes do not need to be reconstructed on a patient-to-patient basis. However, the applicator needs to be correctly aligned with the acquired images. On the other hand, in situations where many implanted seeds need to be localized or when applicators of arbitrary shapes or positions must be reconstructed, image-processing techniques can assist in fast and accurate methods.

7.3.1 Automatic Applicator Reconstruction

Geometrical reconstruction of interstitial applicators can be time consuming and may be related with risks of introducing manual errors, especially in multi-catheter implants (see Chapter 4 for an in-depth discussion). Milickovic et al. (2000) developed an algorithm to automatically detect and reconstruct interstitial applicators in CT images using intensity-based analysis. The centers of 2-mm applicators were detected and reconstructed in 3D images. In a recent study of gynecological applications, Dise et al. (2015) developed a region-growing algorithm to detect and reconstruct interstitial applicators for the Syed–Neblett template in images acquired by CT.

7.3.2 Automatic Seed Localization

During LDR implant procedures for prostate brachytherapy, ^{125}I or ^{103}Pd seeds are inserted under transrectal US guidance. Since implanted seed visibility on US images may be limited, the dose distribution cannot always be calculated during the implantation procedure. This can potentially lead to poor implant quality. To overcome this problem, methods to complement the real-time US imaging with intraoperative x-ray imaging for seed localization have been developed (Nag et al., 2001; Jain et al., 2005; Fallavollita et al., 2010; Dehghan et al., 2011, 2012; Kuo et al., 2012). Methods typically include the use of multi-angle images to automatically reconstruct the seed collection geometry. Further developments have included methods where the reconstructed seed set subsequently has been registered to the real-time 3D US images (Fallavollita et al., 2010; Dehghan et al., 2012). These methods enable an intraoperative evaluation of the target dose coverage with the possibility to add more seeds in underdosed regions.

Several authors have investigated the possibility of using image processing on 3D US images to track the LDR implant needle in real time (Ding et al., 2006; Wei et al., 2006; Qiu et al., 2013). For instance, Wei et al. (2006) suggested using subtraction of images with and without the needles in combination with a needle segmentation algorithm. Once the needle position was found, they searched for seeds in the region along the trajectory of the needle. Image-processing techniques for automatic seed reconstruction have also been developed for post-implant evaluation of LDR implants, mainly using CT imaging (Liu et al., 2003; Holupka et al., 2004; Lu et al., 2013).

7.4 CONCLUSION

In this chapter, we have presented some of the latest applications of image processing in brachytherapy. Investigations of image registration with the purpose of dose accumulation in brachytherapy, with or without the addition of the EBRT dose distribution, have been focused mostly on cervical cancer treatments, although other sites have been studied as well. The challenges for accurate image registration in the pelvic region are substantial due to large intra- and inter-fraction anatomical variations, as well as to the presence of applicators. Several groups are currently addressing these issues. In addition, on-going work involves methods to validate deformable image registration algorithms. Image processing has recently also been used in prostate brachytherapy applications. For US-based brachytherapy, both MR and multi-angle x-ray imaging are used to assess how the presence of the transrectal probe affects target and OAR anatomy. For temporary and permanent US-based prostate brachytherapy, there are newly developed image-processing-based methods that aim at increasing the accuracy of seeds/needle reconstruction and target definition in the intraoperative setting.

REFERENCES

Andersen, E., L. Muren, T.S. Sørensen et al. Bladder dose accumulation based on a biomechanical deformable image registration algorithm in volumetric modulated arc therapy for prostate cancer. *Physics in Medicine and Biology*, 2012; 57(21): 7089.

Andersen, E.S., K.Ø. Noe, T.S. Sørensen et al. Simple DVH parameter addition as compared to deformable registration for bladder dose accumulation in cervix cancer brachytherapy. *Radiotherapy and Oncology*, 2013; 107(1): 52–57.

Berendsen, F., A. Kotte, A. de Leeuw, I. Jürgenliemk-Schulz, M. Viergever, and J. Pluim. Registration of structurally dissimilar images in MRI-based brachytherapy. *Physics in Medicine and Biology*, 2014; 59(15): 4033.

Bondar, L., M.S. Hoogeman, E.M.V. Osorio, and B.J. Heijmen. A symmetric nonrigid registration method to handle large organ deformations in cervical cancer patients. *Medical Physics*, 2010; 37(7): 3760–3772.

Castillo, R., E. Castillo, R. Guerra et al. A framework for evaluation of deformable image registration spatial accuracy using large landmark point sets. *Physics in Medicine and Biology*, 2009; 54(7): 1849.

Christensen, G.E., B. Carlson, K.C. Chao et al. Image-based dose planning of intracavitary brachytherapy: Registration of serial-imaging studies using deformable anatomic templates. *International Journal of Radiation Oncology* Biology* Physics*, 2001; 51(1): 227–243.

Crum, W.R., T. Hartkens, and D. Hill. Non-rigid image registration: Theory and practice. *The British Journal of Radiology*, 2014; 77(suppl. 2): S140–153.

Dehghan, E., A.K. Jain, M. Moradi et al. Brachytherapy seed reconstruction with joint-encoded C-arm single-axis rotation and motion compensation. *Medical Image Analysis*, 2011; 15(5): 760–771.

Dehghan, E., J. Lee, P. Fallavollita et al. Ultrasound–fluoroscopy registration for prostate brachytherapy dosimetry. *Medical Image Analysis*, 2012; 16(7): 1347–1358.

Dice, L.R. Measures of the amount of ecologic association between species. *Ecology*, 1945; 26(3): 297–302.

Ding, M., Z. Wei, L. Gardi, D.B. Downey, and A. Fenster. Needle and seed segmentation in intra-operative 3D ultrasound-guided prostate brachytherapy. *Ultrasonics*, 2006; 44: e331–e336.

Dise, J., X. Liang, J. Scheuermann et al. Development and evaluation of an automatic interstitial catheter digitization tool for adaptive high-dose-rate brachytherapy. *Brachytherapy*, 2015; 14(5): 619–625.

Fallavollita, P., Z.K. Aghaloo, E. Burdette, D. Song, P. Abolmaesumi, and G. Fichtinger. Registration between ultrasound and fluoroscopy or CT in prostate brachytherapy. *Medical Physics*, 2010; 37(6): 2749–2760.

Ghose, S., L. Holloway, K. Lim et al. A review of segmentation and deformable registration methods applied to adaptive cervical cancer radiation therapy treatment planning. *Artificial Intelligence in Medicine*, 2015; 64(2): 75–87.

Godley, A., E. Ahunbay, C. Peng, and X.A. Li. Automated registration of large deformations for adaptive radiation therapy of prostate cancer. *Medical Physics*, 2009; 36(4): 1433–1441.

Graves, Y.J., A.-A. Smith, D. Mcilvena et al. A deformable head and neck phantom with in-vivo dosimetry for adaptive radiotherapy quality assurance. *Medical Physics*, 2015; 42(4): 1490–1497.

Gu, X., H. Pan, Y. Liang et al. Implementation and evaluation of various demons deformable image registration algorithms on a GPU. *Physics in Medicine and Biology*, 2010; 55(1): 207.

Hilts, M., D. Batchelar, J. Rose, and J. Crook. Deformable image registration for defining the post-implant seroma in permanent breast seed implant brachytherapy. *Brachytherapy*, 2015; 14(3): 409–418.

Holupka, E., P. Meskell, E. Burdette, and I. Kaplan. An automatic seed finder for brachytherapy CT postplans based on the Hough transform. *Medical Physics*, 2004; 31(9): 2672–2679.

Jain, A.K., Y. Zhou, T. Mustufa, E.C. Burdette, G.S. Chirikjian, and G. Fichtinger. Matching and reconstruction of brachytherapy seeds using the Hungarian algorithm (MARSHAL). *Medical Physics*, 2005; 32(11): 3475–3492.

Jamema, S.V., U. Mahantshetty, E. Andersen et al. Uncertainties of deformable image registration for dose accumulation of high-dose regions in bladder and rectum in locally advanced cervical cancer. *Brachytherapy*, 2015; 14(6): 953–962.

Jamema, S.V., U. Mahantshetty, K. Tanderup et al. Inter-application variation of dose and spatial location of D_{2cm^3} volumes of OARs during MR image based cervix brachytherapy. *Radiotherapy and Oncology*, 2013; 107(1): 58–62.

Kessler, M.L. Image registration and data fusion in radiation therapy. *The British Journal of Radiology*, 2014; 79(special issue 1): S99–108.

Kirby, N., J. Chen, H. Kim, O. Morin, K. Nie, and J. Pouliot. An automated deformable image registration evaluation of confidence tool. *Physics in Medicine and Biology*, 2016; 61(8): N203.

Kuo, N., A. Deguet, D.Y. Song, E.C. Burdette, J.L. Prince, and J. Lee. Automatic segmentation of radiographic fiducial and seeds from X-ray images in prostate brachytherapy. *Medical Engineering & Physics*, 2012; 34(1): 64–77.

Lang, S., N. Nesvacil, C. Kirisits et al. Uncertainty analysis for 3D image-based cervix cancer brachytherapy by repetitive MR imaging: Assessment of DVH-variations between two HDR fractions within one applicator insertion and their clinical relevance. *Radiotherapy and Oncology*, 2013; 107(1): 26–31.

Liu, D., T. Meyer, N. Usmani et al. Implanted brachytherapy seed movement reflecting transrectal ultrasound probe-induced prostate deformation. *Brachytherapy*, 2015; 14(6): 809–817.

Liu, H., G. Cheng, Y. Yu et al. Automatic localization of implanted seeds from post-implant CT images. *Physics in Medicine and Biology*, 2003; 48(9): 1191.

Lu, H., Z. Cuan, F. Zhou, and B. Liu. *An automatic 3D detection method of seeds on CT images.* In *2013 IEEE International Conference on Medical Imaging Physics and Engineering (ICMIPE)*. Shenyang City, 2013. IEEE. http://www.ieee.org/conferences_events/conferences/conference-details/index.html?Conf_ID=31005

Maintz, J.A. and M.A. Viergever. A survey of medical image registration. *Medical Image Analysis*, 1998; 2(1): 1–36.

Milickovic, N., S. Giannouli, D. Baltas et al. Catheter autoreconstruction in computed tomography based brachytherapy treatment planning. *Medical Physics*, 2000; 27(5): 1047–1057.

Mitra, J., Z. Kato, R. Martí et al. A spline-based non-linear diffeomorphism for multimodal prostate registration. *Medical Image Analysis*, 2012; 16(6): 1259–1279.

Nag, S., J.P. Ciezki, R. Cormack et al. Intraoperative planning and evaluation of permanent prostate brachytherapy: Report of the American Brachytherapy Society. *International Journal of Radiation Oncology* Biology* Physics*, 2001; 51(5): 1422–1430.

Nesvacil, N., K. Tanderup, T.P. Hellebust et al. A multicentre comparison of the dosimetric impact of inter- and intra-fractional anatomical variations in fractionated cervix cancer brachytherapy. *Radiotherapy and Oncology*, 2013; 107(1): 20–25.

Nie, K., C. Chuang, N. Kirby, S. Braunstein, and J. Pouliot. Site-specific deformable imaging registration algorithm selection using patient-based simulated deformations. *Medical Physics*, 2013; 40(4): 041911.

Oliveira, F.P. and J.M.R. Tavares. Medical image registration: A review. *Computer Methods in Biomechanics and Biomedical Engineering*, 2014; 17(2): 73–93.

Osorio, E.M.V., M.S. Hoogeman, D.N. Teguh et al. Three-dimensional dose addition of external beam radiotherapy and brachytherapy for oropharyngeal patients using nonrigid registration. *International Journal of Radiation Oncology* Biology* Physics*, 2011; 80(4): 1268–1277.

Osorio, E.M.V., I.-K.K. Kolkman-Deurloo, M. Schuring-Pereira, A. Zolnay, B.J. Heijmen, and M.S. Hoogeman. Improving anatomical mapping of complexly deformed anatomy for external beam radiotherapy and brachytherapy dose accumulation in cervical cancer. *Medical Physics*, 2015; 42(1): 206–220.

Pötter, R., C. Haie-Meder, E. Van Limbergen et al. Recommendations from gynaecological (GYN) GEC ESTRO working group (II): Concepts and terms in 3D image-based treatment planning in cervix cancer brachytherapy—3D dose volume parameters and aspects of 3D image-based anatomy, radiation physics, radiobiology. *Radiotherapy and Oncology*, 2006; 78(1): 67–77.

Qiu, W., M. Yuchi, M. Ding, D. Tessier, and A. Fenster. Needle segmentation using 3D Hough transform in 3D TRUS guided prostate transperineal therapy. *Medical Physics*, 2013; 40(4): 042902.

Reynier, C., J. Troccaz, P. Fourneret et al. MRI/TRUS data fusion for prostate brachytherapy. Preliminary results. *Medical Physics*, 2004; 31(6): 1568–1575.

Rohlfing, T. Image similarity and tissue overlaps as surrogates for image registration accuracy: widely used but unreliable. *IEEE Transactions on Medical Imaging*, 2012; 31(2):153–163.

Rylander, S., S. Buus, L. Bentzen, E.M. Pedersen, and K. Tanderup. The influence of a rectal ultrasound probe on the separation between prostate and rectum in high-dose-rate brachytherapy. *Brachytherapy*, 2015; 14(5): 711–717.

Serban, M., E. Heath, G. Stroian, D.L. Collins, and J. Seuntjens. A deformable phantom for 4D radiotherapy verification: Design and image registration evaluation. *Medical Physics*, 2008; 35(3): 1094–1102.

Simha, V., F.D. Patel, S.C. Sharma, B. Rai, A.S. Oinam, and B. Dhanireddy. Evaluation of intrafraction motion of the organs at risk in image-based brachytherapy of cervical cancer. *Brachytherapy*, 2014; 13(6): 562–567.

Sparks, R., B.N. Bloch, E. Feleppa, D. Barratt, and A. Madabhushi. Fully automated prostate magnetic resonance imaging and transrectal ultrasound fusion via a probabilistic registration metric. In *SPIE Medical Imaging. 2013. International Society for Optics and Photonics*, Lake Buena Vista.

Teo, B.-K., L.P.B. Millar, X. Ding, and L.L. Lin. Assessment of cumulative external beam and intracavitary brachytherapy organ doses in gynecologic cancers using deformable dose summation. *Radiotherapy and Oncology*, 2015; 115(2): 195–202.

van Dyk, S., S. Kondalsamy-Chennakesavan, M. Schneider, D. Bernshaw, and K. Narayan. Assessing changes to the brachytherapy target for cervical cancer using a single MRI and serial ultrasound. *Brachytherapy*, 2015; 14(6): 889–897.

Watt, E., S. Husain, M. Sia, D. Brown, K. Long, and T. Meyer. Dosimetric variations in permanent breast seed implant due to patient arm position. *Brachytherapy*, 2015; 14(6): 979–985.

Wei, Z., L. Gardi, D.B. Downey, and A. Fenster. Automated localization of implanted seeds in 3D TRUS images used for prostate brachytherapy. *Medical Physics*, 2006; 33(7): 2404–2417.

Wognum, S., S. Heethuis, T. Rosario, M. Hoogeman, and A. Bel. Validation of deformable image registration algorithms on CT images of ex vivo porcine bladders with fiducial markers. *Medical Physics*, 2014; 41(7): 071916.

Zhang, G., T.-C. Huang, V. Feygelman, C. Stevens, and K. Forster. Generation of composite dose and biological effective dose (BED) over multiple treatment modalities and multistage planning using deformable image registration. *Medical Dosimetry*, 2010; 35(2): 143–150.

Zhen, X., H. Chen, H. Yan et al. A segmentation and point-matching enhanced efficient deformable image registration method for dose accumulation between HDR CT images. *Physics in Medicine and Biology*, 2015; 60(7): 2981.

CHAPTER 8

FMEA for Brachytherapy

Susan Richardson, Daniel Scanderbeg,
and Jamema Swamidas

CONTENTS

8.1 INTRODUCTION

A quality high-dose rate (HDR) brachytherapy radiation treatment remains dependent on many intertwined and complex procedures culminating in the single push of a button. Prior to that button push, calculations are double checked; multiple sets of eyes review and confirm planning parameters; checklists are dutifully completed. Barriers are hard wired into the system to prevent failures and morning quality assurance tests to check the mechanics of the radiation delivery system are performed. These were all developed using a variety of quality methodologies to ensure a safe and high-quality treatment for the patient.

In 2012, the American Society for Therapeutic Radiation Oncology (ASTRO) published a document titled "Safety is No Accident: A Framework for Quality Radiation Oncology and Care." The document describes not only the need for patient safety to be of utmost importance in the care and treatment of patients, but also the need for a new type of culture to emerge from departments in which each team member plays an important role in that safety (Safety is No Accident). It also gives examples of tools and methodologies to minimize the frequency of mistakes from reaching the patient, including the use of failure modes and effects analysis (FMEA) as a risk analysis tool.

For medical physicists, the ultimate resource for describing FMEA in a clinical setting is Task Group 100—"A New Paradigm for Quality Management in Radiation Therapy" (2013). While this document is yet unpublished at the time of this writing, it has been a 10-year culmination of work of physicists and collaborators to guide and educate clinical medical physicists in the successes, failures, complexities, and importance of risk analysis in our field. The motivation for TG100 is to transition from a measurement, reactive-based process into a theoretical and anticipatory methodology for predicting potential errors and mistakes. In this way, medical physicists can tailor their measurements and efforts into work that is the most effective in reducing errors that reach the patient and improve overall patient safety.

While it is agreed upon that there is no perfect method to eliminate all errors (Dunscombe, 2012), FMEA is a risk analysis tool used in many industries, including medical device manufacturing, airline travel, nuclear power, etc. At the time of this writing, a quick Google scholar search for "FMEA Radiation Therapy" results in over 1000 different "hits." FMEA has been applied to tomotherapy (Broggi et al., 2013), proton therapy (Cantone et al., 2013), image-guided radiotherapy (Noel et al., 2014), dynamic Multi-Leaf Collimator (MLC) tracking (Sawant et al., 2010), stereotactic radiosurgery (Younge et al., 2015), intraoperative electron therapy (Ciocca et al., 2012), and finally, sparsely in brachytherapy (Venselaar et al., 2012; Wilkinson and Kolar, 2013; Mayadev et al., 2015); essentially it may be used for analysis of any radiation delivery technique. This chapter will describe the techniques and tools needed to perform an FMEA and other safety strategies and give examples when applied to HDR gynecological brachytherapy and low-dose rate (LDR) breast brachytherapy.

8.2 SAFETY STRATEGIES AND TECHNIQUES

Eric Ford and colleagues estimated that there is approximately a 0.2% rate of misadministration in the United States in radiation oncology (Ford and Terezakis, 2010). That works out to about 1 in every 500 patients. However, they compared that to the chance of injury or death on a U.S. commercial airline, which works out to approximately 1 in 10 million. While a misadministration does not directly correlate to patient injury or death in the medical field, this statistic alone should demonstrate the need for formal, prospective methodologies to anticipate and mitigate accidents in radiation oncology; there is room for improvement. In addition, given that there is no "standard" clinic, resources and workflow can differ considerably at each institution, leading to large variability in rankings of risks, or different failure modes altogether. Therefore, it is crucial that each institution perform a prospective risk analysis using currently allocated resources and current workflow to assist in identifying potential areas of risk or hazard.

One of the largest issues with implementation of such a program can be institutional culture. Creating and developing an effective safety culture can be the key component in setting up a program. It is much easier to find errors or potential errors if the staff feels empowered to report unsafe circumstances, so there needs to be a safe and easy mechanism to report information in a nonpunitive environment. However, this takes cooperation from administration leaders and effective communication of the program through all ranks. In addition, it takes a team to collate the reports, analyze them, and disseminate the information, lesson(s) learned, and subsequently implement safety barriers to prevent further mishaps. An excellent example of such a process was reported by Potters and Kapur. Their "no fly policy" program, initiated in 2008 and published in 2012, aimed at improving patient care, but also worked on changing their departmental safety culture (Kapur and Potters, 2012; Potters and Kapur, 2012). After analyzing their department clinical pathways, a "no fly policy" was implemented that delayed the start of a patient's treatment if tasks were not performed at certain predetermined time intervals, instead of rushing the patient's treatment.

There are a variety of patient safety improvement strategies that have been suggested for implementation in medicine, and more specifically, radiation oncology. Recommendations from the American Association of Physicists in Medicine (AAPM) in their Task Group 100 report describe risk analysis such as FMEA and fault tree analysis (FTA) (Thomadsen et al., 2013). Meanwhile, in publication 112 from the International Commission on Radiological Protection (ICRP, 2009), FMEA along with probabilistic safety assessment and risk matrices are recommended. Brief descriptions of some of these strategies are given in the following sections.

8.2.1 Failure Mode and Effects Analysis

FMEA was first developed by the U.S. military and an early adopter of this formalism was the U.S. National Aeronautics and Space Administration (NASA) for a large number of its programs. It was then used by the civil airline industry and adopted by the Society for Automotive Engineers (SAE). Since then, FMEA has seen widespread adoption by a variety of industries such as healthcare, aerospace, motor vehicles, semiconductors, and agriculture.

FMEA is a prospective risk management tool that helps users to identify key failure modes, their cause(s), and the risk or hazard associated with them; with the risk assessed quantitatively based on three parameters: frequency, detectability, and severity. Frequency is how likely the particular failure mode is to occur, detectability is how likely the failure mode would be detected with current procedures in place, and severity is the severity of the outcome such as no effect or death of patient or staff, in the case of healthcare. These three parameters are usually scored on a scale of 1–10 with 10 being the highest score (highest frequency, lowest probability of detection, or most severe outcome) and then the three values are multiplied together to get an RPN score ($=F \times D \times S$). The RPN score is the risk priority number and those failure modes with the highest RPN scores are usually the first to be addressed with mitigation techniques. Usually, a threshold RPN should be established and RPNs that are higher than that number need to be analyzed for the potential causes of failures and feasible and effective ways to improve these processes. This number is entirely arbitrary based on the number of failure modes addressed and the overall distribution of the RPN score. One institution's FMEA may have RPNs over 100 investigated, while another might select 150.

The overall goal for risk management purposes is to reduce the RPN found for each failure. This can be done by several methods: increasing the detectability, reducing the severity, or redesigning the entire process so the failure mode no longer exists. Many resources exist to help the users perform such a task such as employing redundancy, using written documentation and procedures, hard-stops, etc. (Kubo et al., 1998). Any textbook or paper on brachytherapy quality assurance can help the user design a safe program. In addition, the knowledge and incorporation of common errors reported in brachytherapy can be helpful in designing an FMEA or determining failure modes, such as those reported by Thomadsen and Richardson (Thomadsen et al., 2003; Richardson, 2012) or other publications by the ICRP and International Atomic Energy Agency (IAEA) (Valentin, 2004; IAEA Safety Report Series 17, 2000).

8.2.2 Fault Tree Analysis

FTA was first developed at Bell Laboratories by H.A. Watson in the early 1960s. It was initially used to look at the launch control systems of the Minuteman missiles. However, due to its utility and effectiveness, it was adopted by several groups, including the aircraft manufacturer Boeing (Ericson, 2015). It was eventually adopted by the Federal Aeronautics Administration (FAA) and also the Nuclear Regulatory Commission (NRC) and even applied to the Three Mile Island nuclear accident.

FTA is a top-down approach that creates a visual diagram using standard logic symbols, most commonly comprised of blocks for events and connected using AND or OR gates. A fault tree is the complement of process trees (described below) and begins with something that could go wrong or one of the failure modes. The fault tree analyzes what actions or events could cause an error, or the paths that lead to the error. In this way, a person can graphically follow events from their failure back to potential causes. Thus, it is very important when choosing a failure event such that it is not too general to cause an unreasonable or unmanageable number of potential causes and that it is not too specific to yield any useful analysis. The U.S. NRC has an entire handbook devoted to FTA (NUREG-0492, 1981).

8.2.3 Process Trees and Fishbone Diagrams

There are several types of schematic diagrams that are extremely useful in prospective analysis and have been used in a variety of fields. A process tree is a schematic diagram that is typically used in industrial design engineering during the conceptual stages of product design when prospectively thinking of the processes that a product will go through during its lifetime with time "passing" from left to right. For radiation therapy, the process tree could be used from the perspective of the patient and the different interactions, places, and steps the patient (or their data) must go through to be treated. It is referred to as a tree, as it consists of a main trunk (process) and a lot of branches (subprocesses) attached to the trunk. The major processes run in the middle like a trunk of a tree, which will branch out into various subprocesses. The process tree is specific to the institution and the process itself. To generate this process tree, it is recommended that a team is formed consisting of all the members involved in the workflow, for example; administrator, physician, nurse, physicist, radiotherapy technologist, etc.

A fishbone diagram has a variety of names including Ishikawa, herringbone, or cause and effect diagram. These are considered one of the seven basic tools of quality developed

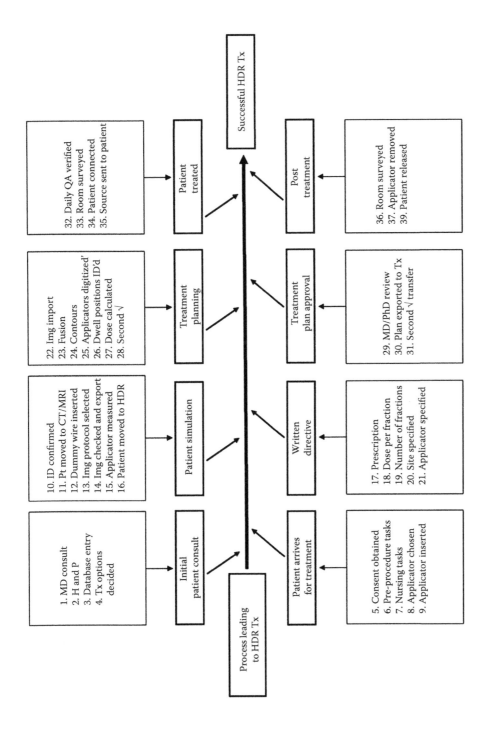

FIGURE 8.1 Fishbone diagram for HDR brachytherapy.

by Professor Kaoru Ishikawa in the 1960s at Tokyo University. This type of diagram is a graphic design grouping causes together for a particular effect. A simple design example that could be applied to HDR brachytherapy is shown in Figure 8.1. Although the patient may have many different processes through their treatment, it may be easier to break each process into a smaller step, and then subdivide into individual steps in the processes.

8.3 IMPLEMENTATION OF FMEA

As previously described, FMEA is a method, which is used to proactively detect risks in a particular process and correct potential errors before adverse events occur. FMEA is carried out in a systematic, multidisciplinary team-based approach (Wilkinson and Kolar, 2013). However, there are valid criticisms of FMEA (Papadopoulos et al., 2004; Bradley and Guerrero, 2011)

- The FMEA is only as good as the team put together to review the system
- FMEA is time consuming
- FMEA only assesses problems; it doesn't eliminate them
- RPN scores are guesses or estimates
- Often severities cannot be reduced and will always be a 10
- Not proven to be cost-effective

Even with these limitations, FMEA is still a tool that can provide insight and guidance for radiation therapy departments. Used in conjunction with other risk analysis tools and methodologies, it can be part of an effective safety program. The objective of the following section is to briefly describe the implementation of FMEA, and the associated challenges for the example of HDR gynecological brachytherapy.

8.3.1 Implementation of FMEA—Example 1: FMEA of HDR Brachytherapy Implemented in Tata Memorial Hospital, Mumbai, India

Cervical cancer is the most common cancer among women in India and Tata Memorial Hospital (TMH) treats a large patient load of gynecological brachytherapy procedures (Dinshaw et al., 1999). Although the procedure is well established, the recent transition from two-dimensional (2D) to three-dimensional (3D)-Computed Tomography (CT)/ Magnetic Resonance (MR) image-based dosimetry, introduction of new applicators, new planning systems/algorithms, multiple planners, and growing patient numbers suggested a more systematic approach for quality control/management. Hence, it was decided to carry out a systematic analysis of FMEA for intracavitary brachytherapy for cervical cancer (Swamidas et al., 2010).

A multidisciplinary team consisting of two physicians, two physicists, two dosimetrists, a medical resident, a nurse, and a secretary was formed. A weekly meeting was held for 4 months. A process tree was created based on the overview of the entire process, with

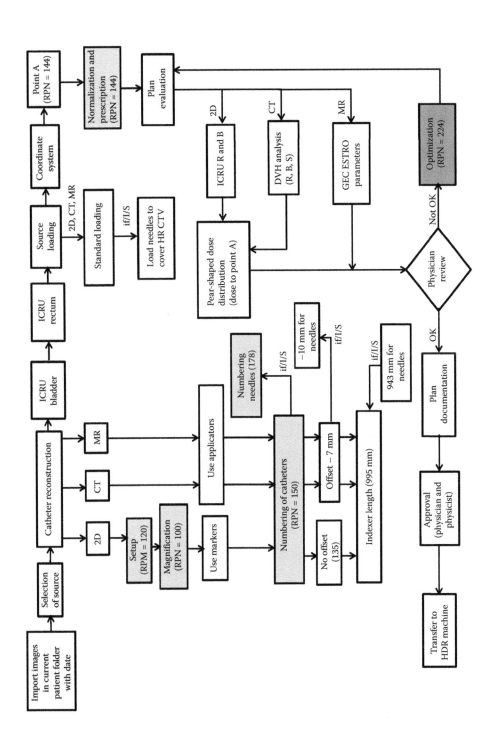

FIGURE 8.2 Marked process map of treatment planning. Dark gray indicates RPN>200 and light gray indicates RPN between 100 and 200.

the main branches as follows: procedure in the operating room, patient imaging, contouring, treatment planning, machine quality assurance, and treatment delivery (Figure 8.2). The process tree starts from the initial consultation, leading to HDR intracavitary brachytherapy treatment for cervical cancer. Both conventional dosimetry based on orthogonal radiographs and CT/MR image-based planning were considered, as both techniques are routinely practiced based on the departmental protocol. Each team member assigned RPNs based on the predefined scoring system (Ford et al., 2009). For a particular failure mode, if the RPN assigned by one member differed from the other, the highest RPN was taken into consideration.

The process tree consisted of 185 modes, with RPNs ranging from 1 to 220, with 77 possible failure modes. Four modes were found with RPN>200, which were considered for immediate process improvements. Twenty four modes were found to be with RPNs ranging from 100 to 200. All 24 processes were considered for process improvement, out of which 12 improvements were found effective and feasible, which includes failure modes with high severity score $S \geq 8$.

Table 8.1 describes the top scoring RPN modes and the corresponding process improvements that were implemented. Table 8.2 lists the failure modes with RPNs ranging from

TABLE 8.1   Failure Modes with High RPNs and the Corresponding Process Improvements

Failure Number	RPNs >200	Process Improvement
1	Patient entry: treating a wrong patient	
	Introduced an electronic medical record system for brachytherapy. Such a system with patient identification (with photograph) details already existing in the department for external beam radiotherapy; however, it was implemented for brachytherapy. The patient wrist also contains a tag with the patient details which can also be used to verify the identity of the patient	
2	Choosing an incorrect plan, wrong dose prescription (6, 7, or 9 Gy)	A checklist (treatment delivery) was introduced with specific details which includes the plan name with a nomenclature, for example: Final_date_dose, which could act as a quick check for selecting a correct plan. This was further facilitated by the radiation oncology information system introduced for brachytherapy
3	Plan optimization: MR image-based interstitial implant planning—prescription and normalization cannot be done with selected catheters	
	The process improvement was done by educating the physicists by familiarizing them with the planning aspects, and also introduction of checklist for treatment planning. Intracavitary and interstitial plans were made only by senior and experienced physicists	

TABLE 8.2   Processes with RPN 100–200

Failure Mode	Process	Subprocess	RPN	F	D	S
1	Application in operations theater	Incorrect review of patient details	130	4	4	8
2		Uterine perforation	135	3	5	9
3		Incorrect external fixation of the applicator	100	2	5	10
4		Enema/preparation not done properly	100	4	5	5
5	Imaging	Incorrect patient registration in simulator	160	5	4	8
6		Bladder protocol	168	6	7	4
7		Fixation of isocenter	120	4	6	5
8	Contouring	Incorrect import of images	150	3	5	10
9		Posterior vaginal wall not seen	120	6	4	5
10		Incorrect OARs	140	7	4	5
11		Incorrect tumor volume	192	6	4	8
12		Physician review	130	4	4	8
13	Treatment planning	Incorrect setup in reconstruction	120	5	6	4
14		Incorrect magnification	100	5	5	4
15		Incorrect numbering of catheters	160	4	5	8
16		Incorrect numbering of needles	160	4	5	8
17		No offset for first dwell position	135	3	9	5
18		Incorrect normalization and prescription	144	3	6	8
19	Machine QA	Incorrect source strength	140	4	5	7
20		Incorrect dwell position accuracy	120	4	5	6
21		Source strength specification	180	5	4	9
22		Applicator transfer tube QA	160	4	5	8
23	Treatment delivery	Check cable run	100	5	5	4
24		Check treatment time	120	4	5	6

100 to 200 with their corresponding RPNs. The processes with high RPNs (>200) were reduced after the introduction of process improvements. For the other processes, standard procedures were modified; for example: checks of the applicators/transfer tubes, rechecking the plan by a second physicist, treatment delivery details documented by the radiotherapy technologist at the treatment console in the radiation oncology information system among others. After the introduction of process improvements for RPNs more than 200, other subprocesses related to those failure modes were found to achieve high detectability and less severity.

Figure 8.2 shows the highest RPN locations in the process map. Upon further analysis, it was found that the common causes of failure, especially the high-ranking RPNs, were due to lack of attention, human error, and work pressure, which concurred with other publications of FMEA such as those by Wilkinson and Mayadev (Wilkinson and Kolar, 2013). Interestingly, human error factors were predominantly discussed in brachytherapy FMEA, as compared to other external beam techniques. This may be attributed to the absence of record and verification systems in most brachytherapy delivery systems as compared to external beam radiotherapy. Or, it could be due to the compressed time frame of imaging, planning, and treatment occurring in a single day. Another important factor related to the high ranking RPNs is the lack of communication among team members, which is also in agreement with Mayadev et al. One such example would be the length of the needles, or size of the ovoid used in a particular procedure in the operating room. A physician might forget to write these details in the file, which could lead to the errors. An introduction of a template sheet consisting of the geometry of the implant and other specific details written by the physician performing the procedure has reduced errors significantly at our center. At the end of the FMEA process, two checklists were introduced, one completed by the physician after the applicator insertion, and the other by the physicists after the treatment planning. It has been observed that this has significantly reduced the errors arising due to lack of communication. The process map was marked and it was found that the failure processes were uniformly distributed throughout the process tree in the present study which is again in agreement with Mayadev et al., where the failure modes were found in insertion, treatment planning, simulation, and treatment delivery (with severity >7). It is also important to consider individual frequency, severity, and detectability rankings instead of solely concentrating on the composite RPN of each failure mode. For example, severities of 10 (possibly patient death) should be examined even if they are extremely unlikely.

On comparison of failure modes, it was found that many of the same potential modes were similar to other studies. These include wrong image data set, incorrect bladder fill, Dose Volume Histogram (DVH) constraints not met, dwell position offset not given, wrong patient orientation, wrong indexer length, wrong dose prescription point, and catheter mislabeling, among others. Although, failure modes are similar, RPNs may vary among institutions because of the different possible implementations and differences in process trees. The current work aims not to suggest specific RPNs or threshold RPNs for action, but instead to act as a guide for the practical implementation of FMEA. The goal of any clinical procedure is to safely and efficiently deliver necessary care to the patient. The FMEA described here represents a model of the HDR brachytherapy process, and has been compared with the published literature. However, other clinics may have different needs, equipment, staff, and work culture, which may have to be adopted and fit to its own specific needs while implementing the FMEA.

It is important to note that the FMEA is limited in its ability to quantify risk. The scoring system is based on certain factors which makes it somewhat subjective. Therefore to get the most meaningful RPN scores, it is critical to include all members of the treatment team. In some cases it may be better to take the worst case scenario (e.g., highest severity and lowest detectability) into account by considering the highest score rather than the average score. There are various other issues which make it subjective and sometimes these tools can be difficult to utilize without proper training and support. Comparing the current work to Mayadev et al., it is interesting to note the two institutions showed similar results. The process trees for each institution were quite similar for the HDR treatments and the failure modes with the highest RPN values were mostly in agreement between the institutions. Three of the top five common failure modes (highest RPN score) at each institution were related to wrong applicator length (either measured or entered into the treatment planning system), wrong connection of the transfer guide tubes to the applicator, or wrong applicator inserted, that is also in agreement with other published results of the same process (Scanderbeg et al., 2012).

8.3.2 Implementation of FMEA—Example 2: FMEA for a New Procedure; LDR Breast Brachytherapy

Starting a new procedure in the clinic can be an intimidating process. Not only must everyone be trained in how to perform the new task at hand, but also often the team must rely on second-hand information rather than first-hand knowledge of how to perform the task. Only one person on the team might be familiar with the procedure or have gone to the manufacturer's training. Therefore, serious potential problems can be overlooked because not everyone on the clinical team has the necessary experience or knowledge to anticipate them. An FMEA can be performed for every new procedure as part of the implementation of the technique. In this example, the new procedure is relatively similar to other procedures done in brachytherapy. Therefore, the knowledge and experience of the team can be used to design the FMEA building upon the processes of other procedures in the department. However, an alternative approach exists in which the team performs a "reverse" FMEA where one can start with the effects instead of the process because the process steps are not known *a priori* (Siochi, 2017). In this method, there is no need for RPN because one does not know the observability or detectability in this circumstance.

A relatively novel procedure in North America is the LDR permanent breast brachytherapy procedure described by Pignol et al. (2006, 2015). This treatment is similar to a classic LDR prostate implant because a fixed template is used to guide needles and stranded seeds are placed in the target. In the breast brachytherapy procedure, an angled template is used to guide ^{103}Pd seeds (in strands) to the breast via an interstitial needle. This procedure is relatively new (started in 2006) and is gaining acceptance with a vendor (Breast Microseed, est. 2015) now providing equipment and training. To date, there are no publications describing an FMEA for this technique. In this procedure, the patient is an early stage breast cancer patient with an existing lumpectomy cavity. To succinctly describe the process, the patient is simulated and a treatment plan is developed by creating a custom CT data set in a plane parallel to a fiducial anchoring needle that is placed through the middle of the planning target volume (PTV). Unlike prostate brachytherapy, the template is not secured or flush against the patient's skin, rather, it is placed at some angle (30–50°) that allows the implant needles to easily cover the PTV. The patient is implanted similarly to a prostate implant with preloaded stranded ^{103}Pd needles, then subsequently receives a CT scan for post implant dosimetry (Keller et al., 2012).

To analyze this new procedure, first a simple process map can be built by breaking down the entire patient treatment into smaller steps. For example, the first step might be the patient coming in for CT simulation, followed by developing a treatment plan. Each logical step in the procedure can be sectioned out into the map flowing from left to right, or top to bottom. An example of a simple process map for the breast brachytherapy process is shown in Figure 8.3. For each step in the process map, the subsequent related procedures for that step can be listed in an itemized fashion. Often, it is helpful to separate it from the original process map so that each step can be considered in isolation from the next step. For

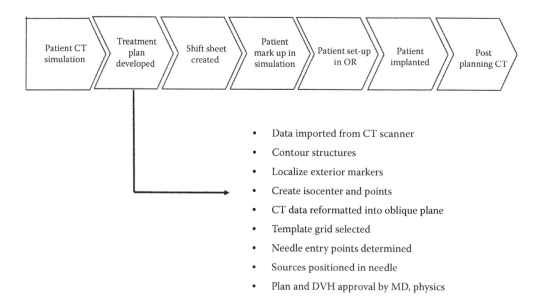

FIGURE 8.3 Treatment plan development process.

example, the second step shown above is "treatment plan developed." Each action, process, or procedure that occurs during that step should be listed.

Depending on the time and effort that is available to spend on the FMEA and process mapping, the options are endless. Each output from the process step under evaluation can be subprocessed in different ways. For example, this can be accomplished by color coding or using various shapes to indicate different types of steps. Data transfer could be indicated by a blue square, critical decisions by a red triangle, and steps where there is a hand off between individuals indicated by an asterisk. Arrows can be used to show the logical flow of time. Events where data are entered manually could be entered in a specific color so the step readily stands out from the rest. Figure 8.4 shows the expanded detail possible from the step "treatment plan developed."

It is entirely up to the team creating the process map to use tools as they see fit. One advantage of being more specific at an early stage is that related events can be quickly and easily seen during the FMEA process—for example, if the network in the department goes down, all blue square processes, which involve data transfer will be halted. Or when Quality Control (QC) steps need to be added, all the manual entry steps can be readily identified and two individuals assigned to the step. Where information is handed off between two individuals, for example, a dosimetrist and a physician, communication must occur. How that communication will be documented (through a quality checklist or planning order) and should be discussed among the individuals that receive information in that step and incorporated in the clinical procedure.

Now, each subprocesses may be analyzed for potential failure modes. In the critical step identified in Figure 8.4 "CT data reformatted into oblique plane," failure modes can be identified and placed in the FMEA. Any and all potential failure modes should be identified. These could range from: "CT data reformatted upside down" to "Dosimetrist forgets

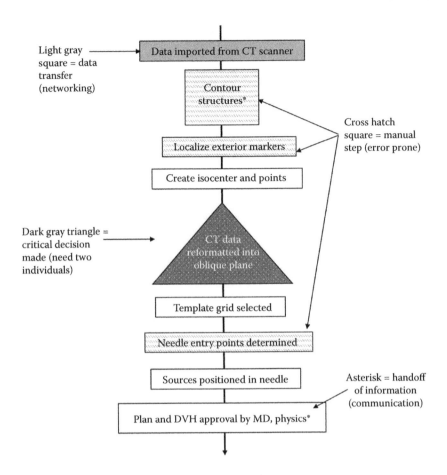

Light gray square = data transfer (networking)

Data imported from CT scanner

Contour structures*

Localize exterior markers

Cross hatch square = manual step (error prone)

Create isocenter and points

Dark gray triangle = critical decision made (need two individuals)

CT data reformatted into oblique plane

Template grid selected

Needle entry points determined

Sources positioned in needle

Asterisk = handoff of information (communication)

Plan and DVH approval by MD, physics*

FIGURE 8.4   Detailed process map of treatment planning with color coordinated levels.

to reformat data." One can imagine these two failure modes having largely different RPN scores due to the first one (being upside down) being less detectable than forgetting to reformat the data entirely. The failure mode of the patient data being upside down may actually be difficult to detect, and the severity of such a failure would be high if it went all the way undetected—it could potentially lead to the dose distribution literally being upside down in the patient if it is not somehow recognized further down the line. Therefore, with this high RPN score, a quality check or checklist item should be added to verify the orientation of the dataset after reformatting. Thus, a high-quality program is built step by step.

In addition, when determining the effects of each failure mode, one should be cognizant that the effects can "happen" to more than just the patient. For example, when deploying the seed/strand into the patient, the needle could potentially jam and fail to deploy. Possibly, the strand itself might fall on the floor, separated from the needle and hub. Not only would this failure "reach" the patient (loss of dose contribution from that needle) but also it would unintentionally expose the staff. This could also be an important failure mode depending on the circumstances. The institution itself might be effected, if a case is suddenly canceled and the institution must pay for the cost of the seeds. In building the FMEA, all effected parties should be considered.

TABLE 8.3 RPN Rankings for Errors for Breast Brachytherapy

Process	Subprocess	Ranking Before (After)				QA Step to Implement
		F	**D**	**S**	**RPN**	
Patient simulation error	Breast board angle not recorded	3 (2)	6	5	90 (60)	Have checklist for sim RTT
	Nipple marker not placed	5 (3)	3	6	90 (54)	Have checklist for sim RTT
	Wrong slice thickness	4 (2)	3	4	48 (24)	Create CT protocol
	Pt breathing irregular/choppy	3	4 (2)	5	60 (30)	Physics review CT scan in control area
Interstitial needle implantation error	Wrong needle announced off plan	4	5 (2)	7	140 (56)	MD to echo/call back needle, physics to check plan
	Wrong needle grabbed from kit	3	5 (2)	7	105 (42)	OR staff to watch MD grab needle
	Needle implanted into improper template location	4 (2)	5 (2)	7	140 (28)	Physics to watch MD implant needle into template
	Needle implanted into correct template hole but does not correlate to correct anatomical location	3 (2)	3 (2)	7	63 (28)	Check needle with ultrasound unit each time, staff training, and education for set-up
	Needle has incorrect seeds or seed locations	2	8 (2)	5	80 (20)	Physics to QA needles prior to going to OR

In Table 8.3, an example of the failure modes association with patient simulation and implantation are examined in accordance with FMEA. Each failure mode is listed with the corresponding F, S, and D. The result assumes there is no quality assurance (QA) or processes in place since this is a new procedure. On the far right is the QA step that can be implemented to either decrease the frequency or increase the detectability (decrease the likelihood the error goes undetected). The scores are given in the format of no QA (with QA). So, each failure mode is approach with the assumption that there is no QA being done on that event. Then, a corrective action or QA step is added to the failure mode to decrease the RPN score. For example, in line 1, once a checklist is created and distributed to the simulation therapists, the RPNs for the frequency of the breast board angle not being recorded or the nipple marker not being placed are subsequently reduced from 90 to 60 and 90 to 54, respectively. To decrease the frequency of the wrong slice thickness being selected, a special breast brachytherapy CT protocol can be created and made available for the therapists. The example of the QA step to have the physicist review the patient CT images on the console requires a new procedure to be developed in which the physicist is present at the brachytherapy simulation.

One of the highest RPN scores for the implantation step is for the needle to be implanted into the improper template location. Based on experience with prostate implants, this can occur quite often as the template letters are small and the lighting in operating rooms is often dim. Once the needle was deployed, there would be little to no chance to detect that the error had occurred at all. Therefore, the F, S, and D values are all larger than average. To add safety steps to this process, the physicist or another member of the implant team could hold a small flashlight and watch the needle insertion into the template. Then, two individuals could be watching the needle and verifying that it went in the proper location. This would decrease the frequency (lower F) and increase the detectability (lower D). The severity would stay the same since the effect of the error would be the same. However, the overall RPN could be lowered from quite significant at 140 down to 28. In this way, the use of FMEA can help create new processes, required checklist items, and training needs. It is worthy to note that the severity of the failure mode is often the most difficult to change, therefore most QA or QC steps implemented effect the frequency or detectability.

The entire FMEA for this procedure will not be given; rather it is the process that is emphasized here. An FMEA can help guide and develop documentation, processes, required personnel, checklists, and other steps in implementing a new procedure. The FMEA should be applied to every step in the patient's journey, from scheduling to follow up. It is important to examine the failure modes not just from the perspective of the patient alone, but also the patient's data and any communication that goes to or from the patient. An FMEA provides excellent documentation of required resources, including new equipment that might be necessary, time for existing employees to perform certain tasks, or even justify the benefit of hiring new employees.

There are dozens of excellent publications on the application of FMEA to clinical radiation therapy. However, none of these publications can help assess the risk in one's own facility; they can only guide one based on the clinical processes and steps that occur at the author's own institution. Some institutions' results may be similar; however, the proposed mechanism for improvement might be different. A large institution's solution to a failure mode might be to have two different people examine the same piece of data. A small institution's solution might be to have an excel spreadsheet, look-up table, or nomogram which contains information that will verify the integrity of the original data. No matter what the solution, it cannot be implemented in a scientific and data-driven environment without the use of FMEA or other quantitative risk analysis tools.

8.4 CONCLUSION

This chapter sought to explain the methodology for implementing FMEA for brachytherapy applications. FMEA can be a useful tool, which uses a systematic approach for quality management of a specific process. However, it must be done properly with an appropriate team. If FMEAs are done correctly, they should reduce errors and improve the efficiency of the process. However, one item to keep in mind: FMEA gives limited insight into the *cause* of an error. It takes teamwork and analysis to mitigate that error by understanding how and why that error occurs. Also, FMEA ranks RPNs under the assumption that all steps in the process occur—if a QA step is missed, then the FMEA will be invalid. Therefore, rigorous quality assurance practices and checks are vitally important. This is true for patient level QA all the way up to systematic machine QA. Finally, the FMEA should be reevaluated every year, or with any major system change (e.g., software upgrade, hardware upgrade, etc.) so that it can be updated to new steps or potential failures. While proper FMEA takes hard work, collaboration, and effort, the FMEA should redistribute workloads toward more efficient safety checks and distribute time and energy on processes that need it the most for the highest patient safety.

REFERENCES

AAPM TG-100. A new paradigm for quality management in radiation therapy. 2013; Available at: http://chapter.aapm.org/pennohio/2013Fall SympPresentations/SI11_Saiful_Huq.pdf

American Society for Radiation Oncology. Safety is No Accident. 2012; Available at: https://www.astro.org/Clinical-Practice/Patient-Safety/Safety-Book/Safety-Is-No-Accident.aspx

Bradley, J. R. and Guerrero, H. H. An alternative FMEA method for simple and accurate ranking of failure modes. *Decision Sciences*. 2011; 42(3): 743–771.

Broggi, S. et al. Application of failure mode and effects analysis (FMEA) to pretreatment phases in tomotherapy. *Journal of Applied Clinical Medical Physics*. 2013; 14(5): 265–277.

Cantone, M. C. et al. Application of failure mode and effects analysis to treatment planning in scanned proton beam radiotherapy. *Radiation Oncology*. 2013; 8(1): 127.

Ciocca, M. et al. Application of failure mode and effects analysis to intraoperative radiation therapy using mobile electron linear accelerators. *International Journal of Radiation Oncology* Biology* Physics*. 2012; 82(2): e305–e311.

Dinshaw, K. A., Rao, D. N., and Ganesh, B. Tata Memorial Hospital Cancer Registry Annual Report. Mumbai, India: Department of Biostatistics and Epidemiology, Tata Memorial Hospital, 1999; 52.

Dunscombe, P. Recommendations for safer radiotherapy: What's the message? *Frontiers in Oncology*. 2012; 2: 129. doi: 10.3389/fonc.2012.00129

Ericson, C., II. *Hazard Analysis Techniques for System Safety*. 2nd edn. Wiley, Hoboken, NJ, 2015.

Fault Tree Handbook (NUREG-0492), *Systems and Reliability Research*, Office of Nuclear Regulatory Research, US Nuclear Regulatory Commission, Washington, DC 20555-0001, 1981.

Ford, E. and Terezakis, S. How safe is safe? Risk in radiotherapy. *International Journal of Radiation Oncology Biology Physics*. 2010; 78: 321.

Ford, E. C. et al. Evaluation of safety in a radiation oncology setting using failure mode and effects analysis. *International Journal of Radiation Oncology* Biology* Physics*. 2009; 74(3): 852–858.

IAEA Safety Report Series 17. *Lessons Learned from Accidental Exposures in Radiotherapy*. Vienna, Austria, IAEA. IAEA Safety Reports Series, 2000; 17.

International Commission on Radiological Protection (ICRP). Preventing accidental exposures from new external beam radiation therapy technologies. ICRP Publication 112, *Annals of the ICRP*. 2009; 39(4): 1–86.

Kapur, A and Potters, L. Six sigma tools for a patient safety-oriented, quality-checklist driven radiation medicine department. *Practical Radiation Oncology*. 2012; 2: 86.

Keller, B. M. et al. Permanent breast seed implant dosimetry quality assurance. *International Journal of Radiation Oncology* Biology* Physics*. 2012; 83(1): 84–92.

Kubo, H. D. et al. High dose-rate brachytherapy treatment delivery: Report of the AAPM Radiation Therapy Committee Task Group No. 59. *Medical Physics*. 1998; 25(4): 375–403.

Mayadev, J. et al. A failure modes and effects analysis study for gynecologic high-dose-rate brachytherapy. *Brachytherapy*. 2015; 14(6): 866–875.

Noel, C. E. et al. Process-based quality management for clinical implementation of adaptive radiotherapy. *Medical Physics*. 2014; 41(8): 081717.

Papadopoulos, Y., David, P., and Christian, G. Automating the failure modes and effects analysis of safety critical systems. In: *Proceedings of the Eighth IEEE International Symposium on High Assurance Systems Engineering*, Tampa, FL, 2004. IEEE, 2004.

Pignol, J.-P. et al. First report of a permanent breast 103 Pd seed implant as adjuvant radiation treatment for early-stage breast cancer. *International Journal of Radiation Oncology* Biology* Physics*. 2006; 64(1): 176–181.

Pignol, J.-P. et al. Report on the clinical outcomes of permanent breast seed implant for early-stage breast cancers. *International Journal of Radiation Oncology* Biology* Physics*. 2015; 93(3): 614–621.

Potters, L. and Kapur, A. Implementation of a "No Fly" safety culture in a multicenter radiation medicine department. *Practical Radiation Oncology*. 2012; 2: 18.

Richardson, S. A 2-year review of recent Nuclear Regulatory Commission events: What errors occur in the modern brachytherapy era? *Practical Radiation Oncology*. 2012; 2, 157.

Sawant, A. et al. Failure mode and effect analysis-based quality assurance for dynamic MLC tracking systems. *Medical Physics*. 2010; 37(12): 6466–6479.

Scanderbeg, D., Richardson, S., and Pawlicki, T. Critical analysis of FMEA implementation for brachytherapy at two institutions. OC-115, *Radiotherapy and Oncology*. 2012; 103(2): S46.

Siochi, Alfred. Failure modes and effects failure modes and effects analysis (FMEA) for radiation medicine. *Presentation at the Radiological and Medical Physics Society of New York*, New York. 2017.

Swamidas, J. V., Sharma, S., Mahantshetty, U. M., Khanna, N., Somesan, V., Deshpande, D. D., and Shrivastava, S. K. Risk assessment in intracavitary brachytherapy based on failure mode and effective analysis. *Brachytherapy*. 2010; 9(1): S4.

Thomadsen, B., Brown, D., Ford, E., Huq, M. S., and Rath, F. Risk assessment using the TG-100 methodology. In: *Quality and Safety in Radiotherapy: Learning the New Approaches to Task Group 100 and beyond, Medical Physics Monograph*, edited by B. Thomadsen, P. Dunscombe, E. Ford, S. Huq, T. Pawlicki, and S. Sutlief, Medical Physics Publishing, Madison, WI, 2013; 36: 95–112.

Thomadsen, B., Lin, S. W., Laemmrich, P., Waller, T., Cheng, A., Cladwell, B., Rankin, R., and Stitt, J. Analysis of treatment delivery errors in brachytherapy using formal risk analysis techniques. *International Journal of Radiation Oncology Biology Physics*. 2003; 57: 1492.

Valentin, J. Prevention of high-dose-rate brachytherapy accidents. ICRP Publication 97. *Annals of the ICRP*. 2004; 35(2): 1–51.

Venselaar, J., Meigooni, A. S., Baltas, D., and Hoskin, P. J. *Comprehensive Brachytherapy: Physical and Clinical Aspects*. CRC Press, Boca Raton, FL, 2012.

Wilkinson, D. A. and Kolar, M. A. Failure modes and effects analysis applied to high-dose-rate brachytherapy treatment planning. *Brachytherapy*. 2013; 12(4): 382–386.

Younge, K. C. et al. Practical implementation of failure mode and effects analysis for safety and efficiency in stereotactic radiosurgery. *International Journal of Radiation Oncology* Biology* Physics*. 2015; 91(5): 1003–1008.

CHAPTER 9

Real-Time In Vivo Dosimetry

Luc Beaulieu, Rick Franich, and Ryan L. Smith

CONTENTS

9.1 INTRODUCTION

The pursuit of *in vivo* dosimetry (IVD) in brachytherapy has been the subject of investigation by prominent research groups around the world. The diversity of approaches numbers almost as many as the research groups themselves. An impressively diverse range of detectors and dosimeters has been brought to bear on the task in the quest to exploit various attributes of the many detection and dosimetry tools available.

The characteristic features of high-dose rate (HDR) brachytherapy treatments are in many cases also the drivers of the need for a solution: HDRs, few fractions, highly conformal dose distributions, with steep dose gradients often in close proximity to organs at risk (OARs). The potential consequences of incorrectly delivered treatments are well recognized. These same features contribute to the difficulty of the task. The dose distribution to be verified is highly localized, and measurement within the target volume may not be easily accessible. The dose distribution can be highly spatially modulated. Thus, even when anatomical cavities or implanted catheters provide an opportunity for internal measurement, high spatial resolution and positional precision become important. IVD conducted beyond the treatment volume risks a lack of sensitivity to discrepancies in treatment dose arising from changes in the delivered plan.

Approaches to IVD in brachytherapy typically seek to combine the attributes of dosimetric precision, high spatial resolution, and ready interpretation of results, with more pragmatic goals such as real-time output and ease of integration into workflow.

What is perhaps most surprising is that the plethora of systems and strategies described in the literature to date has not yet yielded a system that has been adopted into routine clinical use. With one recent exception (described in Section 9.2), this even extends to a lack of routine use by the research groups themselves. This highlights dramatically both the complexity of the problem and the high expectations that the brachytherapy community have of such a system.

9.2 LESSONS LEARNED FROM PAST IN VIVO STUDIES

A PudMed (National Center for Biotechnology Information) search with keywords "*in vivo* dosimetry" and "radiation therapy" yields close to 530 manuscripts, 172 of them in the 2010–2015 period. Less than 10% of those are related to brachytherapy and a smaller number report actual patient measurements during treatments. Table 9.1 provides a portrait of the most significant clinical *in vivo* studies in the field. They are presented by the dosimeter system used and detailed action level if reported or key comments from each studies. If an uncertainty budget study was conducted, the expected uncertainty for $k = 1$ (i.e., 1σ) is also indicated.

It is important to distinguish the intrinsic accuracy and reproducibility of each dosimeter and the task of *in vivo* dose measurements that further involve an extended number of related tasks, such as accurate detector positioning, accurate dose calculation, and others. For example, Raffi et al. were able to obtain an uncertainty budget for thermoluminescent dosimetry (TLD) measurements of less than 3%, yet discrepancies as large as 20% or more were seen and related to the inaccuracy of the TG43 dose calculation method (Raffi et al. 2010). This leads to a few important observations: (1) contrary to external beam radiation therapy, where all clinical physicists can make a dose measurement in standard field conditions within 1% or better, dose measurement in brachytherapy is much more complicated. (2) It is important to understand the strengths and limitations of the dosimetry system chosen to perform *in vivo* measurements; however, some can require more expertise than others. The reported uncertainty budget for TLD in the above-mentioned study should be seen in the context of usage in the hands of experts. It is likely that this uncertainty budget would be higher in more casual users of TLD (Baltas et al. 2006). (3) In performing IVD, it is crucial to question all aspects related to the type of clinical cases to be measured. The shortcoming of the TG43 formalism in this particular example is one of many issues (Rivard et al. 2009, Beaulieu et al. 2012). These will be tackled in the next section, as they are critical in performing more accurate IVD measurements in brachytherapy.

The general picture that can be extracted from the bulk of the studies reported in Table 9.1 is that dose differences below 10% should normally not be acted upon. This is due to the combination of the measurement tool's uncertainty budget and the overall dose calculation chain

TABLE 9.1 A Summary of Clinical Brachytherapy IVD Studies

Dosimeter	Study	Site	Action Level/Comment	Uncertainty (1σ)
TLD	Brezovich et al. (2000)	Prostate, urethra, and rectal dose	Action level: 20% generally	8–10%
	Anagnostopoulos et al. (2003)			
	Das et al. (2007)			
	Toye et al. (2008)			<3% (TLD uncertainty budget)
	Raffi et al. (2010)	Skin (breast)		
MOSFET	Cygler et al. (2006)	Urethra (prostate seed implants)	Action level: 16%	8%
	Bloemen-van Gurp et al. (2009a)			
Alanine/ESR	Schultka et al. (2006)	GYN (^{137}Cs)	Detector volume too large; difference with planning 10+%	None provided
	Anton et al. (2009)	Urethra (prostate HDR)		5% (excl. source strength uncertainty)
Diodes	Alecu and Alecu (1999)	Cervix	Agreement with TPS within 15%	None provided
	Waldhäusl et al. (2005)	Cervix	Action level: 10% (36/55 cases need further investigation)	7%
Glass dosimeters	Seymour et al. (2011)	Rectum (prostate HDR)	95% measurements within 20%	9.8% (meas. only)
	Takayuki et al. (Nose et al. 2008) Hsu et al. (2008)	Prostate	Deviations of more than 20% seen	None provided
	Takayuki et al. (Nose et al. 2008)	GYN		
	Takayuki et al. (Nose et al. 2005)	H&N		
OSLD "NanoDot"	Sharma and Jursinic (2013)	GYN, breast	−4.4% to 6.5% difference to AcurosBV	None provided
Real-time OSL	Andersen et al. (2009)	Cervix (PDR)	Errors detection are distance dependent; time-resolved measurements are better	5%
Plastic scintillation dosimeters	Suchowerska et al. (2011)	Urethra (prostate HDR)	Maximum deviation without imaging 67%; maximum deviation with imaging 9%	None provided
MOSkin	Carrara et al. (2016)	Rectum (prostate HDR)		None provided
	Qi et al. (2012)	Nasopharynx	Action level: 20%	2.5% (MOSkin uncertainty budget)

uncertainty budget. Even for clinical cases perfectly described by TG43 conditions (Rivard et al. 2004), the AAPM–GEC-ESTRO TG138 showed that the uncertainty budget from National Institute of Standards and Technology (NIST) to the output of the treatment planning system (TPS) is of the order of 6.8% for ^{192}Ir and 9% for seed implants ($k = 2$) (DeWerd et al. 2011). Of course, there are also clinical considerations such as the accuracy of applicator reconstruction and so on that adds to the overall uncertainty chain (Kirisits et al. 2014).

In many studies, 10% is used as a sort of caution level or yellow light, which might not mandate immediate response from the clinical team but a posttreatment investigation, that is, an actual clinical error might be occurring but the true positive detection rate for the dosimetry system use, in combination with all other potential uncertainties in the clinical processes to be measured, is too low to mandate immediate action. In the study of Waldhäusl et al., a 10% action level would flag 65% of the cases (Waldhäusl et al. 2005). Most studies will use 2σ or approximately 20%, which included the measurements device uncertainty as well as other components such as dose calculation chain and so on in the overall uncertainty budget, as a direct action threshold. Again, referring to the study of Waldhäusl et al., 5% of the treatments would need immediate attention or reported differently, 95% of the measurements fall within 2σ or ±20%, which makes detection of true error beyond this boundary more certain.

It is further interesting to look at the study by Andersen et al. who used a real-time optically stimulated luminescence (OSL) dosimeter for an *in vivo* study of cervix patients treated using pulsed-dose rate (PDR) brachytherapy (Andersen et al. 2009). They showed that ability to detect an error is strongly dominated by positioning uncertainty between the source and the sensor in the first few centimeters (due to the dose gradient) while measurement uncertainty dominates at larger distances (due to decreased dose rate). Thus, estimation of the overall uncertainty budget might be a more complicated task than previously thought. Another key finding of this particular study is that integral dose measurement encompassing the complete treatment could be mute to significant errors, such as a switch in transfer tube connection. However, they eloquently demonstrated that time-resolved measurements allow monitoring the dose deposition on a catheter-per-catheter basis, even dwell-position specific delivery within a catheter as well as providing a quality assurance (QA) of source transit in and out of the patient. Other groups further confirmed this from phantom studies. Thus, time-resolved (real-time) IVD provides improved actionable data to clinicians.

The primary motivation for real-time IVD is treatment verification—clinical practitioners desire to know that the planned treatment is being delivered as intended. This goal encompasses two objectives: (i) error detection and prevention and (ii) monitoring differences or changes between planning and treatment. Each objective imposes its own demands on the tools and techniques employed.

Error detection guards against errors in the delivery of the planned treatment caused by either operator/procedural mistakes, or possible equipment malfunction. Accidents of this nature are probably rare, but the true incidence is largely unknown as routine monitoring during treatment is not common. Thus, most occurrences would go undetected and therefore unreported. In the modern era, a record of evidence that a treatment was not maladministered could be quite important and valuable. Real-time monitoring offers the additional advantage of the potential to interrupt a treatment if an error is detected. This both protects the patient from injury, and creates an opportunity to recover the prescription dose in a subsequent fraction, via an adapted plan.

The second objective perhaps plays a larger part, if we maintain that accidents and errors are rare, but acknowledge that there will always be some changes between plan-imaging and delivery. Our aim is to be aware of any differences between planned and delivered dose distributions, or their location with respect to anatomy. These changes can arise from anatomical changes, which can in turn cause geometric changes in the implant or applicator positioning. There exists the potential to either underdose regions of the target volume or to over-irradiate OARs, or both.

This means that up to now, IVD in brachytherapy might enable the detection of gross errors (tens of % differences between planned and delivered dose) but could also very well let significant errors, resulting in global or local under or overdosage, go undetected.

9.3 REQUIREMENTS FOR ACCURATE IVD

Dose measurement in brachytherapy is particularly challenging. On one hand, the inverse square dependence of the dose rate dictates strong dose gradients, that is, 10% or more per millimeter for the first few centimeters from the source. As such, a positioning error of 1–2 mm could lead to an *in vivo* measurement above the action threshold solely based on a positioning error, either due to an insufficiently accurate estimation of the sensor-to-source distance or to an actual displacement occurring between the detector placement and the treatment delivery.

It is also interesting to look at the various components of the dose deposited by a brachytherapy source. Figure 9.1 shows the contribution of the primary photons, the total scatter photons as well as the single and multiple scatter components for the MicroSelecton® V2 ^{192}Ir source (Elekta Brachytherapy, Veenendal, the Netherlands). Here, the inverse square dependence has been removed. The primary photons clearly dominate the total dose deposited until about 6.5 cm from the source. However, as depth increases the scatter photons become dominant and beyond 10 cm, the multiple scatter component is the major contributor to the dose. This is important because the spectral composition of the radiation fluence will evolve with the distance from the source. In most clinical *in vivo* measurement situations, at least for ^{192}Ir brachytherapy, one might expect the primary photons would dominate the dose. However, if dose measurements were conducted at larger depth, the detector response might need further corrections. Note that for low-energy seeds, such as ^{103}Pd or ^{125}I, the scatter component becomes dominant at about 2 cm from the source.

Returning to the PubMed search for IVD in radiation therapy, diodes, TLD, electronic portal imaging device (EPID), and metal oxide semiconductor field effect transistor (MOSFET) are the four most commonly found dosimeters when searching for "*in vivo* dosimetry." On the other hand, restricting the search to brachytherapy, MOSFET becomes the most common dosimeter, followed by diodes, TLD, and glass dosimeters. A number of other dosimeters, some not commercially available, such as real-time OSL and plastic scintillation detectors (PSD), are also used in a limited number of studies. The following sections address specific issues related to the intrinsic dosimeter performance, positioning

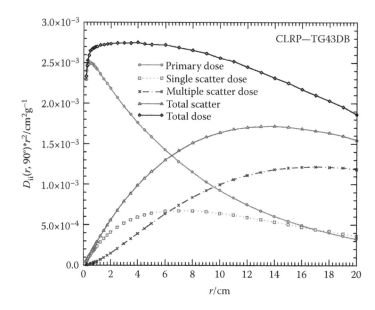

FIGURE 9.1 Primary and scatter dose separation for a MicroSelectron V2 ^{192}Ir source. This figure is reproduced from the Carleton Laboratory for Radiotherapy Physics (CLRP) Database of TG-43 brachytherapy dosimetry parameters (http://www.physics.carleton.ca/clrp/seed_database). (Image courtesy of Taylor R E P, Yegin G and Rogers D W O 2007. *Med Phys* 34, 445–7.)

and monitoring, and how these translate into action thresholds and other clinical requirements that enable better decision trees resulting from *in vivo* measurements.

9.3.1 Dosimeter-Dependent Parameters

Table 9.2 provides an overview of the characteristics of key dosimeters used for IVD in brachytherapy, including those that have the potential to be used for time-resolved IVD (Tanderup et al. 2013). The interested reader will want to consult recent IVD review papers as well as reference textbooks for an in-depth description and discussion of each dosimeter (Baltas et al. 2006, Rogers and Cygler 2009, Tanderup et al. 2013, Kertzscher et al. 2014). The data presented in Table 9.2 is important in that it provides an overview of the effort needed to either carefully obtain correction factors for the dosimeter before usage (e.g., energy dependence) or how much information will be needed during measurements. For example, a dosimeter with a strong angular dependence will necessitate at least accounting for the angle between the source and the sensor for each dwell position correctly intepreting the measurement with respect to the expected dose to be measured. Similarly, temperature dependence might require independent *in vivo* monitoring or allowing for stabilization time if the effect is sufficient to induce a large reading difference from room temperature.

It is interesting to note from Table 9.2 that time-resolved dosimetry has been demonstrated for four detector technologies. Of these, three have (various levels of) energy dependence issues and/or show minor to strong angular dependence (diodes, MOSFET, and radioluminescence [RL]). All have some degree of temperature dependence. For PSDs, it has recently been shown that the temperature dependence can be treated as a stem-like effect

Dosimetry, Real-Time In Vivo

TABLE 9.2 Intrinsic Detectors Characteristics and Features of Importance for Precise Routine IVD in Brachytherapy

	TLD	Diode	MOSFET	Alanine	RL	PSD
Size	+	+/−	+/++	−	++	++
Sensitivity	+	++	+	−	++	+/++
Energy dependence	+	−	−	+	−	+/++
Angular dependence	++	−	+	+	++	++
Dynamic range	++	++	+	−	++	++
Temperature dependence	++	+	+	+	+	+
Time-resolved dosimetry	−	++	+	−	++	++
Commercial availability	++	++	++	++	−	+
Main advantages	No cables, well-studied system	Commercial systems at reasonable price, well-studied system	Small size, commercial system at reasonable price	Limited energy dependence, no cables	Small size, high sensitivity	Small size, no angular and no energy dependence (at ^{192}Ir energy), sensitivity
Main disadvantages	Tedious procedures for calibration and readout, not online dosimetry	Angular and energy dependence, minor temperature dependence (<0.6%/K)	Limited life of detectors, energy dependence, temperature dependence for non-dual bias MOSFET, possible dose/LET dependence	Not sensitive to low doses, tedious procedures for calibration and readout, not online dosimetry, expensive readout equipment not available in clinics	Needs frequent recalibration, stem effect, minor temperature dependence (<0.2%/K) not commercially available	Stem effect, minor temperature dependence (0.05%/K to 0.09%/K), not commercially available for IVD

Source: Adapted from Tanderup K et al. 2013. In vivo dosimetry in brachytherapy. *Med. Phys.* 40, 070902 and supplemented by Kertzscher G et al. 2014. In vivo dosimetry: Trends and prospects for brachytherapy. *Br. J. Radiol.* 87, 20140206.

The items are rated according to advantageous (++), good (+), and inconvenient (−).

and extracted using the previously presented hyperspectral approach (Therriault-Proulx et al. 2015). At least, three technologies (MOSFET, RL, and PSD) were shown to be small enough for insertion into catheters and applicator channels. However, two (RL and PSD) are still not yet available commercially for IVD in brachytherapy. Finally, for large fractionation scheme, such as 1×15 Gy or 1×19 Gy use for prostate HDR brachytherapy, MOSFET responses could vary by as much as 5–6% due to dose or linear energy transfer dependence.

It is crucial for the end users, the clinical physicists, to understand these effects and also to be able to establish a rigorous uncertainty budget for the IVD system to be used. Many recent publications highlight this aspect as described in Table 9.1.

9.3.2 Positioning and Monitoring

The placement of one or more dosimeters is of particular interest. Firstly, because, depending on the location, the dose calculation geometry might not reflect the TG43 conditions (which requires at least 5 cm of tissue for low-energy seeds (Rivard et al. 2004) or at least 20 cm of tissues for high-energy ^{192}Ir source (Perez-Calatayud et al. 2012)). In this particular case, differences will appear between the TPS calculated dose and the measured dose due to the TPS failure to accurately take into account heterogeneities. This is nicely illustrated in the study of Raffi et al. on surface dose in the breast, which highlighted large discrepancies when comparing the measurements to TG43 but within 5% when Monte Carlo or AcurosBV were used (Raffi et al. 2010). Therefore, the placement of a dosimeter in air, close to an air pocket, bone, shield, or any other structures which have interaction cross sections significantly different from water can have significant implication for TPS calculated dose and lead to differences in dose measurements that are not due to a delivery error per se (Beaulieu et al. 2012).

The second element is that the steep dose gradients in brachytherapy requires an exact knowledge of the source-to-dosimeter distance (SDD) to ensure that any difference measured is actually due to a treatment error and not only to a bad estimation of this parameter. When exploring discrepancies between planned and measured doses a number of studies were able to trace them back not to a delivery error but to an incorrect estimate of the SDD (Table 9.1).

We can usually relate SDD problems to a limited number of overarching scenarios

1. Displacement of the dosimeter relative to plan position
   a. Organ-induced displacement
   b. Manipulation error (digitization, displacement before measurements, etc.)
2. Displacement of source position(s) relative to plan position(s)
   a. Displacement of one or more catheters or of an applicator, including rotation for certain applicators
   b. Organ-induced displacement
   c. Manipulation error (wrong transfer tube connection, etc.)

3. Combination of the above two, that is, source and sensor displacements

   a. Perfectly in sync: no effect on dose measured but effect on dose delivered

   b. Out of sync

4. Organ-related change that does not impact the relative distances but organ dosimetry (e.g., swelling, deformation, etc.)

The use of imaging before and during treatment could certainly address some of these issues. The work of Suchowerska et al. demonstrated a strong benefit in reducing the SDD uncertainties (67% without imaging vs. 9% with imaging) (Suchowerska et al. 2011). However, the detection accuracy of the most widely used imaging devices might not be better than 1–2 mm based on clinical data for seed detection or catheter tip detection accuracy (De Brabandere et al. 2006, Kirisits et al. 2014). As such, both the source positions and the dosimeter-digitized position should include this limitation in the measurement uncertainty budget. Moreover, unless the SDD can be monitored in real-time, it is possible that for some clinical situations, the distance could change not only before but also during treatments. Therefore, the stability of the dosimeter should also be part of the consideration when choosing its placement. A possible solution is to accurately track the SDD distance at all times and for all dwell/source positions. Technologies, such as electromagnetic (EM) or radiofrequency (RF) tracking (Wood et al. 2005), are readily available to accomplish this kind of task (Bharat et al. 2014, Damato et al. 2014, Poulin et al. 2015), but have not been widely adopted and integrated with current brachytherapy and dosimeter devices. The RadPos is an example of a tracked MOSFET dosimeter (Cherpak et al. 2009). These technologies will be discussed in the next section.

Item 4 is a special case where the SDD does not change but the relation between target and OARs does. In such a case, the *in vivo* measurement might not reveal any difference. In this case, only frequent or real-time imaging can account for these changes. It is interesting to note that at least one new technology has tried to incorporate IVD (based on MOSFET technology) with real-time transrectal ultrasound (TRUS) providing a common reference frame for the images used by the TPS and the *in vivo* dose measurements (Carrara et al. 2016).

9.3.3 Action Thresholds and Clinical Requirements

The knowledge of the uncertainty budget associated with the measurement tools sets the stage for the minimal detectable dose difference, and it should be obvious by now that it is probably not sufficient to generate clinically efficient action thresholds. Figure 9.2 from Andersen et al. brings together elements discussed in the previous two sections, that is, the interplay between the dose gradient, the SDD and the dose rate versus detector response (Andersen et al. 2009). This figure illustrates that the ability to detect an error (e.g., displacement) is highly dependent on the nominal SDD (named source-to-probe distance on that figure).

In the first few centimeters, the positioning uncertainty dominates over the dosimeter uncertainty budget (a real-time radioluminescent dosimeter in this case). This illustrates

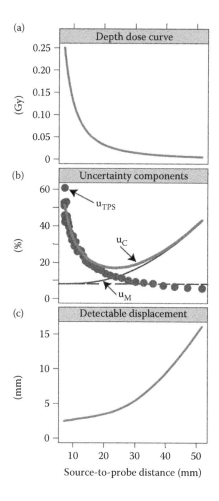

FIGURE 9.2 Relationship between uncertainty, ability to detect displacements, and source-to-probe distance (or SDD) for a 10 s irradiation period using ^{192}Ir. (a) Depth-dose curve, (b) U_{TPS} representing a 1 mm positioning uncertainty relative to TPS reference dose values, U_M the measurement uncertainty associated with the dosimeter (uncertainty budget), and U_C which is the combination of both uncertainties. (c) Approximate displacement distances that can be detected. (Image courtesy of Andersen et al. 2009. Time-resolved in vivo luminescence dosimetry for online error detection in pulsed dose-rate brachytherapy. *Med. Phys.* 36, 5033–43.)

the strong effect of dose gradient, where a small SDD error leads to a large dose difference. At large distance, the dose rate decreases significantly and the dosimeter ability to detect a signal becomes an important characteristic. Thus, at large SDD, the measurement uncertainty dominates the overall combined uncertainty. There is a "sweet-spot" that minimizes U_C. The bottom panel indicates that small source displacements can be detected when the SDD is small while much larger displacements, sometimes above 1 cm, are needed to generate a detectable signal at larger SDD. Kertzscher et al. pushed this analysis further by looking at a wide range of positioning uncertainties (up to 4 mm) and their effect on the combined uncertainty and detectable displacements as a function of SDD (Kertzscher et al. 2011). They concluded that both the source and dosimeter positions should be known to better than 1.5 mm to be able to detect a source displacement error of 5 mm (one dwell position for most standard HDR/PDR treatments).

Based on the above work, using a single action threshold value (such as 10%, 15%, or 20%) does not accurately represent the breadth of configurations that have to be measured by a dosimeter during a full treatment delivery (Tanderup et al. 2013). Instead, Kertzscher et al. argued for the adoption of the statistical discrepancy criterion, where prior knowledge of the plan (e.g., SDD for all source positions) and all known uncertainties are taken into account and applied as the delivery proceeds (Kertzscher et al. 2011). However, implementation of such an approach is limited without the ability to perform time-resolved IVD. At the same time, Andersen et al. demonstrated that integrated *in vivo* measurements are severely limited in detecting certain types of errors such as two interchanged transfer tubes.

Finally, one important aspect that should not be forgotten, we are not treating a catheter, an applicator, or a dosimeter but physical target volumes. It is ultimately the dose to those volumes (target or OARs) that are important. The discussion above tackled the detection of a source displacement error and our ability to measure it. Yet, even when no such error is detected, the delivered dose to the target or OARs could still be wrong if significant changes in volume or shape (deformation) remained undetected. Therefore, imaging as close as possible to treatment delivery or even real-time imaging should be considered.

From the above, a number of key prerequisites emerge for accurate IVD

- Use a dosimeter that possesses the maximum number of good characteristics (Table 9.2).

- Make a careful determination of the measurement uncertainty for the specific detector chosen.

- Determine the uncertainty associated with the extraction of the dosimeter and source positions. Ideally, knowledge in real-time of each SDD would be the best-case scenario.

- Monitor those positions and volume of interest (ideally in real time).

- Generate statistical displacement criteria that are specific to each treatment plan.

9.4 TREATMENT VERIFICATION AND THE ROLE OF IVD

For the goal of *treatment verification*, we would ideally seek to know the dose everywhere in the patient. It is clearly not possible to measure this, even just in the target volume, and IVD for treatment verification is limited to a point-wise sampling exercise. Most IVD systems are capable of only one, or very few, point measurements, and this sparse sampling combined with typical positioning uncertainty, imposes limits on the magnitude of the errors that are identifiable, and even a probability of detecting an error when it occurs.

An alternative approach to treatment verification is source tracking such as in the system of Smith et al. 2013, 2015, 2016). In this approach, the source position is monitored during the treatment and compared to the dwell positions and times in the plan. If any changes are detected the treatment can be interrupted or modified. The consequential dose distribution can be calculated from the known delivered source dwells. The effect of known anatomical changes can also be assessed by recalculation.

IVD lends itself very well to dosimetric assessment of changes between plan and delivery. IVD at a point within the target can confirm the administration of the prescription dose, while IVD monitoring in an OAR can protect against exceeding an OAR dose constraint. For example, to the rectal wall or urethra where such organs are amenable to intracavitary IVD probes. An OAR dose measurement that differs from the corresponding plan dose may not indicate a delivery error, rather an anatomical change that has altered the source-detector separation.

Treatment delivery verification in brachytherapy should provide an assessment of the overall treatment delivery parameters, consider anatomical changes, and report an error if/when it occurs. Assessment of organ dose has been a focus of IVD approaches in brachytherapy, and provides a measure of the dose received by the organ. When applied as a real-time dosimeter, the approach in principle can indicate a problem and the treatment can be interrupted before possible organ overdose occurs.

Recent developments in brachytherapy have concentrated on methods to overcome the detector positioning uncertainty (source to detector geometry) at treatment delivery. Some methods include dosimeters integrated into ultrasound imaging devices, position sensitive dosimeters, and EM catheter tracking. Other approaches to treatment verification do not use IVD, but track the position of the source within the patient in order to verify correct treatment delivery. The introduction of these new technologies aims to reduce measurement uncertainty, create confidence in error detection mechanisms, and establish a reliable approach to overall treatment verification.

9.5 NEW TECHNOLOGIES

Recent years have seen considerable innovation and development in technology specifically aimed at treatment verification in brachytherapy. Although currently the commercially available options remain limited, the developments that have been made within the research groups are commendable in tackling such a difficult problem. This section will briefly look at some new technologies that have the potential to improve treatment verification techniques in brachytherapy.

9.5.1 Position-Sensitive Dosimeters

IVD can provide valuable information about the brachytherapy treatment, but the detector response can also be difficult to interpret if the position of the dosimeter is not well known. Modern three-dimensional (3D)-based treatment planning techniques allow identification of the detector at imaging, but organ motion between planning and treatment, or the inability to accurately reposition the detector at treatment will contribute to the uncertainty in detector position. One novel approach to overcome this problem is to continuously monitor the position of the dosimeter using 3D EM position tracking technology.

The technology combines a microMOSFET dosimeter coupled to a positioning sensor (Ascension 3D-Guidance medSafe EM Tracker, Ascension Tech Corp, Burlington,

California). The guidance system uses a DC magnetic field transmitter that generates a pulsed 3D magnetic field. The response of the position sensor to this magnetic field is monitored by a position tracker, which can determine the 3D position of the sensor. The radiation positioning system (RADPOS) is small enough it can be inserted into brachy-therapy treatment catheters or urethral Foley catheter (Cherpak et al. 2009).

Reniers et al. (2012) evaluated the performance of the RADPOS for the possibility of using it for HDR gynecological brachytherapy. The system is able to accurately determine the position of the dosimeter, in most cases to a resolution of better than 1 mm, which is acceptable for HDR brachytherapy clinical applications. The presence of metal can cause perturbations in the position sensors, potentially reducing the accuracy of the position readings. For HDR gynaecological brachytherapy applications, modern applicators that incorporate materials such as titanium or nonmagnetic stainless steel generally present no position disturbances.

The RADPOS system has also been used as a real-time dosimetric tool for low-dose rate (LDR) prostate brachytherapy (Cherpak et al. 2014). Often the quality of a LDR pros-tate implant is not fully known until post implant imaging is performed usually after the patient is removed from the operating theater. Deviations from an ideal implant can be difficult to address postoperatively. Implementation of real-time dosimetry during the prostate seed implant procedure enables evaluation of the dosimetric coverage and organ sparing. The RADPOS dosimeter was inserted into the urethra to provide real-time dosi-metric data. The position information provided by the system was also able to evaluate the swelling of the prostate over the period of the implant, along with a measure of the defor-mation induced with the TRUS probe inserted.

Tracking of the spatial position of needles, catheters, and applicators *in vivo* is a recent and interesting development in brachytherapy configuration. In theory, it should allow for real-time monitoring of positions and angles of the above-named brachytherapy devices. Various approaches have been proposed to accomplish this task, namely EM tracking (Seiler et al. 2000), optical (infrared) tracking (Khadem et al. 2000, Kral et al. 2013) and more recently, a MR-compatible alternative using embedded shape sensing optical fibers within needles using fiber Bragg grating technology (Taffoni et al. 2013), including 3D reconstruction of needle shape (Roesthuis et al. 2014, Abayazid et al. 2013) as well as navi-gation and tracking (West and Maurer 2004; Waltermann et al. 2014). Already, demonstra-tion of certain quality control (QC) or QA tasks such as automated catheter reconstruction (Poulin et al. 2015) and identifying error in transfer tube connections (Damato et al. 2014) has been demonstrated. A concept demonstrated by Elekta, called Vision 20/20 was shown in 2014 that builds upon these results and incorporates an EM tracking sensor into an afterloader drive wire. Chapter 4 discusses some of these technologies in detail.

9.5.2 Dosimeters Integrated into Ultrasound Imaging

Ultrasound-based HDR prostate brachytherapy has the advantages of imaging, planning, and treating the patient in the same location, eliminating possible anatomical uncertainties associated with moving the patient. The relative short-time period between imaging and treatment delivery is also an advantage as it minimizes unavoidable anatomical changes

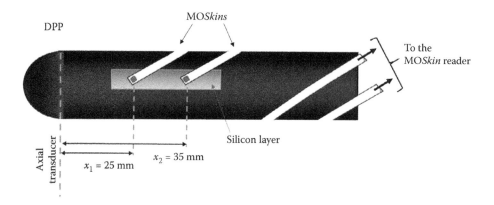

FIGURE 9.3 Schematic representation of the DPP, used both for imaging with the transversal transducer and IVD with the two integrated MO*Skin* dosimeters. (Image courtesy of Carrara M et al. 2016. In vivo rectal wall measurements during HDR prostate brachytherapy with MOSkin dosimeters integrated on a trans-rectal US probe: Comparison with planned and reconstructed doses. *Radiother. Oncol.* 118, 148–53.)

due to swelling from prostatic trauma as needles are inserted. A novel approach to perform IVD measurements during TRUS-based HDR prostate brachytherapy has been introduced by Carrara et al. (2016) with a recent clinical study. The device known as a dual purpose probe (DPP), integrates two MO*Skin* detectors, with a BK TRUS probe, as illustrated in Figure 9.3.

The advantage of this IVD approach is that the position of the dosimeters is well known as they can be located relative to the TRUS images and therefore to the positions of the implanted catheters and surrounding anatomy. A treatment plan is created and the predicted dose values are determined based on the US images. Treatment is delivered and the measured values, representing rectal wall dose, are compared to the predicted values. After treatment, to account for possible further anatomical changes due to prostate swelling, a second set of TRUS images are acquired, catheter positions established, and a revised treatment plan calculated. Although the time between planning and treatment is relatively small (average time was 90 min), the anatomy still changes and therefore impacts the actual delivered dose. The revised predicted rectal wall values were then compared to the measured values and in most cases, the agreement between predicted and measured dose improved.

9.5.3 Source Tracking Verification

An alternative approach to treatment verification is to track the position of the HDR brachytherapy source during the treatment delivery, identify where these source positions are relative to the implanted catheters or applicator and where they occur in relation to the surrounding anatomy. From this, combined with the time the source dwells at each position, it is possible to verify that the overall treatment was delivered as planned and reconstruct the dose that was delivered to the tumor volume and OARs.

Previous studies have investigated methods to perform source position measurements, but these were mostly focused at afterloader system quality assurance measurements (Espinoza

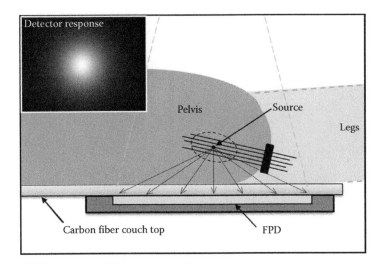

FIGURE 9.4   An illustration showing the source tracking approach applied during treatment delivery of HDR prostate brachytherapy. The FPD is integrated into the treatment couch and captures patient exit radiation (insert), which is processed to locate the source position and verified against the plan.

et al. 2015). At least one study explored the use of an x-ray imager for ^{192}Ir source imaging (Verhaegen et al. 2007). A more recent approach using a flat panel detector (FPD) (Smith et al. 2013, 2016) embedded in a brachytherapy treatment couch has been used as a clinical *in vivo* device (Smith et al. 2015). This method is a two-step approach to verification; initially pretreatment imaging is performed with the FPD and an external x-ray source to verify that the implanted catheters are in the correct position relative to the anatomy, and then source position verification follows by tracking the position of the source during treatment delivery. This source tracking process is illustrated in Figure 9.4 and shows the arrangement of the FPD in relation to the patient, enabling measurement of exit radiation during treatment delivery. This novel method, combined with the pretreatment imaging, allows verification of treatment delivery parameters not easily confirmed with other approaches.

9.5.4 Linear Scintillation Array Dosimeters

Miniature array dosimeters have numerous advantages over single dosimeter apparatus for IVD. Of course, they offer in a single package multiple dose measurement points simultaneously. They can also be inserted inside a catheter (including a Foley catheter) or an applicator commonly used in brachytherapy. Furthermore, Nakano et al. demonstrated that by using three dosimeters, one can triangulate the source position within 2 mm for up to 12 cm from the source (Nakano et al. 2003, 2005). Until recently, real-time *in vivo* array dosimeters were limited to MOSFETs. An array of 5 MOSFETs (Best Medical) and the 10 MOSFET RadFET array have been reported in the literature (Price et al. 2004, Bloemen-van Gurp et al. 2009a, b).

In 2012, Archambault et al. developed a generalized formalism called hyperspectral decomposition, which allowed for multiple scintillating elements to be connected to a single optical light guide (Archambault et al. 2012). In this formalism, the previously

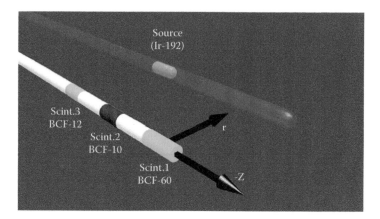

FIGURE 9.5 Example of a three-point plastic scintillation dosimeter using the shelf components. (Image courtesy of Therriault-Proulx F et al. 2013. On the use of a single-fiber multipoint plastic scintillation detector for 192Ir high-dose-rate brachytherapy. *Med. Phys.* 40, 062101.)

described chromatic Cerenkov removal technique as well as multiple spectral decomposition constitutes special cases (Fontbonne et al. 2002, Therriault-Proulx et al. 2012, Darafsheh et al. 2015). This approach has been applied by Therriault-Proulx et al. to build two- and three-point plastic scintillation dosimeters (called mPSD) (Therriault-Proulx et al. 2013). An example of a three-point mPSD is shown in Figure 9.5. Using the hyperspectral formalism, they were able to extract the dose at each scintillator position as well as eliminate the Cerenkov stem effect. For the three-point mPSD, they further demonstrated that the information can be pooled and the source position estimated. Finally, the same mathematical formalism can further be used to extract the temperature in PSD IVD due to small (~0.05%/C) but measurable change in light collection relative to room temperature (Therriault-Proulx et al. 2015).

mPSDs are still not available commercially. However, the technology combines all the "good" properties enumerated in Table 9.2 for scintillation dosimeters to multiple points of measurements, efficient Cerenkov subtraction, and potentially temperature monitoring. At this time, experimental prototypes have been built and tested on phantoms (Therriault-Proulx et al. 2013).

9.6 LOOKING FORWARD

Using the knowledge acquired from the past of *in vivo* studies in brachytherapy, and the new technologies described in this chapter, as well as the tracking technologies presented in Chapter 4, it becomes relevant to ask which clinical issues or challenges are best answered by IVD and for which a more appropriate technology could be used? To answer these questions, we have reframed the table presenting the quality items and typical quality tests presented by Tanderup et al. (2013) by also listing for each the most appropriate technology. Here, "appropriate" is defined as either more precise and/or more efficient. For example, to validate source calibration, a well chamber is the most appropriate technology. It is a robust measurement tool, its calibration is traceable to a standard, it is precise, and

TG138 recommends using the measured source strength as input to the planning system (e.g., ^{192}Ir sources). The result of this exercise is presented in Table 9.3. It is important to notice that the only quality item where IVD is the sole technology that can perform exactly the task described is recording of the dose as delivered in the patient. IVD is relevant to at least six more quality items, however, other technologies such as tracking (EM or optical)

TABLE 9.3   Recast in Terms of Potentially Appropriate Technology for the Test to be Performed

Quality Item	Typical Quality Test	Appropriate Technology
Source calibration	Independent source calibration in the department	Well chamber
Afterloader source positioning and dwell time (nonpatient specific)	Autoradiography, commissioning of applicators, other source stepping and dwell time QA	• Films (position) • Well chamber (time) • Flat panel (possible for both)

Afterloader malfunction	Unpredictable afterloader malfunction is difficult to target with general QA	• Real-time feedback
		  • IVD
		  • Flat panel

Patient identification	Manual check	Not relevant
Correct treatment plan	Manual check	Checksum and similar independent software validation tools is the more efficient
Intra- and interfraction organ/applicator movement	Reimaging performed just before treatment delivery can in some cases be used to assess organ or tumor dose in image-guided brachytherapy	• For organ motion: real-time imaging if possible • Applicator motion: real-time tracking of the applicator or 3D source tracking within the applicator (EM, optical, flat panel) • IVD could be relevant depending on decoupling between applicator and dosimeters
Applicator reconstruction and fusion errors	Manual check	• IVD relevant to catch error during treatment • Pre-delivery or real-time imaging • Independent applicator/channel reconstruction technology (e.g., EM tracking)
Applicator length/ source-indexer length	Manual check	IVD relevant to catch error during treatment Flat panel technology EM tracking (if link to the afterloader drive wire)
Source step size (patient specific)	Manual check	• IVD relevant to catch error during treatment • Flat panel technology
Interchanged guide tubes	Manual check	• IVD relevant to catch error during treatment • Flat panel technology • EM tracking (if link to the afterloader drive wire)
Recording of dose	General QA related to dose calculation in the TPS	• IVD only direct mean to measure the real delivered dose • Flat panel (dose engine needed, not direct)

Source: Adapted from Tanderup K et al. 2013. *Med. Phys.* 40, 070902.

and flat panel imaging could perform these tasks with greater accuracy. Finally, for some quality items, IVD is irrelevant. Of course, incorporating tracking technology with IVD would increase the overall accuracy of the IVD process significantly but would not alleviate the need for greater use of real-time imaging to ensure that any organ related motions are accounted for.

9.7 CONCLUSION

To date, brachytherapy has lagged behind external-beam radiation therapy (EBRT) in the development and adoption of IVD into routine practice. This is largely a consequence of the demanding requirements specific to brachytherapy including variable source–target–detector geometry, high-dose gradients, and internalized delivery of radiation. These challenges, combined with the use of isotopic sources—which inherently lack the ability to be "switched off"—are the very reasons why the development of IVD and treatment monitoring are imperative.

Currently, emerging technologies are addressing the particular challenges of source and detector position monitoring and show great promise in the diversity and richness of novel solutions being brought to bear on the problem. Clearly, some quality-related tasks will be better served by technologies other than IVD. However, with the specific objective of radiotherapy dosimetry within and/or near target volumes and OARs, IVD treatment verification remains the gold standard.

REFERENCES

Abayazid M, Kemp M and Misra S 2013. 3D flexible needle steering in soft-tissue phantoms using fiber Bragg grating sensors. In *Proceedings of the IEEE International Conference on Robotics and Automation*, ICRA 2013, 6–10 May 2013, Karlsruhe, Germany. pp. 5843–9.

Alecu R and Alecu M 1999. In-vivo rectal dose measurements with diodes to avoid misadministrations during intracavitary high dose rate brachytherapy for carcinoma of the cervix. *Med. Phys.* 26, 768–70.

Anagnostopoulos G, Baltas D, Geretschlaeger A, Martin T, Papagiannis P, Tselis N and Zamboglou N 2003. In vivo thermoluminescence dosimetry dose verification of transperineal 192Ir high-dose-rate brachytherapy using CT-based planning for the treatment of prostate cancer. *Int. J. Radiat. Oncol. Biol. Phys.* 57, 1183–91.

Andersen C E, Nielsen S K, Lindegaard J C and Tanderup K 2009. Time-resolved *in vivo* luminescence dosimetry for online error detection in pulsed dose-rate brachytherapy. *Med. Phys.* 36, 5033–43.

Anton M, Wagner D, Selbach H-J, Hackel T, Hermann R M, Hess C F and Vorwerk H 2009. In vivo dosimetry in the urethra using alanine/ESR during (192)Ir HDR brachytherapy of prostate cancer—A phantom study. *Phys. Med. Biol.* 54, 2915–31.

Archmabault L, Therriault-Proulx F, Beddar S and Beaulieu L 2012. A mathematical formalism for hyperspectral, multipoint plastic scintillation detectors. *Phys. Med. Biol.* 57, 7133–45.

Baltas D, Sakelliou L and Zamboglou N 2006. *The Physics of Modern Brachytherapy for Oncology.* Boca Raton, Florida: Taylor & Francis.

Beaulieu L, Carlsson Tedgren A, Carrier J-F, Davis S D, Mourtada F, Rivard M J, Thomson R M, Verhaegen F, Wareing T A and Williamson J F 2012. Report of the Task Group 186 on model-based dose calculation methods in brachytherapy beyond the TG-43 formalism: Current status and recommendations for clinical implementation. *Med. Phys.* 39, 6208–36.

Bharat S, Kung C, Dehghan E, Ravi A, Venugopal N, Bonillas A, Stanton D and Kruecker J 2014. Electromagnetic tracking for catheter reconstruction in ultrasound-guided high-dose-rate brachytherapy of the prostate. *Brachytherapy.* 13, 640–50.

Bloemen-van Gurp E J, Haanstra B K C, Murrer L H P, van Gils F C J M, Dekker A L A J, Mijnheer B J and Lambin P 2009a. In vivo dosimetry with a linear MOSFET array to evaluate the urethra dose during permanent implant brachytherapy using iodine-125. *Int. J. Radiat. Oncol. Biol. Phys.* 75, 1266–72.

Bloemen-van Gurp E J, Murrer L H P, Haanstra B K C, van Gils F C J M, Dekker A L A J, Mijnheer B J and Lambin P 2009b. In vivo dosimetry using a linear Mosfet-array dosimeter to determine the urethra dose in 125I permanent prostate implants. *Int. J. Radiat. Oncol. Biol. Phys.* 73, 314–21.

Brezovich I A, Duan J, Pareek P N, Fiveash J and Ezekiel M 2000. In vivo urethral dose measurements: A method to verify high dose rate prostate treatments. *Med. Phys.* 27, 2297–301.

Carrara M, Tenconi C, Rossi G, Borroni M, Cerrotta A, Grisotto S, Cusumano D et al. 2016. In vivo rectal wall measurements during HDR prostate brachytherapy with MOSkin dosimeters integrated on a trans-rectal US probe: Comparison with planned and reconstructed doses. *Radiother. Oncol.* 118, 148–53.

Cherpak A, Ding W, Hallil A and Cygler J E 2009. Evaluation of a novel 4D *in vivo* dosimetry system. *Med. Phys.* 36, 1672.

Cherpak A J, Cygler J E, E Choan and Perry G 2014. Real-time measurement of urethral dose and position during permanent seed implantation for prostate brachytherapy. *Brachytherapy.* 13, 169–77.

Cygler J E, Saoudi A, Perry G, Morash C and E Choan 2006. Feasibility study of using MOSFET detectors for *in vivo* dosimetry during permanent low-dose-rate prostate implants. *Radiother. Oncol.* 80, 296–301.

Damato A L, Viswanathan A N, Don S M, Hansen J L and Cormack R A 2014. A system to use electromagnetic tracking for the quality assurance of brachytherapy catheter digitization. *Med. Phys.* 41, 101702.

Darafsheh A, Zhang R, Kanick S C, Pogue B W and Finlay J C 2015. Separation of Čerenkov radiation in irradiated optical fibers by optical spectroscopy. *Proc. SPIE 9315, Design and Quality for Biomedical Technologies* VIII, 93150Q (March 11, 2015). doi:10.1117/12.2079441.

Das R, Toye W, Kron T, Williams S and Duchesne G 2007. Thermoluminescence dosimetry for in-vivo verification of high dose rate brachytherapy for prostate cancer. *Australas. Phys. Eng. Sci. Med.* 30, 178–84.

De Brabandere M, Kirisits C, Peeters R, Haustermans K and Van den Heuvel F 2006. Accuracy of seed reconstruction in prostate postplanning studied with a CT- and MRI-compatible phantom. *Radiother. Oncol.* 79, 190–7.

DeWerd L A, Ibbott G S, Meigooni A S, Mitch M G, Rivard M J, Stump K E, Thomadsen B R and Venselaar J L M 2011. A dosimetric uncertainty analysis for photon-emitting brachytherapy sources: Report of AAPM Task Group No. 138 and GEC-ESTRO. *Med. Phys.* 38, 782–801.

Espinoza A, Petasecca M, Fuduli I, Howie A, Bucci J, Corde S, Jackson M, Lerch M, and Rosenfeld, AB 2015. The evaluation of a 2D diode array in "magic phantom" for use in high dose rate brachytherapy pretreatment quality assurance. *Med. Phys.* 42, 663–73.

Fontbonne J M, Iltis G, Ban G, Battala A, Vernhes J C, Tillier J, Bellaize N et al. 2002. Scintillating fiber dosimeter for radiation therapy accelerator. *IEEE Trans. Nucl. Sci.* 49, 2223–7.

Hsu S-M, Yeh C-Y, Yeh T-C, Hong J-H, Tipton A Y H, Chen W-L, Sun S-S and Huang D Y C 2008. Clinical application of radiophotoluminescent glass dosimeter for dose verification of prostate HDR procedure. *Med. Phys.* 35, 5558–64.

Kertzscher G, Andersen C E and Tanderup K 2011. Identifying afterloading PDR and HDR brachytherapy errors using real-time fiber-coupled Al(2)O(3):C dosimetry and a novel statistical error decision criterion. *Radiother. Oncol.* 100, 456–62.

Kertzscher G, Rosenfeld A, Beddar S, Tanderup K and Cygler J E 2014. In vivo dosimetry: Trends and prospects for brachytherapy. *Br. J. Radiol.* 87, 20140206.

Khadem R, Yeh C C and Tehrani M S 2000. Comparative tracking error analysis of five different optical tracking systems. *Comp. Aid. Surg.* 5, 98–107.

Kirisits C, Rivard M J, Baltas D, Ballester F, De Brabandere M, van der Laarse R, Niatsetski Y et al. 2014. Review of clinical brachytherapy uncertainties: Analysis guidelines of GEC-ESTRO and the AAPM. *Radiother. Oncol.* 110, 199–212.

Kral F, Puschban E J, Riechelmann H and Freysinger W 2013. Comparison of optical and electro-magnetic tracking for navigated lateral skull base surgery. *Int. J. Med. Robot.* 9, 247–52.

Nakano T, Suchowerska N, Bilek M M, McKenzie D R, Ng N and Kron T 2003. High dose-rate brachytherapy source localization: Positional resolution using a diamond detector. *Phys. Med. Biol.* 48, 2133–46.

Nakano T, Suchowerska N, McKenzie D R and Bilek M M 2005. Real-time verification of HDR brachy-therapy source location: Implementation of detector redundancy. *Phys. Med. Biol.* 50, 319–27.

Nose T, Koizumi M, Yoshida K, Nishiyama K, Sasaki J, Ohnishi T and Peiffert D 2005. In vivo dosimetry of high-dose-rate brachytherapy: Study on 61 head-and-neck cancer patients using radiophotoluminescence glass dosimeter. *Int. J. Radiat. Oncol. Biol. Phys.* 61, 945–53.

Nose T, Koizumi M, Yoshida K, Nishiyama K, Sasaki J, Ohnishi T, Kozuka T et al. 2008. In vivo dosimetry of high-dose-rate interstitial brachytherapy in the pelvic region: Use of a radio-photoluminescence glass dosimeter for measurement of 1004 points in 66 patients with pelvic malignancy. *Int. J. Radiat. Oncol. Biol. Phys.* 70, 626–33.

Perez-Calatayud J J, Ballester F F, Das R K R, Dewerd L A L, Ibbott G S G, Meigooni A S A, Ouhib Z Z, Rivard M J M, Sloboda R S R and Williamson J F J 2012. Dose calculation for photon-emitting brachytherapy sources with average energy higher than 50 keV: Report of the AAPM and ESTRO. *Med. Phys.* 39, 2904–29.

Poulin E, Racine E, Binnekamp D and Beaulieu L 2015. Fast, automatic, and accurate catheter reconstruction in HDR brachytherapy using an electromagnetic 3D tracking system. *Med. Phys. Int. J.* 42, 1227–32.

Price R A, Benson C, Joyce M J and Rodgers K 2004. Development of a RadFET linear array for intracavitary *in vivo* dosimetry during external beam radiotherapy and brachytherapy. *IEEE Trans. Nucl. Sci.* 51, 1420–6.

Qi Z-Y, Deng X-W, Cao X-P, Huang S-M, Lerch M and Rosenfeld A 2012. A real-time *in vivo* dosi-metric verification method for high-dose rate intracavitary brachytherapy of nasopharyngeal carcinoma. *Med. Phys.* 39, 6757–63.

Raffi J A, Davis S D, Hammer C G, Micka J A, Kunugi K A, Musgrove J E, Winston J W, Ricci-Ott T J and DeWerd L A 2010. Determination of exit skin dose for 192Ir intracavitary accelerated partial breast irradiation with thermoluminescent dosimeters. *Med. Phys.* 37, 2693–702.

Reniers B B, Landry G G, Eichner R R, Hallil A A and Verhaegen F F 2012. In vivo dosimetry for gynaecological brachytherapy using a novel position sensitive radiation detector: Feasibility study. *Med. Phys.* 39, 1925–35.

Rivard M J, Coursey B M, DeWerd L A, Hanson W F, Saiful Huq M, Ibbott G S, Mitch M G, Nath R and Williamson J F 2004. Update of AAPM Task Group No. 43 Report: A revised AAPM protocol for brachytherapy dose calculations. *Med. Phys.* 31, 633–74. Online: http://link.aip. org/link/MPHYA6/v31/i3/p633/s1&Agg=doi

Rivard M J, Venselaar J L M and Beaulieu L 2009. The evolution of brachytherapy treatment plan-ning. *Med. Phys.* 36, 2136–53.

Roesthuis R J, Kemp M, van den Dobbelsteen J J and Misra S 2014. Three-dimensional needle shape reconstruction using an array of fiber Bragg grating sensors. *IEEE/ASME Trans. Mechatron.* 19, 1115–26.

Rogers D W O and Cygler J E 2009. *Clinical Dosimetry Measurements in Radiotherapy*. Madison, Wisconsin: Medical Physics.

Schultka K, Ciesielski B, Serkies K, Sawicki T, Tarnawska Z and Jassem J 2006. EPR/alanine dosimetry in LDR brachytherapy—A feasibility study. *Radiat. Prot. Dosimetry* 120, 171–5.

Seiler P G, Blattmann H, Kirsch S, Muench R K and Schilling C 2000. A novel tracking technique for the continuous precise measurement of tumour positions in conformal radiotherapy. *Phys. Med. Biol.* 45, N103–10.

Seymour E L, Downes S J, Fogarty G B, Izard M A and Metcalfe P 2011. In vivo real-time dosimetric verification in high dose rate prostate brachytherapy. *Med. Phys.* 38, 4785.

Sharma R and Jursinic P A 2013. In vivo measurements for high dose rate brachytherapy with optically stimulated luminescent dosimeters. *Med. Phys.* 40, 071730.

Smith R L, Haworth A, Millar J, Matheson B, Hindson B, Taylor M and Franich R 2015. Clinical implementation of *in vivo* source position verification in high dose rate prostate brachytherapy. *Radiother. Oncol.* 115, S87–8.

Smith R L, Haworth A, Panettieri V, Millar J L and Franich R D 2016. A method for verification of treatment delivery in HDR prostate brachytherapy using a flat panel detector for both imaging and source tracking. *Med. Phys.* 43, 2435–42.

Smith R L, Taylor M L, McDermott L N, Haworth A, Millar J L and Franich R D 2013. Source position verification and dosimetry in HDR brachytherapy using an EPID. *Med. Phys.* 40, 111706.

Suchowerska N, Jackson M, Lambert J, Yin Y B, Hruby G and McKenzie D R 2011. Clinical trials of a urethral dose measurement system in brachytherapy using scintillation detectors. *Int. J. Radiat. Oncol. Biol. Phys.* 79, 609–15.

Taffoni F, Formica D, Saccomandi P, Di Pino G and Schena E 2013. Optical fiber-based MR-compatible sensors for medical applications: An overview. *Sensors (Basel).* 13, 14105–20.

Tanderup K, Beddar S, Andersen C E, Kertzscher G and Cygler J E 2013 In vivo dosimetry in brachytherapy. *Med. Phys.* 40, 070902.

Taylor R E P, Yegin G and Rogers D W O 2007. Benchmarking brachydose: Voxel based EGSnrc Monte Carlo calculations of TG-43 dosimetry parameters. *Med. Phys.* 34, 445–7.

Therriault-Proulx F, Archambault L, Beaulieu L and Beddar S 2012. Development of a novel multi-point plastic scintillation detector with a single optical transmission line for radiation dose measurement. *Phys. Med. Biol.* 57, 7147–59.

Therriault-Proulx F, Beddar S and Beaulieu L 2013. On the use of a single-fiber multipoint plastic scintillation detector for 192Ir high-dose-rate brachytherapy. *Med. Phys.* 40, 062101.

Therriault-Proulx F, Wootton L and Beddar S 2015. A method to correct for temperature dependence and measure simultaneously dose and temperature using a plastic scintillation detector. *Phys. Med. Biol.* 60, 7927–39.

Toye W, Das R, Kron T, Franich R, Johnston P and Duchesne G 2008. An *in vivo* investigative protocol for HDR prostate brachytherapy using urethral and rectal thermoluminescence dosimetry. *Radiother. Oncol.* 91, 243–8.

Verhaegen F, Palefsky S, Rempel D and Poon E 2007. Imaging with Iridium photons: An application in brachytherapy. *Proc. SPIE 6510, Medical Imaging 2007*: Physics of Medical Imaging, 65103S (March 14, 2007). doi:10.1117/12.707983.

Waldhäusl C, Wambersie A, Pötter R and Georg D 2005. In-vivo dosimetry for gynaecological brachytherapy: Physical and clinical considerations. *Radiother. Oncol.* 77, 310–7.

Waltermann C, Koch J, Angelmahr M, Schade W, Witte M, Kohn N, Wilhelm D, Schneider A, Reiser S and Feußner H 2014. Femtosecond laser aided processing of optical sensor fibers for 3D medical navigation and tracking (FiberNavi). In *Proc. SPIE 9157, 23rd International Conference on Optical Fibre Sensors*, 91577G (June 2, 2014), doi:10.1117/12.2059599.

West J B and Maurer C R 2004. Designing optically tracked instruments for image-guided surgery. *IEEE Trans Med Imaging.* 23, 533–45.

Wood B J, Zhang H and Durrani A 2005. Navigation with electromagnetic tracking for interventional radiology procedures: A feasibility study. *J. Vasc. Interv. Radiol.* 16, 493–505.

CHAPTER 10

Quality Assurance Technologies

Antonio L. Damato, Dorin A. Todor, Laura
Cervino, and Robert A. Cormack

CONTENTS

10.1 INTRODUCTION

New technologies may enter the clinic to improve treatments but may exceed the capabilities of existing quality assurance (QA) paradigms. An example is the introduction of multi-leaf collimator technologies in external beam radiotherapy, which resulted eventually in the widespread adoption of intensity modulated radiation therapy (IMRT). The use of the available secondary independent dose calculations was not applicable to IMRT dose distributions. The result was the advent of measurement-based IMRT plan verification, and the effective and efficient method of performing IMRT QA remains an active point of discussion in the field. In brachytherapy, we are seeing the emergence of technologies that have the potential to disrupt the current treatment paradigms, such as direction modulated brachytherapy (DMBT) (Han et al., 2014) and rotating shield brachytherapy (RSBT) Yang et al. (2013) allowing fluence modulation. While this particular development is still in its infancy, and future clinical use remains uncertain, there are technologies that may be considered established in external beam planning are being newly applied to brachytherapy. Two recent innovations for brachytherapy, the use of deformable registration (Chapter 7) and inhomogeneity corrected dose calculations (Chapter 5) may seem less consequential, but in practice they can influence clinical decisions. For instance, a voxel-by-voxel summation of the dose received during the external beam course will likely significantly alter dose optimization of a brachytherapy boost. In another example, correct calculation of dose will affect applicator selection and design of commercial or custom shields for normal organs. In the next two sections of this chapter we will discuss the QA implications of these two innovations in brachytherapy practice.

New technologies should not be considered only as an overall burden to a clinic QA budget. Technological developments may also provide the ability to improve a QA program. A section of this chapter is dedicated to the use of electromagnetic tracking technology in QA of brachytherapy implantations, plans, and treatments. In another case, technological changes in clinical practice provide the means to facilitate and expand QA practice. In the last section, we will discuss how electronic medical record (EMR) initiative, can be used to design tools and software that mine patient-specific and clinic-specific information to automate and enhance some aspects of QA.

10.2 DEFORMABLE IMAGE REGISTRATION

Traditionally, dose summation from different brachytherapy fractions has been based on point doses. In every fraction, the point dose for target and for organs at risk (OARs) is calculated from the treatment planning system or estimated on radiographs. The use of Point A in cervical cancer brachytherapy is one example. Accurate reproducibility in every fraction of Point A relative to the anatomy cannot be guaranteed. In the same manner, the rectum point dose may be in a different position within the rectum in every fraction. Doses are then added up as if it were the same anatomical point every day. As volumetric images (e.g., CT, MRI) became available, dose volume histograms provide doses to target (CTV_{HR}) and OARs (rectum, bladder, sigmoid, etc.). Still, most clinics will use the Point A for dose prescription and dose recording. As for the OARs, the maximum dose to 2 cm^3 is usually treated now as the OAR point dose, and added up among different fractions counting for radiobiological equivalent dose.

Recent advancements in deformable image registration (DIR) methods have allowed us to calculate a better match for the anatomical position of target and OAR in different fractions, which in turn, leads to better dose accumulation estimation. Although commercial software is available to perform DIR, none are specific to brachytherapy. Several research teams are working on developing methods that deal with the added challenge that we face in brachytherapy, that is, the presence of the applicator (Zhen et al., 2015). More details on efforts on DIR are provided in Chapter 7. It is not well understood yet how the uncertainties associated with DIR methods affect the dose calculations and resulting clinical metrics, and efforts are now being put into that task (Salguero et al., 2011; Samavati et al., 2016).

Evaluating the accuracy of deformable registration algorithms is challenging due to the inexistence of a ground truth for patient images. Deformable phantoms are used instead for both commissioning and QA. Physical deformable phantoms aim at providing the ground truth while not affecting the deformable registration algorithm (one such phantom is depicted in Figure 10.1) (Kashani et al., 2006, 2007; Serban et al., 2008; Kirby et al., 2011; Graves et al., 2015). However, these phantoms are rare and not specific to brachytherapy. Commissioning of the image registration algorithm and QA upon software upgrades need to evaluate results in digital phantoms. Digital phantoms are created

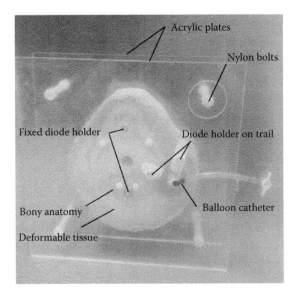

FIGURE 10.1   A deformable phantom that can serve the purpose of validating a deformable image registration algorithm. (Image courtesy of Graves YJ et al. *Med Phys* 2015;42:1490–1497.)

or modified from patient images in software with a known deformation field applied, as is done in Zhen et al. (2015) and shown in Figure 10.2, or independently created images based on clinical scenarios (such as the XCAT phantom). These latter phantoms are limited by the realism with which they represent human anatomy and many features that are seen in human images are impossible to reproduce in phantoms. It is recommended that the algorithm used to create the deformation in the images is not the same as the algorithm used in the deformable registration algorithm. Digital phantoms can be developed in-house or purchased from vendors. The purpose of using these images is to then recover the deformation vector field with the algorithm under evaluation. Tests should

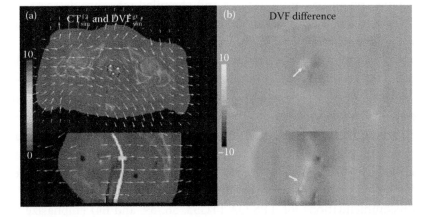

FIGURE 10.2   (a) Artificially created deformable vector fields (ground truth). (b) Difference of the deformation obtained with a deformable registration algorithm by Zhen et al. (2015) with the ground truth. (Image courtesy of Zhen X et al. *Phys Med Biol* 2015;60:2981–3002.)

be done during commissioning and on an annual basis. Phantoms should model basic deformation and sliding deformation, as recommended by TG132. The recommended tolerance in TG132 is 95% of voxels to be within 2 mm and the maximum error to be less than 5 mm.

Deformable registration accuracy should be reviewed for each patient as well upon application of the software. Visual confirmation of relevant boundaries and features should be within 1–2 voxels, and any additional error should feed into margins.

10.3 MODEL-BASED DOSE CALCULATION ALGORITHMS IN BRACHYTHERAPY

The TG43 dose calculation algorithm (in its updated version, TG43U1) is likely the most clinically used dose formalism in brachytherapy today. Slowly emerging in the clinical practice is a new paradigm which replaces the infinite "water world" with actual patient data, taking into account tissue heterogeneities, interfaces and scatter conditions, shielded applicators, etc. These new dose computation algorithms are based on Monte Carlo, collapsed cone, and the linear Boltzmann transport equations and generically referred to as "model-based dose calculation algorithms" (MBDCAs) (Beaulieu et al., 2012). More in-depth details are provided in Chapter 5. While many methods to account for material inhomogeneity and scatter conditions were developed in the last two decades, two are currently integrated in commercially available high-dose rate (HDR) treatment planning systems (TPS): Oncentra® Brachy with advanced collapsed-cone engine (ACE™) and BrachyVision™ Acuros BV™. Planning in brachytherapy essentially relies on two prerequisites: a correct description of the implant/applicator geometry and an accurate calculation of dose deposited by the sources into the targeted tissues. While simplistic in its assumptions, the TG43 formalism for dose calculation, together with a standard for source strength specification, created a coherent and unified practice of brachytherapy across institutions worldwide. Our challenge today is to preserve this coherence while transitioning to MBDCA by offering societal guidance, and an infrastructure for commissioning and QA activities: phantom configurations and compiled reference datasets. In the context of TG43, water was the reference medium and, while an oversimplification of complex clinical settings, it removed the need for tissue/material assignment and boundary description but facilitated inter-institutional comparisons of clinical results. MBDCA dose computation adds a complexity layer in that interaction cross sections need to be specified at voxel level for tissues, sources, and applicators. The possibility that tissue composition, source, and applicator models used in TPS might not accurately describe reality (including actual applicators used for treatment) creates new and significant error pathways requiring new layers of QA. Under TG43, both TPS source parameters and dose computation could be verified by performing simple calculations in easy to reproduce geometries containing either single sources or combinations of sources. Most of the AAPM Task Group containing recommendations for TPS QA (TG53, 56, 59, and 64) emphasize the overall functionality of the system or simple to test items like source decay or treatment time for given source distributions. Patient-specific QA is typically performed using nomograms derived based on clinical experience from academic centers with extensive practice (e.g., prostate LDR nomograms from the Memorial Sloan Kettering Cancer Center (MSKCC)). All these well developed and widely spread practices should become the minimum test for a MBDCA TPS: at a minimum, MBDCA should reproduce the results TG43 would produce in an equivalent water phantom. While trivial in appearance, this statement contains the implicit assumption that the source definitions for TG43 and MBDCA are in fact equivalent in the clinical range of distances (0.1–10 cm). The particularities of TPSs and HDR sources associated with remote afterloaders from different manufacturers make commissioning and comparisons very difficult. An interesting idea to alleviate these difficulties is presented in a recent paper "A generic high-dose rate ^{192}Ir brachytherapy source for evaluation of model-based dose calculations beyond the TG43 formalism" (Ballester et al., 2015). The authors propose and present the development of a hypothetical HDR ^{192}Ir source and a virtual water phantom in DICOM (Digital Imaging and Communications in Medicine) format. Full MC-based calculation employing a host of currently used codes (Taylor et al., 2007; Afsharpour et al., 2012) are used to generate TG43 and MBDCA source parameters. MC-based dose in a phantom (20.1 cm size water cube with 1 mm voxels surrounded by 51.1 cm side air cube) is then compared with the two MBDCA commercially available TPSs, both currently implemented for ^{192}Ir HDR brachytherapy only. Without getting into the details of the study, a test HDR source and afterloader was created and imported into both Oncentra® Brachy and BrachyVision™, together with a DICOM phantom in which two scenarios were created: source centered and source displaced. Dose distributions in both systems were compared with TG43 and MC-based methods. Agreement, when compared with MC, was for both single source cases investigated, within (−1%,+1%) range for Acuros™ and (−2%,+3%) range for ACE™. This type of comparison using reference data seems to be a suitable method to be used for commissioning and annual QA for MBDCA-based TPS.

Plan optimization is a difficult subject since a lot of practitioners in brachytherapy (physicians and physicists) have preconceived ideas on what optimization should or can do. MBDCA-based TPSs, even though fast enough to compute a "real" dose distribution in a few minutes are still too slow to include such a calculation in an optimization loop. The obvious question follows: is a plan using TG43 in the optimization loop and MBDCA in the final calculation still optimal? The literature on QA for optimization is virtually nonexistent. There is one exception (Deufel and Furutani, 2014) even though the authors tackle the issue of whether optimization results are "reasonable," rather than optimal. They show that the "simple optimization" method they are proposing as a QA tool produces results which "…are surprisingly similar to the results from a more complex, commercial optimization for several clinical applications." They also conclude that "the improvements expected from sophisticated linear optimizations, such as PARETO methods, will largely be in making systems more user-friendly and efficient, rather than in finding dramatically better source strength distributions." Our own opinion is that nomograms can be a valuable QA tool in capturing consistency (and consequently outliers) for a particular treatment site and modality, provided consistency in the planning structures and optimization parameters. An unfortunate "legacy" constraint in dose optimization seems to be, for many practitioners, the "smoothness" of dwell times distributions along a given applicator. A "band aid" created early in the development of optimization algorithms in order to "temper" rather unstable behavior of these algorithms, it proved to be very resilient and it is still used today, when its only achievement is to create less than optimal plans by producing eye pleasing dwell times distributions.

10.4 ELECTROMAGNETIC TRACKING FOR BRACHYTHERAPY QA

In a HDR brachytherapy failure mode analysis published in 2013 (Wilkinson and Kolar, 2013), 3 of the 6 highest failures with the highest risk probability number were related to catheter reconstruction (also see Chapter 8). In the era of image guided brachytherapy, any discrepancy between the digitization of an available source path and its actual location compared to the surrounding anatomy may result in large differences between the planned dose and the delivered dose. Discrepancies may arise from reconstruction errors (Figure 10.3), such as inverting the connector-tip orientation of a catheter, or assigning the incorrect channel number to a catheter. These errors are not noticeable in the dose distribution showed in the planning system, and are therefore mostly invisible to plan review by the physician authorized user. Detecting these errors may be also challenging during a secondary physics check. Another source of discrepancy can arise when a change in the geometry of the catheters compared to each other may have occurred between simulation and treatment. An independent physical measurement of the catheters' paths is necessary to perform an effective check. Recently, interest has increased in the emerging use of electromagnetic tracking technology in brachytherapy (Zhou et al., 2013; Damato et al., 2014; Mehrtash et al., 2014; Poulin et al., 2015). See Chapter 4 for an in-depth review. A particular challenge for the application of this technology is the registration of the electromagnetic frame of reference to the frame of reference of the images where the plan is generated. Under this respect, the application of this technology to QA seems particularly promising, as the comparison of reconstructed dwell positions from electromagnetic data and from image analysis (manual or automatic) allow for very accurate registration of the two spaces. A procedure for the QA of catheter digitization based on this approach has been described (Damato et al., 2014), and it shows promise as having good sensitivity and

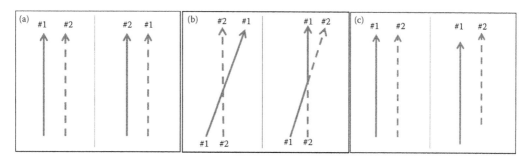

FIGURE 10.3 Possible reconstruction errors during catheter digitization in HDR brachytherapy. (a) "Swap" error, in which catheter #1 (#2) is incorrectly identified as catheter #2 (#1). (b) "Mix" error, in which two catheters coming in close proximity inside the patient can be mistakenly identified due to an overlap of their image signature. (c) "Shift" error, in which a catheter tip is incorrectly digitized.

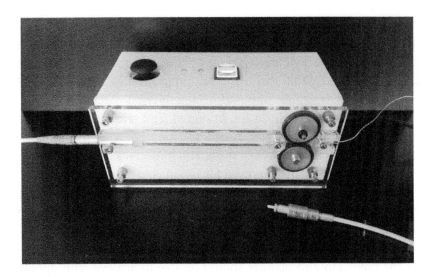

FIGURE 10.4 A prototype of a robot for the automatic delivery of an electromagnetic sensor through a brachytherapy catheter.

specificity in clinical scenarios. A similar technique can also be implemented to monitor the changes over time in the geometry of an interstitial implant (Damato et al., 2014).

The main roadblock in the widespread implementation of this QA technique is the need to build custom systems from raw components (e.g., electromagnetic sensors, field emitters, and data processing) and developing the software for routine use. Preliminary reports of clinical applications suggest that robotic delivery of the electromagnetic sensor is very desirable to perform fast, reliable, and programmable data acquisition. A prototype of such a system was recently presented at an Annual Meeting of the AAPM, and it is depicted in Figure 10.4. Ideally, this technology may one day be implemented in commercial afterloaders, with deployment of the electromagnetic sensor controlled by the robotic system in the afterloader and data analysis and comparison to the reconstructed dwell locations handled directly by the treatment console.

10.5 SEMI-AUTOMATIC PLAN VERIFICATION

The advent of EMRs has provided opportunities and challenges. In radiation oncology departments, an increasing number of institutions have abandoned paper charts and adopted electronic-only records. A variety of solutions were found for ad-hoc problems, such as prompt communication of set-up instructions to therapist in a linac vault, or efficient recording and propagation of changes to a treatment plan. In some cases, hybrid solutions were embraced in which some information was still produced in printed form, only to be scanned and ported into the EMR later. Brachytherapy has probably seen a higher number of these hybrid solutions, due to historic disconnect between brachytherapy treatment units and record and verify systems. Moreover, regulatory uncertainty about electronic approvals of a treatment directive may have slowed the adoption of paperless formats. These approaches detract much from the potential benefits of electronic records, such as data mining. A recent report (Damato et al., 2014) has underscored the potential of using the data mining capabilities inherent in EMRs to increase the efficiency and efficacy of the secondary physics check of brachytherapy plans, potentially resulting in an increase in safety. This is performed with the use of ad-hoc programs that access the EMR and planning information, comparing data, and performing internal checks on the consistency of plans with clinic practices and built-in conditions. Once a reasonably thorough transition to an EMR environment is performed, such a software has shown to decrease the time required to perform a secondary physics check and to decrease the number of errors and inconsistencies in treatment plans approved for treatment.

Aside from its use to detect errors, automatic analysis of electronic planning information can also be used to verify the quality of a plan. Access to old plans and patient contours allow the development of training sets for knowledge-based algorithms capable of predicting, given a new patient anatomy and applicator placement, what are reasonable plan metrics that the physician and planner should be able to achieve. This approach, which has been demonstrated for external beam planning (Appenzoller et al., 2012) and preliminarily reported for brachytherapy plans (Damato et al., 2013), would allow the detection of suboptimal plans. This capability can enhance QA practices providing previously unavailable quantitative tools for assessing plan quality. In institutions with multiple physicists and radiation oncologists, potentially separated on multiple sites, knowledge-based algorithms provide a quantitative approach at harmonizing plan quality among different users. The emerging opportunity to develop software capable of performing such analysis will likely be available, in the foreseeable future, only to centers with the infrastructure to develop and commission custom software for medical applications. Guidelines exist (Lillicrap, 2000) on ways to validate the custom software. It has been shown that in some instances, an increase in the technological sophistication of the QA process can result in an increase in errors (Patton et al., 2003). Extreme care should be used to avoid the introduction of systematic problems in the QA practice due to poor coding or poor architecture.

10.6 CONCLUSION

Brachytherapy has evolved greatly from the days of planar films and point dose characterization of treatment plans. Innovations in radiation sources (Mobit et al., 2015), applicators (Klopp et al., 2013), insertion assistance (Viswanathan et al., 2013; Podder et al., 2014), and computational techniques have produced a field of brachytherapy that is significantly different from existing published QA protocols (Cormack, 2008). While technology provides new QA challenges, it also provides new means to facilitate QA efforts. Professionals in the field of brachytherapy should develop standardized protocols appropriate for modern methodologies in clinical practice.

REFERENCES

Afsharpour H, Landry G, D'Amours M et al. ALGEBRA: ALgorithm for the heterogeneous dosimetry based on GEANT4 for BRAchytherapy. *Phys Med Biol* 2012;57:3273–3280.
Appenzoller LM, Michalski JM, Thorstad WL et al. Predicting dose-volume histograms for organs-at-risk in IMRT planning. *Med Phys* 2012;39:7446–7461.

Ballester F, Carlsson Tedgren A, Granero D et al. A generic high-dose rate (192)Ir brachytherapy source for evaluation of model-based dose calculations beyond the TG-43 formalism. *Med Phys* 2015;42:3048–3061.

Beaulieu L, Carlsson Tedgren A, Carrier JF et al. Report of the Task Group 186 on model-based dose calculation methods in brachytherapy beyond the TG-43 formalism: Current status and recommendations for clinical implementation. *Med Phys* 2012;39:6208–6236.

Cormack RA. Quality assurance issues for computed tomography-, ultrasound-, and magnetic resonance imaging-guided brachytherapy. *Int J Radiat Oncol Biol Phys* 2008;71:S136–S141.

Damato AL, Cormack RA, Viswanathan AN. Characterization of implant displacement and deformation in gynecologic interstitial brachytherapy. *Brachytherapy* 2014;13:100–109.

Damato AL, Devlin PM, Bhagwat MS et al. Independent brachytherapy plan verification software: Improving efficacy and efficiency. *Radiother Oncol* 2014;113:420–424.

Damato AL, Viswanathan AN, Cormack RA. Validation of mathematical models for the prediction of organs-at-risk dosimetric metrics in high-dose-rate gynecologic interstitial brachytherapy. *Med Phys* 2013;40:101711.

Damato AL, Viswanathan AN, Don SM et al. A system to use electromagnetic tracking for the quality assurance of brachytherapy catheter digitization. *Med Phys* 2014;41:101702.

Deufel CL, Furutani KM. Quality assurance for high dose rate brachytherapy treatment planning optimization: Using a simple optimization to verify a complex optimization. *Phys Med Biol* 2014;59:525–540.

Graves YJ, Smith AA, McIlvena D et al. A deformable head and neck phantom with in-vivo dosimetry for adaptive radiotherapy quality assurance. *Med Phys* 2015;42:1490–1497.

Han DY, Webster MJ, Scanderbeg DJ et al. Direction-modulated brachytherapy for high-dose-rate treatment of cervical cancer. I: Theoretical design. *Int J Radiat Oncol Biol Phys* 2014;89:666–673.

Kashani R, Balter JM, Hayman JA et al. Short-term and long-term reproducibility of lung tumor position using active breathing control (ABC). *Int J Radiat Oncol Biol Phys* 2006;65:1553–1559.

Kashani R, Lam K, Litzenberg D et al. Technical note: A deformable phantom for dynamic modeling in radiation therapy. *Med Phys* 2007;34:199–201.

Kirby N, Chuang C, Pouliot J. A two-dimensional deformable phantom for quantitatively verifying deformation algorithms. *Med Phys* 2011;38:4583–4586.

Klopp AH, Mourtada F, Yu ZH et al. Pilot study of a computed tomography-compatible shielded intracavitary brachytherapy applicator for treatment of cervical cancer. *Pract Radiat Oncol* 2013;3:115–123.

Lillicrap S. Physics aspects of quality control in radiotherapy (Report No. 81). *Phys Med Biol* 2000;45:815.

Mehrtash A, Damato A, Pernelle G et al. EM-navigated catheter placement for gynecologic brachytherapy: An accuracy study. *Proc SPIE Int Soc Opt Eng* 2014;9036:90361F.

Mobit PN, Packianathan S, He R et al. Comparison of Axxent-Xoft, (192)Ir and (60)Co high-dose-rate brachytherapy sources for image-guided brachytherapy treatment planning for cervical cancer. *Br J Radiol* 2015;88:20150010.

Patton GA, Gaffney DK, Moeller JH. Facilitation of radiotherapeutic error by computerized record and verify systems. *Int J Radiat Oncol Biol Phys* 2003;56:50–57.

Podder TK, Beaulieu L, Caldwell B et al. AAPM and GEC-ESTRO guidelines for image-guided robotic brachytherapy: Report of Task Group 192. *Med Phys* 2014;41:101501.

Poulin E, Racine E, Binnekamp D et al. Fast, automatic, and accurate catheter reconstruction in HDR brachytherapy using an electromagnetic 3D tracking system. *Med Phys* 2015;42:1227–1232.

Salguero FJ, Saleh-Sayah NK, Yan C et al. Estimation of three-dimensional intrinsic dosimetric uncertainties resulting from using deformable image registration for dose mapping. *Med Phys* 2011;38:343–353.

Samavati N, Velec M, Brock KK. Effect of deformable registration uncertainty on lung SBRT dose accumulation. *Med Phys* 2016;43:233.

Serban M, Heath E, Stroian G et al. A deformable phantom for 4D radiotherapy verification: Design and image registration evaluation. *Med Phys* 2008;35:1094–1102.

Taylor RE, Yegin G, Rogers DW. Benchmarking brachydose: Voxel based EGSnrc Monte Carlo calculations of TG-43 dosimetry parameters. *Med Phys* 2007;34:445–457.

Viswanathan AN, Szymonifka J, Tempany-Afdhal CM et al. A prospective trial of real-time magnetic resonance-guided catheter placement in interstitial gynecologic brachytherapy. *Brachytherapy* 2013;12:240–247.

Wilkinson DA, Kolar MD. Failure modes and effects analysis applied to high-dose-rate brachytherapy treatment planning. *Brachytherapy* 2013;12:382–386.

Yang W, Kim Y, Wu X et al. Rotating-shield brachytherapy for cervical cancer. *Phys Med Biol* 2013;58:3931–3941.

Zhen X, Chen H, Yan H et al. A segmentation and point-matching enhanced efficient deformable image registration method for dose accumulation between HDR CT images. *Phys Med Biol* 2015;60:2981–3002.

Zhou J, Sebastian E, Mangona V et al. Real-time catheter tracking for high-dose-rate prostate brachytherapy using an electromagnetic 3D-guidance device: A preliminary performance study. *Med Phys* 2013;40:021716.

CHAPTER 11

Additive Manufacturing (3D Printing) in Brachytherapy

Ananth Ravi, Lior Dubnitzky, and Harry Easton

CONTENTS

11.1 CONFORMING PATIENTS TO APPLICATORS

Historically, brachytherapy treatment delivery has been based on standard applicators that have been designed to conform to population-based approximations of patient anatomy. Unfortunately, not all patients are the same, and some have unique anatomies that require customized approaches to treat optimally. These patient-specific considerations may arise from varying body habitus, surgical sequelae altering the normal appearance of the anatomy, or due to medical contraindications preventing the use of general anesthetic necessitating less invasive approaches. With conventional manufacturing techniques tailoring applicators or needle guides to the needs of individuals can be impractical, resulting in the patient's anatomy having to conform to the shapes of these applicators.

Additive manufacturing (AM) in the context of brachytherapy attempts to address shortcomings of standardized applicators with the ability of creating customized, robust, sterilizable applicators that aim to conform the applicators to suit the patient's needs.

11.2 WHAT IS AM?

AM is a process that has been in development for the last 30 years and began with the invention of stereolithography by, Charles W. Hull (1986, 1990). AM is also known as 3D printing and refers to seven distinct process types. These processes create components by growing or adding materials one layer at a time with each subsequent layer being fused to the previous layer, repeating until the desired component is built or complete. AM is able to create parts that cannot be made with conventional machining, which typically uses various methods of subtracting material to create subcomponents that are fixed together to create the final part. Drawing an example from the aerospace industry, a complex fuel injector would normally be manufactured as an assembly of as many as 20 components which need to be individually cast, molded, and machined often utilizing specialized alloys, and then fixed together to form the finished article. AM can now be used to form these injectors as one component with all the internal channels and structures printed layer by layer. This method allows for significant optimization of power, efficiency, and weight.

Prototypes and end-use components created using AM are now in use in various industries from consumer products, automotive and aerospace, to medicine utilizing a multitude of materials including plastics, ceramics, metals, and composites. As the field of AM advances, investment in AM equipment is becoming more affordable, low-end consumer machines are available such as the Cube® 3D (3D systems, South Carolina, USA) retailing in the thousands of dollars to high end Direct Metal Printing machines which can cost over a million dollars and require special infrastructure.

AM starts with the generation of a 3D model created using a computer aided design (CAD) software program with the finished model being saved as a standard tessellation language (STL) file which breaks the model into tiny triangles. Further software is utilized to slice the digital data into thin individual layers, which are then sent to the AM unit. The AM unit interprets the slice data and creates the part by joining layers of material one on top of the other until the part is complete. This process allows for design freedom and manufacturing of extremely complex components, which using a subtraction style of manufacturing would be impossible to achieve. Some secondary processes may be needed

FIGURE 11.1 Sunnybrook Odette Physics and Mechanical Design and Engineering services vat photopolymerization unit on site at Sunnybrook Research Institute (SRI).

to finish the component or directly to finish high definition parts and can be achieved depending on the AM machine type. Other methods of obtaining a 3D model are available through reverse engineering, via hand-held scanning devices using laser or light to capture topography and output the resulting data as an STL file. Volumetric medical imaging data such as computed tomography (CT), magnetic resonance imaging (MRI), and 3D ultrasound (US) can be segmented and then converted into STL files that can be printed.

The following subsections present a discussion on the various methods of AM (Carter, 2001).

11.2.1 Vat Photopolymerization

During vat photopolymerization, liquid photopolymer in a vat is selectively cured by light-activated polymerization. The process is also referred to as light polymerization. Figure 11.1 is a photograph of a commercially available vat photopolymerization unit.

11.2.2 Material Jetting

Material jetting is a process by which a print head selectively deposits material to the build area. These droplets are comprised of photopolymers with secondary materials (e.g., wax) to create support structures during the build process. An ultraviolet (UV) light solidifies the photopolymer material to form cured parts. Post build processing will remove the support material.

11.2.3 Material Extrusion

Material extrusion involves a thermoplastic material that is fed through a heated nozzle and deposited on the build platform. The nozzle then melts the material and extrudes it to create each object layer. Continuation of this process completes the part.

11.2.4 Powder-Bed Fusion

During powder-bed fusion, particles of material (plastic or metal) are selectively fused together using a thermal energy source, for example, a laser. Once a layer is fused, a new layer is created by spreading powder over the top of the object and repeating the process. Object support is achieved using unfused material thus reducing the need for support systems.

11.2.5 Binder Jetting

In binder jetting, material particles are joined together selectively using a liquid binding agent (e.g., glue). Colored parts can be achieved by the addition of inks. After each layer is formed, a new one is created by spreading powder over the top of the object and repeating the process. Repetition of this action continues until the object is formed. Unbound material is used to support the object being produced, thus reducing the need for support systems.

11.2.6 Sheet Lamination

During sheet lamination, thin sheets of material, metal, or plastic are bonded together using a variety of methods gluing or ultrasonic welding to form an object. Each new sheet of material is positioned over previous layers. A laser or knife is used to cut around the required part and unneeded material is removed. This process is repeated until the part is completed.

11.2.7 Directed Energy Deposition

During directed energy deposition, focused thermal energy fuses metal as it is being deposited. Directed energy deposition systems may employ either wire or powder-based approaches.

Given the variety of AM methodologies and apparatus, it is advisable to consult AM professionals available through the vendors. By providing detailed end-use requirements of the intended product the best AM process can be selected. Many factors will be taken into account such as surface finish, density, clarity, porosity, biocompatibility, and material. These details will guide which process is the most suitable to achieve the best results. To date, the use of AM to create brachytherapy devices has used vat photopolymerization as the mainstay methodology, where the finalized devices are suitable for clinical use. Vat photopolymerization suits medical needs because very fine features can be achieved and the end product can be fabricated from medical-grade materials that can be sterilized. The uncured liquid easily drains out of the printed hollow channels that will later be used for the catheters to transport the high-dose rate source. The device has low CT and MR signature and a high clarity finish, which allows the viewing of the internal structure for precise positioning on the patient and visual confirmation of catheter positioning. The next section will discuss some of the clinical applications of AM in brachytherapy.

11.3 AM AND PATIENT-SPECIFIC APPLICATORS

The current commercially available fleet of applicators is discretized in their shapes and sizes. These shapes and sizes have been designed to treat the majority of patients based on anatomical approximations. However, instances arise where standard applicators are suboptimal for a patient's unique anatomy. Despite efforts in improving traditional applicators for brachytherapy to add customization based on unique patient geometry, conventional manufacturing methods have a variety of limitations that make them impractical for rapidly creating customized brachytherapy devices on-demand. AM on the other hand benefits from being able to rapidly manufacture prototypes with high accuracy and specificity while still being relatively inexpensive. With AM brachytherapy devices, brachytherapy can be delivered without compromising clinical objectives from target coverage to critical organ avoidance due to anatomical and physical constraints imposed by standard applicators.

In the following subsections, the various innovative uses of AM in brachytherapy will be discussed, from custom interstitial needle guides to avoid unnecessary trauma to normal tissues or sensitive organs to intricate surface applicators where the catheters are accurately positioned from the skin surface.

11.3.1 Plesiobrachytherapy, Superficial Therapy

The bulk of plesiobrachytherapy is done today using either the Freiburg flap, a flexible planar applicator with linear catheter trajectories or wax molds. In instances where standard Freiburg flap approaches do not adequately conform to the treatment area, a wax mold is typically used. However, wax molds are cumbersome and laborious requiring skilled staff to create the mold and cast in the channels. The positions of the channels are rarely optimized but rather are a direct result of the casting process. As such the catheter trajectories are limited to straight paths, and may not optimally follow the curvature of the target surface. Below are some examples of how AM has overcome some of these disadvantages.

11.3.1.1 Scalp Plesiobrachytherapy

Traditionally, lesions on a patient head and scalp have not been ideally treated by the set of standardized applicators as they do not conform to the curvature of the surface. Harris et al. (2015) designed a scalp brachytherapy applicator that accurately placed catheters such that they conformed to the surface of the scalp while maintaining a fixed offset. The study used a 3D Touch v5.4.1 printer (3D Systems Inc., SC, USA) with ABS plastic photopolymer. Dosimetric qualities of ABS plastic were validated using gafchromic film. The quality of the print was evaluated and deemed to be suitable using mechanical verification of the catheter channel dimensions using a Vernier caliper. Error in catheter positioning was found to be less than 1.0 mm, which was deemed to be suitable for clinical use.

11.3.1.2 Penile Plesiobrachytherapy

Penile brachytherapy has been delivered predominantly using interstitial means (Crook et al., 2002, 2009); however, due to the invasiveness of the treatment and resource utilization with hospital admission, a plesiobrachytherapy approach is attractive. The challenges with conventional methods of creating an applicator are that it is difficult to ensure apposition with the penis surface and secure the catheters at an optimal distance to the skin surface. D'Alimonte et al. (2015) and Helou et al. (2015) explored the use of AM to fabricate clear penile applicators (Figures 11.2 and 11.3). Negative impressions of the patient's penis were cast in dental alginate. The impressions were imaged using CT, the voids were then segmented and a virtual model of the patient's penis was created in 3D Slicer, the virtual

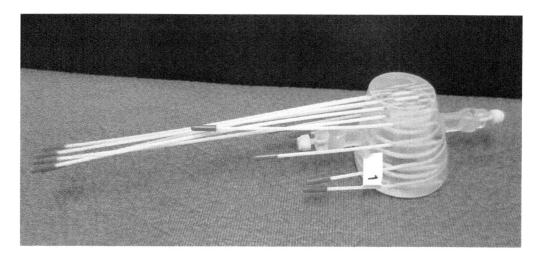

FIGURE 11.2 Photograph of plesiobrachytherapy applicator for penile brachytherapy. The clarity of the applicator aids in a visual verification of its placement on the lesion.

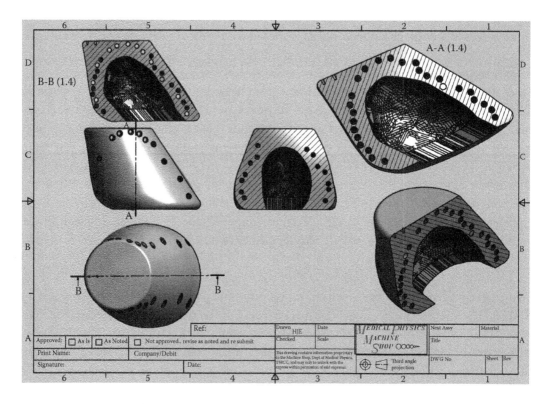

FIGURE 11.3 CAD of the plesiobrachytherapy applicator. This diagram depicts the uniform spacing of the catheters from the surface of the penis. Additionally, it highlights the level of detail captured in the applicator of the penile surface.

penis was converted into an STL file and transferred to the CAD software where an applicator was designed around it. The applicator positioned catheters that ran circumferentially around the penis at a distance of 5 mm. After the applicator was printed, it was re-imaged using CT and a treatment plan was generated using anatomy-based inverse optimization. Patient satisfaction and acute skin toxicities were favorably reported (D'Alimonte et al., 2015; Helou et al., 2015).

11.3.2 Intracavitary Therapies

Intracavitary brachytherapy has been premised on the principle that rigid applicators, when placed in potential spaces within the body, force the tissue to conform to the applicator. Unfortunately, this is sometimes not the case in vaginal vault therapy where postsurgical patients have scar tissue, which is less pliable. Alginate molds have been used in the past to create customization; however, there is a risk of the mold getting stuck if improper technique is used while they are being created. Also, there are limited options for catheter track optimization. The next subsections discuss some of the novel applications of AM in intracavitary brachytherapy.

11.3.2.1 Vaginal Intracavitary Brachytherapy

Vaginal vault brachytherapy is typically treated using standardized cylindrical applicators which are designed to force the vaginal cavity to conform to the applicator. In cases where the patient's vaginal cuff does not conform to a hemispherical dome geometry, treatment objectives can suffer (Richardson et al., 2010). Wiebe et al. (2015) and Cunha et al. (2015) explored AM techniques to create custom 3D printed applicators based on individualized patient anatomy for gynecological intracavitary brachytherapy. Cunha et al. replicated standard designs of traditional cylindrical applicators out of PC-ISO polymer (Sratasys, MN, USA) using AM techniques and added customized catheter trajectories which created dosimetric advantages (Cunha et al., 2015). Due to water equivalency, PC-ISO was difficult to see on CT imaging, and thus made it difficult to confirm orientation and positioning. Coating the applicators in radiopaque dye and condom pre-insertion may help solve this problem (Cunha et al., 2015). Both Weibe et al. (2015) and Cunha et al. (2015) confirmed the radiological equivalency of the polymers they used to water. Weibe et al. reported on clinical case study that used a customized vaginal applicator made with AM techniques. Given the patient's narrow introitus, only a 2.5 cm cylinder applicator could have been used, which would have underdosed the target. A customized, two-part (Figure 11.4a–c) applicator was created provided sufficient coverage while simultaneously limiting dose to sensitive organs at risk. The results of the study showed an improvement in target coverage of 13.2%, and a reduction of doses to the organs at risk (Wiebe et al., 2015). This was largely because the multi-catheter applicator allowed for dispersion of the dose laterally at the apex independently of the anterior or posterior directions (Wiebe et al., 2015).

11.3.2.2 Nasopharyngeal Intracavitary Brachytherapy

Surgical treatment of nasopharyngeal carcinoma often requires alteration or removal of bony anatomy during the procedure. Standard applicators for adjuvant radiation therapy

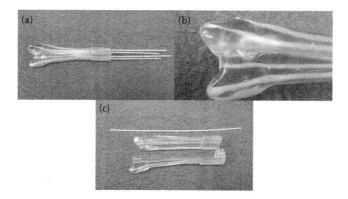

FIGURE 11.4 (a) Photograph of custom AM fabricated vaginal vault applicator with catheters in place. (b) Magnified photograph of AM vaginal vault applicator depicting the clarity of the applicator, in addition to the surface finish and the optimized paths of the catheters. (c) Photograph of the vaginal vault applicator in two pieces such that it could be placed through a narrow introitus. It can also be locked together using a mechanical key to ensure stability.

are often inadequate in that they do not account for the altered anatomy Scwaderer et al. (2000). used a stereolithographic model of a patient's nasopharyngeal cavity to design a custom applicator made of Silastic MDC-4-421 (Dow Corning Corporation, MI, USA) elastomer. A spiral CT scan of the postoperative cavity and surrounding margin was obtained and segmented using the Mimics application package (Materialise, Leuven, Belgium). The applicator was separated into two parts to facilitate insertion of the applicator through the narrow introitus. The model was built using AM techniques and the cavity was then used to create a mold in "stone plaster." Afterloading catheters were inserted into the molded applicator made of vulcanized elastomer. On post-implant CT imaging it was demonstrated that the custom mold optimally fit into the patient's nasocavity.

11.3.3 Interstitial Brachytherapy

Interstitial brachytherapy is one of the most effective methods of profound dose escalation in radiation therapy. With the accurate placement of needles in the target volume, large hypofractionated doses can be delivered while dramatically sparing organs at risk. However, for the most part conventional needle templates/guides restrict the needle guidance to only one direction that is orthogonal to the template face. There may be instances where oblique needle trajectories are desirable where tumor implantation may be possible without perforating adjacent normal tissues or sensitive organs. Areas where this may have the greatest impact include a transvaginal approach to treat cervical cancer interstitially, head and neck brachytherapy, orbital brachytherapy, and breast brachytherapy. The subsequent subsections will discuss applications of AM to create customized needle guides for these clinical sites.

11.3.3.1 Orbital Brachytherapy

Interstitial orbital brachytherapy is a viable alternative for the treatment of orbital hamangiopericytoma to total exenteration of the orbit. Due to the proximity of critical structures in the orbit, interstitial brachytherapy is technically challenging and difficult. Poulsen et al. (1999) reported a clinical case study where they created an orbital brachytherapy template using stereolithographic modeling. CT scans of the patient's skull were obtained and segmented. A physical model of the patient's bony anatomy, tumor, and optic nerve were created using vat photopolymerization techniques. This biomodel was then used to create a custom needle guide that robustly fixed to the patient's orbit and allowed the insertion of interstitial needles into the tumor while sparing the optic nerve. Patient reported outcomes indicated that local control was achieved.

11.3.3.2 Breast Brachytherapy

Traditional interstitial breast brachytherapy is predominantly performed with the parallel needle templates such as the Kuske applicator (Nucletron, Veenendal, The Netherlands) (Vicini et al., 2001; Kuske and Bolton, 2004; Vicini and Arthur, 2005). This applicator has recognized limitations in that it provides limited immobilization of the breast upon compression making catheter insertion challenging. Also, there are suggestions that fat necrosis and cosmetic outcomes are a function of the number of catheters implanted in the patient, and reducing the number of catheters implanted may be desirable. Pompeu-Robinson et al. (2012) evaluated the feasibility of using 3D printed catheter guides for improved tumor targeting and non-orthogonal needle implantation. CT images of breast phantoms with simulated tumor volumes were obtained and segmented using Mimics (Materialise, Leuven, Belgium). 3D templates catheter guides were created in Magics® (Materialise, Leuven, Belgium) and exported for printing. The templates were printed using AM fabrication and the accuracy of needle guidance was evaluated. The study reported a trajectory placement error of less than 2.57 mm. Poulin et al. (2015) applied the technique developed by Pompeu-Robinson et al. (2012) and improved upon it by incorporating 3DUS (3D ultrasound) guidance along with a freehand catheter optimization algorithm. Treatment plan quality was analyzed comparing the Kuske applicator and the AM fabricated template. Poulin et al. (2015) used the Replicator™ 2X printer (Makerbot, NY, USA) to create their template out of polylactide (PLA). Poulin et al. (2015) demonstrated that the personalized template were able to achieve optimal clinical objectives while reducing the number of catheters compared to the Kuske applicator.

11.3.3.3 Head and Neck Brachytherapy

Brachytherapy for the treatment of head and neck carcinomas has had limited acceptance in the broader radiation oncology community due to the complexity of the implantation technique. Huang et al. (2012) have clinically evaluated the use of custom needle guides to enable the accurate implantation of I-125 radioactive seeds on 31 patients with malignant tumors of the head and neck. The authors remark that conventional needle templates do not account for the flexibility of patient's head and neck resulting in significant discrepancies between planning and treatment positions. Huang et al. (2012) used the combination of CAD and AM techniques with the aim of reducing the impact of patient positioning errors on implant quality. In this instance, AM was used to create a fitted template which aligned to physical landmarks on the patient such as the patient's noses, lower borders of mandible, and the zygomas of the ears (Huang et al., 2012). CT scans of each patient were acquired to generate the treatment plan. Needles were virtually positioned to avoid bones, blood vessels, and important organs using the BTPS (Beijing Atom and High Technique Industries Inc., Beijing, China) treatment planning software. The Mimics® (Materialise, Belgium) image processing software was used to convert the CT and the needle positions into a virtual needle guide. The virtual model was then fabricated using a vat photopolymerization technique on an Eden250™ (Objet, Israel) AM system. Huang et al. (2012) demonstrated that in using a custom template they were able to significantly reduce set-up error and maintain template stability over the course of the implantation procedure. Post-implant dosimetry was evaluated revealing that clinical objectives for target coverage and organs at risk sparing were achieved.

11.3.3.4 Gynecological Brachytherapy

Typical delivery of interstitial brachytherapy is relegated to the Vienna/Utrecht interstitial/intracavitary applicators (Kirisits et al., 2006; Nomden et al., 2012) and transperineal templates (Martinez et al., 1984). The intracavitary approaches do not ideally target the lateralized tumors as they are limited in their lateral reach. Transperineal approaches have the consequence of perforating large amounts of normal tissue in order to reach the tumor volume. Garg et al. (2013) have capitalized on an interesting opportunity presented by the use of AM, which is the optimization of the catheter trajectories in three dimensions. They have developed an automated algorithm that calculates the ideal positions of the needles and constrains them to their minimum radius of curvature within intracavitary applicators. In phantom evaluations they successfully directed needles in to the tumor volume through the vaginal canal, sparing sensitive organs and normal tissues.

Lindegaard et al. (2015) have developed a novel method to produce patient-specific 3D printed transvaginal intracavitary–interstitial templates that couple with a conventional ring and tandem applicator (Figure 11.5). The template was designed around a Varian 26 mm ring and tandem applicator set (Varian Medical Systems Inc, CA, USA). AM was accomplished using a ProJet® 3510 SD (3D Systems Inc, SC, USA) with biocompatible Visijet® M3 Crystal material (3D Systems Inc, SC, USA). Parallel and oblique needle trajectories were planned on pre-brachytherapy MR images to adequately target the tumor while minimizing the perforation of normal tissues. Transabdominal US guidance was used to confirm correct placement of the intracavitary applicator and provided image guidance for subsequent needle insertions. Lindegaard et al. (2015) demonstrated that the use of a transvaginal brachytherapy template customized to the particular geometry of a patient's tumor is able to achieve optimal target coverage while limiting OAR doses and soft-tissue damage through perforation.

11.4 MATERIAL SELECTION AND QUALITY ASSURANCE

Complex customized brachytherapy devices can be manufactured using volumetric medical imaging, treatment planning, and CAD systems. This section deals with some of the important quality assurance considerations one ought to incorporate into their AM processes to ensure safe medical applications of the medical device.

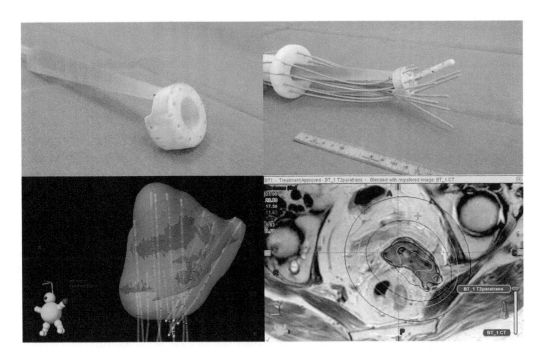

FIGURE 11.5   Application of an individualized 3D printed applicator in a stage IIIB cervical cancer patient (Lindegaard et al., 2015). The applicator was designed based on tumor extension at the time of brachytherapy as assessed in a treatment planning MRI acquired 1 week prior to brachytherapy. Upper panels: 3D printed ring without (left) and with (right) needles. Lower panels: treatment plan with HR CTV (red) and 85 Gy EQD2 isodose curve (cyan) in a 3D reconstruction (left) and para-transversal MRI (right).

11.4.1 Material Selection

When designing a brachytherapy applicator with the intention of manufacturing it using AM, material selection is paramount. In the case of internal use where the device will come into contact with tissue and/or blood, biocompatibility is of vital importance; as such, a very specific set of conditions have to be met. Government and institutional regulations, for example, the United States Pharmacopeia (USP) classification system will determine which materials are suitable for the device. In addition, special considerations should also be made to account for the type of volumetric medical imaging that will be performed with the applicator in place. It is prudent to select materials that are visible on the specific imaging modality but that do not create excessive artifacts. The material selection is also dependent on the type of processing to be done on the completed device to ensure that it is minimally clean or if interstitial applications are expected, appropriately sterilized for clinical use. The choice of the sterilization process will have an impact on material selection; such as, whether steam sterilization or a less aggressive alternative like low-temperature plasma or ethylene oxide sterilization is used. Hundreds of materials are available across all the processes and new materials are constantly being developed. There is a small group of materials that have been certified with the (USP) CLASS V1 classification and can be processed by AM satisfying the above constraints.

Two examples of such materials or composite resins would be polyether ether ketone (PEEK) which up until recently was only available for large-scale industrial SLS AM machines. However, only in 2015, PEEK filament been introduced to the desktop printer category utilizing the FDM process. PEEK can be repeat steam sterilized opening up the possibility of repeat-use applicators, templates, and guides that are MR- and bio-compatible. Another USP class V1 AM material is Accura® ClearVue™ (3D Systems Inc., SC, USA). Accura® ClearVue™ is a rigid, non-metallic, clear (post curing) plastic with a density of 1.17 g/cm³ that can be steam sterilized to (USP) class V1 guidelines.

11.4.2 Post-Processing and Quality Assurance

After the AM process is complete, typically most parts require a period of time for the polymer to cure and harden, often this process is accelerated using UV light. This process may leave uncured resin in and around the new device, which must be rigorously cleaned away during a rinsing step after curing. It is important to verify that AM devices meet the mechanical specification laid out in the applicator design file. Due to this concern, it is imperative that the source not directly travels within the applicator; instead, it should travel within approved catheters inserted in the applicator.

Some post AM work may be required to finish the device; such as, further cleaning based on the material supplier's instructions to comply with USP class VI conditions to ensure the device is free from contaminants. The flexible channels are now inserted and fixed in place (e.g., Figure 11.2).

Upon final fabrication of the applicator the device will undergo quality assurance. This typically involves physical inspection of the device followed by a high-resolution CT scan of the applicator. The CT image of the applicator is evaluated to ensure that the geometry of the applicator is what was planned, the density and consistency of the applicator are uniform, and finally the catheter paths are free of debris, kinks, and occlusions. The CT scan can also be used to develop a pretreatment plan prior to the brachytherapy application. After QA is complete, the applicator is submitted for processing to ensure that it is sterilized appropriately for clinical use. Depending on the material used in the applicator, an appropriate sterilization process must be selected. For most materials used in applicator fabrication, low-temperature sterilization processes are favorable such as low-temperature plasma if available and ethylene oxide.

Generally, if a patient has consented to the use of AM applicator, and it is not being sold commercially, additional regulatory approval is not required. However, it would be prudent for brachytherapy practitioners to verify that there are no local and/or national regulatory restrictions to the use of institutionally created applicators. Furthermore, it is also advised to check with institutional legal counsel whether the added liability of using in-house developed applicators is acceptable by the institution prior to beginning an AM brachytherapy program.

11.5 CONCLUSION

AM is a disruptive technology that is being applied extensively in all areas in brachytherapy. AM enables truly personalized medicine by creating customized applicators that address a specific patient's needs. While this technology is still in its infancy the intuitive benefits have already manifested in improved oncologic endpoints such as optimized targeting of disease and sparing of adjacent organs at risk. As advances in AM continue and costs continue to decrease, accessibility to this technology will only improve. AM is currently an emerging technology, but will become ubiquitous in the not-too-distant future. When this occurs the landscape of brachytherapy will be forever altered, from one where patients are manipulated to fit stock sets of applicators to one where applicators are purposefully designed and created to address a patient's needs.

REFERENCES

P.W. Carter, Advances in rapid prototyping and rapid manufacturing, *Proceedings of the Electrical Insulation and Electrical Manufacture Coil Winding Conference (Cat. No.01CH37264)*, Cincinnati, OH, pp. 107–114, 2001.

J. Crook, L. Grimard, J. Tsihlias, C. Morash, and T. Panzarella, Interstitial brachytherapy for penile cancer: an alternative to amputation, *J. Urol.*, 167(2), 506–511, 2002.

J. Crook, C. Ma, and L. Grimard, Radiation therapy in the management of the primary penile tumor: An update, *World J. Urol.*, 27(2), 189–196, 2009.

J.A.M. Cunha et al., Evaluation of PC-ISO for customized, 3D printed, gynecologic 192 Ir HDR brachytherapy applicators, *J. Appl. Clin. Med. Phy.*, 16(1), 246–253, 2015.

L. D'Alimonte, J. Helou, A. Ravi, H. Easton, L. Jurincic, and G. Morton, Customized surface moulds for the treatment of penile cancer with high dose rate (HDR) brachytherapy, *J. Med. Imaging Radiat. Sci.*, 46(1), S4, 2015.

A. Garg et al., An algorithm for computing customized 3D printed implants with curvature constrained channels for enhancing intracavitary brachytherapy radiation delivery, in *2013 IEEE International Conference on Automatic Science and Engineering* (IEEE, 2013), pp. 466–473.

B.D. Harris, S. Nilsson, and C.M. Poole, A feasibility study for using ABS plastic and a low-cost 3D printer for patient-specific brachytherapy mould design, *Australas. Phys. Eng. Sci. Med.*, 38(3), 399–412, 2015.

J. Helou, G. Morton, H. Easton, L. Jurincic, L. D'Alimonte, and A. Ravi, Customized penile plesio-brachytherapy using latest stereolithography techniques, *Brachytherapy*, 14(2015), S99–S100, 2015.

M.-W. Huang et al., A digital model individual template and CT-guided 125I seed implants for malignant tumors of the head and neck, *J. Radiat. Res.*, 53(6), 973–977, 2012.

C.W. Hull, Apparatus for production of three-dimensional objects by stereolithography, US 4575330 A 1986.

C.W. Hull, U.S. Patent, Method for production of three-dimensional objects by stereolithography, US Patent 4,929,402, 1990.

C. Kirisits, S. Lang, J. Dimopoulos, D. Berger, D. Georg, and R. Pötter, The Vienna applicator for combined intracavitary and interstitial brachytherapy of cervical cancer: Design, application, treatment planning, and dosimetric results, *Int. J. Radiat. Oncol. Biol. Phys.*, 65(2), 624–630, 2006.

R.R. Kuske and J.S. Bolton, A phase I/II trial to evaluate brachytherapy as the sole method of radiation therapy for stage I and II breast carcinoma, *Radiat. Ther. Oncol. Gr. Publ.*, 13(1), 95–17, 2004.

J. Lindegaard et al., Individualised 3D printed vaginal template for MRI guided brachytherapy in locally advanced cervical cancer, *J. Radiother. Oncol.*, 118(1), 173–175, 2015.

A. Martinez, R.S. Cox, and G.K. Edmundson, A multiple-site perineal applicator (MUPIT) for treatment of prostatic, anorectal, and gynecologic malignancies, *Int. J. Radiat. Oncol. Biol. Phys.*, 10, 297–305, 1984.

C.N. Nomden, A.A.C. de Leeuw, M.A. Moerland, J.M. Roesink, R.J.H.A. Tersteeg, and I.M. Jürgenliemk-Schulz, Clinical use of the Utrecht applicator for combined intracavitary/interstitial brachytherapy treatment in locally advanced cervical cancer, *Int. J. Radiat. Oncol. Biol. Phys.*, 82(4), 1424–30, 2012.

a Pompeu-Robinson, M. Kunz, C.B. Falkson, L.J. Schreiner, C.P. Joshi, and G. Fichtinger, Immobilization and catheter guidance for breast brachytherapy, *Int. J. Comput. Assist. Radiol. Surg.*, 7(1), 65–72, 2012.

E. Poulin, L. Gardi, A. Fenster, J. Pouliot, and L. Beaulieu, Towards real-time 3D ultrasound planning and personalized 3D printing for breast HDR brachytherapy treatment, *Radiother. Oncol.*, 114(3), 335–338, 2015.

M. Poulsen, C. Lindsay, T. Sullivan, and P. D'Urso, Stereolithographic modelling as an aid to orbital brachytherapy, *Int. J. Radiat. Oncol. Biol. Phys.*, 44(3), 731–735, 1999.

S. Richardson, G. Palaniswaamy, and P.W. Grigsby, Dosimetric effects of air pockets around high-dose rate brachytherapy vaginal cylinders, *Int. J. Radiat. Oncol. Biol. Phys.*, 78(1), 276–279, 2010.

E. Schwaderer et al., Soft-tissue stereolithographic model as an aid to brachytherapy, *Medica Mundi*, 44(1), 48–51, 2000.

F. a. Vicini and D.W. Arthur, Breast brachytherapy: North American experience, *Semin. Radiat. Oncol.*, 15, 108–115, 2005.

F.A. Vicini et al., Accelerated treatment of breast cancer, *J. Clin. Oncol.* 19(7), 1993–2001, 2001.

E. Wiebe, H. Easton, G. Thomas, L. Barbera, L. D'Alimonte, and A. Ravi, Customized vaginal vault brachytherapy with computed tomography imaging-derived applicator prototyping, *Brachytherapy*, 14(3), 380–384, 2015.

CHAPTER 12

Robotics in Brachytherapy

Tarun K. Podder and Aaron Fenster

CONTENTS

12.1 INTRODUCTION

Literature reviews indicate that biochemical and clinical relapse rates are very low with current brachytherapy procedures (Peschel and Colberg, 2003; Martinez et al., 2015). Nevertheless, there is evidence that improvements are still required in order to reduce the rate of complications, which include erectile dysfunction, urinary retention, incontinence, and rectal injuries, to provide better quality assurance and to treat patients consistently. Brachytherapy complications have been associated with high-dose loads to large portions of the prostate (Wust et al., 2004), or the pelvis for gynecological cases, to high urethral dose (Zelefsky et al., 2003) and to needle trauma.

Poor visualization of the prostate, prostate motion and swelling and the need to re-register preoperative to intraoperative images during the procedure cause seed insertion errors and make prostate brachytherapy a procedure that is difficult to learn and perform. Indeed, seed placement errors can be significant and complication rates are seen to decline with training. During each needle insertion, needle forces move the prostate and cause seed targeting errors. Prostate swelling and deformation caused by edema, and motion due to respiration also generate errors (Yamada et al., 2003). Treatment plan execution problems are also caused by the interference of the pubic arch with the Z-direction trajectories generated by the planner and guided by the XY template, which often require re-positioning of the guiding template.

In prostate brachytherapy, positional accuracy of the radioactive seeds or sources is very important for optimizing the dose delivery to the targeted tissues sparing the critical organs and structures. But accurate steering and placement of surgical needles in soft tissues are challenging because of several reasons. Some of them are tissue heterogeneity and elastic stiffness, unfavorable anatomy, needle bending, inadequate sensing or imaging, tissue/organ deformation and movement, poor maneuverability, and change in the dynamics of various organs. In currently practiced procedures, the fixed grid holes in the template allow the surgeon to insert the needle at specified fixed positions. Very little can be done to steer the needle to a place other than straight passing through the hole in the template. Change in needle insertion position may be required based on intraoperative dynamic planning. Sometimes, especially for larger prostates, the needle needs to be angulated to avoid pubic arch interference (PAI) and get access to the desired target position in the prostate for seed deposition or high-dose rate (HDR) source dwelling. In current brachytherapy procedure with fixed template the needle angulation is almost impossible. However, the surgeon can use a hook to bend the needle and get to the desired position after several trials. Whereas a robotic system, with sufficient degree-of-freedom (DOF), can provide flexibility in positioning and orientating (angulating) along with improved accuracy of needle insertion and seed deposition. Additionally, with the assistance of robotic systems, less skillful surgeons will be able to treat patients with higher quality.

12.1.1 Objectives and Requirements for Robot-Assisted Brachytherapy

Robotic assistance in brachytherapy is attractive for several reasons. The main advantages are increased accuracy in needle/applicator placement and seed/source positioning, reduced human variability, and possible reduction of operation time. The robotic systems designed and developed for brachytherapy are expected to meet the following functional requirement and objectives and satisfy the following functional requirements (Podder et al., 2014).

12.1.1.1 Objectives
The main objectives are to

- Improve accuracy of needle placement and seed/source delivery/positioning

- Improve consistency of seed implant or source positioning

- Improve avoidance of critical structures (urethra, pubic arch, rectum, and bladder) during needle placement

- Reduce clinician's fatigue

- Reduce radiation exposure, and

- Reduce learning curve

12.1.1.2 Requirements

The expected functional requirements of the system are as follows:

- Quick and easy disengagement in case of emergency

- Provision for reverting to conventional manual brachytherapy method at any time

- Method for the clinician to review and approve the motion plan before needle placement

- Visual (and/or haptic force) feedback during needle insertion

- Visual confirmation by the chosen imaging technique of each seed deposition or the needle tip at the resting position

- Ease of operation in the operating room (OR) environment

- Ease of cleaning and decontamination

- Compatible for sterilization of the required components

- Highly robust and reliable, and

- Safe for the patient, clinician, and the OR environment

In addition to the above objectives and requirements, the following objectives and functional requirements may be incorporated in an advanced robotic system.

Additional objectives for an advanced robot:

- Update dosimetry after each needle is implanted (automatic seed localization) or HDR source dwelling is completed for each individual needle/applicator

- Detect tissue heterogeneities and deformation via force sensing and imaging feedback

- Reduce trauma and edema

Additional functional requirements for an advanced robot

- Improvement of organ/target immobilization techniques

- Provision for periodic quality assurance checking

- Perform adaptive planning considering on implanted seeds or dwell position are completed and current shape and size of the target and OARs

- Ability to modulate velocity and needle rotation by automatic feedback control

- Steering of the needle by automatic feedback control, and

- A teaching mode to simulate force/velocity patterns of expert practitioners

There may not be any currently developed robotic system that can successfully execute all the above requirements. However, these requirements can provide good guidelines for robotic system designers and developers.

12.1.2 Considerations for Clinical Implementation

The above-mentioned objectives and requirements encompass both robotic system development and clinical implementations. So far, fifteen or more robotic systems have been developed for performing various brachytherapy procedures. However, of these robotic systems, very few are marginally successful in clinical implementations. Some of the main considerations for clinical implementation for a transrectal ultrasound (TRUS)-guided prostate brachytherapy (radioactive seed implantation) are discussed below.

12.1.2.1 Clinical Workflow

After initializing the robot, the patient information should be entered into the computer by the authorized user. Then the TRUS images are acquired and used in delineating the appropriate anatomical structures, such as prostate boundary, urethra, pubic bone, rectum, and seminal vesicles. Then, three-dimensional (3D) models of the prostate and adjacent required structures are generated automatically.

This 3D model of the prostate is used for dosimetric planning to obtain the desired coordinates of the radioactive seed distribution. The planning software should be able to display the planned iso-dose contours, needle position, and seed locations in 3D. This would provide the clinicians a useful visualization of the whole treatment plan, and if required, the clinicians can edit the plan.

Once the clinicians approve the plan, the needle insertion is performed in a sequential order, either by the physician or by the robot (e.g., Euclidian). After the needle insertion is completed, the physician or the robot can deliver the seeds according to the dosimetric plan. On completion of the seed implant, the needle is withdrawn. Some advance software may have dynamic validation of the seed and dose distribution. However, researchers are also considering insertion of multiples needles simultaneously (Podder et al., 2010). A detailed flowchart for an advanced robotic brachytherapy procedure is shown in Podder et al. (2014).

12.1.2.2 Interactions

12.1.2.2.1 Robot–Clinician Interaction   In the current generation of brachytherapy robots, two main types of interactions are implemented: (1) autonomous robots execute subtasks under human supervision in an autonomous way, (2) semi-autonomous robots position a needle guide which is used by the clinician to manually insert a needle and delivery of the seeds. Robots that are capable of operation in both autonomous and semi-autonomous modes must be capable of displaying their current operational mode, with an emergency reset to default to semi-autonomous or resting mode. Due to the complexity and constraints of the clinical environment where the clinician is an expert with high abilities to detect, analyze, and react to unwanted critical situations, the clinician and the robot have to work together in a collaborative synergistic way.

12.1.2.2.2 Robot–Patient Interaction Interaction between the robot and the patient is very important because of the nature of tasks intended to be performed by a brachytherapy robot. The motion of the robot is controlled by the instructions (generated automatically or given by the clinician) that are required to execute the tasks as per the dosimetric plan. The patient is inactive, and does not interact with robot as such. However, the robot may have provisions for detecting excessive movement of the patient that might endanger the placement of seeds or otherwise alter the dose delivery to the patient. The clinicians supervise all the tasks of the robot and/or execute some the tasks for the robot. Both robot–clinician and robot–patient interactions should be designed to ensure the robustness, reliability, and safety of the correct dose, in the correct site, for the correct patient.

12.1.2.3 Safety and Reliability Issues

12.1.2.3.1 Radiation Safety Radioactive seeds need to be handled with extreme care. The seeds are to be put in a protective seed cartridge to reduce radiation exposure. If the seeds are expelled from the cartridge using a motorized stylet, precaution must be taken for not exerting excessive force causing damage to the seeds resulting in spilling the radioactive material. Confirmation delivery of required number of seeds is to be maintained through one or multiple methods such as visual feedback, counting manually, or sensory feedback. Clinical staff must be vigilant to avoid issues, such as seed jamming, undesired seed delivery, or lack of seed delivery.

12.1.2.3.2 Safety of the Patient, Clinical Staff, and the OR Environment Safety of the patient is one of the most critical criteria. Since the robot is moving in close proximity of the patient and clinical personnel, all the movements must be verified/checked for avoiding potential physical injury as well as collision with the OR environment. These are more critical for advanced robots, which carry needles and radioactive seeds as well as delivering them automatically.

12.1.2.3.3 Cleaning and Decontamination Cleaning and decontamination are two highly effective infection prevention measures that can minimize the risk of transmission of microorganisms to healthcare workers, and in breaking the infection transmission cycle for patients. For robotic systems, the cleaning and decontamination must be done in the similar manner as in the standard operation room procedures for cleaning and decontamination of ultrasound steppers, brachytherapy stands, etc. used in conventional techniques. There may be several parts (which come in contact with needle and/or seeds) in the robot, which need sterilization before the start of each procedure.

12.1.2.3.4 System Robustness and Reliability Before introducing any new robotic system in a clinical environment, the robustness and reliability of the system should be evaluated (Dhillon, 2000). Accelerated testing methodology mimicking the real operating procedure can be used for this purpose (Podder et al., 2010).

12.2 ROBOTIC SYSTEMS FOR BRACHYTHERAPY

12.2.1 Design and Development of Robotic System for Brachytherapy

12.2.1.1 Workspace Analysis

The available workspace for the robotic system is quite limited while the patient is in the lithotomy position for transperineal prostate brachytherapy (Figure 12.1). Thus, most of the industrial robots may lose dexterity or encounter a singularity (lose a DOF) while working in such a severely constrained workspace (may be less than 120 mm in lateral direction) in OR. Moreover, the robotic system should not occupy too much space, so that the clinicians get enough space to reach or work with the patient. Successful clinical implementation of a robotic system critically depends upon the shape and size of the robot. Therefore, it is very important to analyze the robotic workspace in conjunction with available OR–patient workspace before designing or selecting any robotic system for brachytherapy.

12.2.1.2 System Design and Development

Currently, it is difficult to find any commercial or industrial robot that can readily be used for brachytherapy. Therefore, several custom-made robotic systems have been developed and some are being developed at different research institutes and hospital settings. Since these robotic systems carry a surgical tool (needle) and radioactive seeds and come in contact with patient as well as close proximity to the clinical staff, the robots have to be at least partly cleanable, sterilizable (needle and seed/source passage), robust, reliable, and safe. These impose critical considerations on design choices. Several hardware and software engineering solutions can contribute to solving them. Safety, accuracy, user-friendliness, and reliability (SAUR) are the critical criteria that need to be satisfied for any robotic system to be implementable in prostate seed implantation.

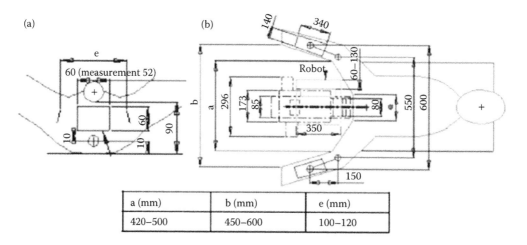

a (mm)	b (mm)	e (mm)
420–500	450–600	100–120

FIGURE 12.1 Workspace for robotic insertion of brachytherapy needle: (a) front view, (b) top view. (Image courtesy of Podder et al. 2014. *Med Phys* 41(10):101501.)

12.2.2 Overview of Currently Available Robotic Systems

During the past decade, investigators and companies have developed robotic systems for prostate brachytherapy. These image-guided brachytherapy (IGBT) systems made use of transrectal ultrasonography (TRUS), computed tomography (CT), and magnetic resonance image (MRI) guidance. An overall summary of the systems is listed in Table 12.1. The majority of these robotic systems are based on TRUS guidance (Meltsner et al., 2005; Rivard et al., 2005; Fichtinger et al., 2007; Yu et al., 2007; Salcudean et al., 2008; Hungr et al., 2009; Podder et al., 2010; Bax et al., 2011). Next most popular image guidance modality is MRI (Stoianovici et al., 2007; van den Bosch et al., 2010; Tokuda et al., 2012). Although some of these systems still use a template for guiding the brachytherapy needles, many incorporated sufficient flexibility allowing oblique trajectories into the prostate to avoid PAI. Tracking and verification of the needle and seed locations have also been explored through both mechanical and software approaches. The robotic systems make use of motors to manipulate the needle guide for the selected needle trajectory, but needle insertion is primarily performed manually. However, a few systems incorporate mechanisms to insert the needles and deposit the seeds using the robot, that is, motorized needle insertion where the needles as well as the seeds can be deposited by the robot (Yu et al., 2007; Podder et al., 2010). In the following section, we review the robotic systems that have been developed primarily for prostate brachytherapy and a few for other applications.

12.2.2.1 TRUS-Based Robotic Prostate Brachytherapy Systems

12.2.2.1.1 Elekta's FIRST™ (SeedSelectron®) System The Fully Integrated Real-time Seed Treatment (FIRST™) developed by Nucletron (Veenendaal, Netherlands) combines a computer-controlled 3D TRUS system, a robotic source delivery, and needle retraction device called the seedSelectron®, and the SPOT-PRO treatment planning system. This integrated product received FDA and Health Canada approvals in 2001 and CE approval in 2002 and has been available for over a decade. The seedSelectron® is the robotic component and is mounted on the same support unit as the TRUS transducer. It builds and loads the needle in real-time while allowing modification at any point during the delivery process, based on previously placed needles or delivered sources. Using the source and spacer cartridges the seedSelectron® builds a non-stranded source–spacer sequence. The sources and spacers are expelled from the cartridges by a drivewire and a detector is used to measure source strength as the sources pass it. The source train is delivered by the drivewire into a needle through a transfer tube connected to the needle. The system delivers the sources and spacers train with the first source at its reference depth. After delivery of the sources, the needle is retracted automatically by a small robotic arm, keeping the drivewire in place to avoid source suction from the retraction of the needle. The drivewire is then removed after the needle retraction allowing the needle to be disconnected from the transfer tube and removed. The system is then ready for the next insertion or further plan modification. Extensive testing of the seedSelectron® accuracy has been performed and the clinical performance of the complete system has been reported (Rivard et al., 2005; Beaulieu et al., 2007).

12.2.2.1.2 Thomas Jefferson University: EUCLIDIAN A team at Thomas Jefferson University, Philadelphia, developed a robotic system called the Endo-Uro Computer Lattice for Intra-tumoral Delivery, Implantation, and Ablation with Nanosensing (EUCLIDIAN). This system consists of three modules: (1) a fully motorized surgical module with a 2 DOF TRUS, a 3 DOF gantry, and

TABLE 12.1   Summary of the Currently Available Robotic Brachytherapy Systems

Features	TRUS-based Robotic Prostate Brachytherapy Systems								MR-based Robotic Prostate Brachytherapy Systems				US-based Robotic Brachytherapy Systems for Other Organs		
	First	Euclidian	MIRAB	UW robot	JHU-robot1 (1)	UBC	RRI	CHUG	UMCU	JHU-MrBot (2)	JHU-MR (3)	JHU-MR (4)	MIRA-V	PARA-BRACHYROB	DMBT
Institute/Lab	Elekta-Nucletron	TJU	TJU	UW	JHU	UBC	RRI	CHUG	UMCU	JHU	JHU	JHU	UWO	TUCN	UCSD
Year (approx.)	2001–2004	2005–2010	2007–2012	2005–2008	2002–2008	2007–2009	2005–2011	2007–2011	2006–2010	2003–2008	2005–2008	2007–2011	2005–2009	2013–2016	2011–2016
RIA Class	2	3	3	2	2	2	3	2	2	3	2	2	3	2	3
Brachy Class	II	III	III	II	II	II	II	II	II	III	II	II	II	II	II
Application	PSI	PSI	PSI/HDR	PSI/HDR	PSI	PSI	PSI	PSI	PSI/HDR	PSI	PSI	PSI	LSI (lung)	Seed Implantation	HDR (rectum/breast)
Imaging modality	U/S (auto & manual)	U/S (auto & manual)	U/S (auto & manual)	U/S (manual)	U/S (manual)	U/S (manual)	U/S (auto & manual)	U/S	MRI	MRI	MRI	MRI	U/S		
Degrees-of-freedom (DOF)	2 DOF	5 DOF surgical, 2 DOF U/S, 6 DOF positioning, 3 DOF cart	5 DOF surgical, 2 DOF U/S, 6 DOF positioning, 3 DOF cart	6 DOF	4 DOF surgical	4 DOF surgical	5 DOF	5 DOF	5 DOF	4 DOF	3 DOF	6 DOF	5 DOF	5 DOF	3 DOF (including HDR source movement)
Number of channel/needle	Single	Single	16 needles	Single	Single	Single	Single	Single	Single	Single	Single	Single	Single	Single	8–16 channels
Needle insertion	Manual	Autonomous	Autonomous	Auto and/ or Manual	Manual	Manual	Manual	Autonomous	Autonomous tapping	Autonomous			Manual	Autonomous	N/A
Needle rotation	No	Yes	Yes	Yes	No	No	Manual	Yes	No	No	No	No	No	No	N/A
Angled Insertion	No	Yes	Yes	Yes	Yes	Yes	Yes	Yes	Yes	No	Yes	Yes	Yes	Yes	N/A
Source delivery/positioning	Autonomous	Autonomous	Autonomous	Manual (auto in research)	Manual	Manual	Manual	Manual	Manual	Autonomous	Manual	Manual	Manual	N/A	Autonomous
Needle/source withdraw	Autonomous	Autonomous	Autonomous	Auto and/ or Manual	Manual	Manual	Manual	Manual	Manual	Autonomous	Manual	Manual	Manual	Autonomous	Autonomous
Physical template	Yes	No	Yes	No	No	No	No	No	No	No	No	No	No	No	No
Template/perineum area coverage	Conventional	62 mm × 67 mm	60 mm × 60 mm	250 mm × 250 mm	50 mm × 50 mm	150 mm × 150 mm	60 mm × 60 mm	105 mm × 105 mm	N/A	40 mm × 40 mm	N/A	50 mm × 50 mm	N/A	N/A	N/A

(Continued)

TABLE 12.1 (*Continued*) Summary of the Currently Available Robotic Brachytherapy Systems

Features	TRUS-based Robotic Prostate Brachytherapy Systems								MR-based Robotic Prostate Brachytherapy Systems				US-based Robotic Brachytherapy Systems for Other Organs		
	First	Euclidian	MIRAB	UW robot	JHU-robot1 (1)	UBC	RRI	CHUG	UMCU	JHU-MrBot (2)	JHU-MR (3)	JHU-MR (4)	MIRA-V	PARA-BRACHYROB	DMBT
Depth movement	Conventional	312 mm	240 mm	250 mm	120 mm	150 mm	70 mm	N/A	150 mm	40 mm	N/A	120 mm	N/A	N/A	45 mm
TPS	Oncentra Seeds	In-house, FDA-IDE approved	In-house	N/A	FDA approved Interplant	N/A	In-house	N/A	N/A	N/A	N/A	none	N/A	N/A	N/A
Needle tip positioning accuracy in air	N/A	<0.2 mm	<0.2 mm	N/A	N/A	<0.3 mm	0.2 mm	N/A	N/A	0.32 mm	N/A	0.94 mm	<0.5 mm	N/A	N/A
Needle tip positioning accuracy in phantom	<0.5 mm	<0.5 mm	<0.5 mm	N/A	1.04 mm	N/A	0.9 mm	1.0 mm	N/A	<0.5 mm	2.0 mm	3.0 mm	0.9 mm	N/A	N/A
Accuracy in source/seed deposition	<1 mm (tested)	<1 mm (tested)	<1 mm	<1 mm	N/A	1.2 mm	1.6 mm	N/A	N/A	<1 mm	N/A	N/A	N/A	N/A	1 mm
Emergency stop	Yes	Yes	Yes	Yes	Yes	Yes	Yes	Yes	Yes	Yes	Yes	Yes	Yes	N/A	N/A
Provision for reverting to conventional mode	Yes	Yes	Yes	Yes	Yes	Yes	Yes	Yes	No	No	No	No	No	N/A	N/A
Force-torque sensor	No, but motor stops if too much force needed	Yes	No	Yes	No	No	No	No	No	No	No	No	No	No	N/A
FDA, CE approval	Yes, also CE	IDE	No	No	No	No	No	No	No	No	No	No	No	No	No

Note: PSI – prostate source implantation; LSI – lung source implantation; HDR – high-dose-rate; N/A – not available (or not applicable).

a 2 DOF needle inserter capabilities; (2) a 6 DOF positioning module; and (3) a 3 DOF cart with electronic housing (Figure 12.2). The 2 DOF TRUS transducer capability is achieved with servo motors allowing maximum translation of about 185 mm and a maximum rotation of about 180°. Translation and rotation can also be operated manually using two manual knobs. A template holder at the end of the TRUS transducer driver enables manual takeover if required. The prostate can be stabilized with a needle guide that can orient the needle in the sagittal and coronal planes. The multi-DOF capability of the gantry is achieved with two translation motions in the X- and Y-direction with a range of 62 mm × 67 mm, and one rotational motion in vertical plane used to avoid PAI or to reach closer to the rectum.

The most complex subsystem is the needle driver, which incorporates three force–torque sensors (stylet, cannula, and whole needle sensors) and autonomous needle insertion and seed delivery. This robot has the capability of needle rotation used to reduce needle insertion force, resulting in reduction of organ/target deformation and displacement. Although needle insertion and seed delivery is fully automatic, it can be interrupted and its movements can be manipulated manually. Manual operation of the system can also be achieved in the case of motor failure by inserting a conventional template, and continuing the procedure using the same dosimetric plan. The 6 DOF of the positioning platform allows translation and orientation of the whole surgical module to accommodate for the patient's position and orientation (Figure 12.2). The computer, system electronics and cable junctions are housed in an electronics housing (Figure 12.2). Software modules provide contouring, 3D anatomical model generation, dosimetric planning, 3D visualization, and needle tracking capabilities. Detailed description of this system can be found in Yu et al. (2007).

12.2.2.1.3 Thomas Jefferson University: MIRAB The team at Thomas Jefferson University, Philadelphia, also developed the modularized Multichannel Image-guided Robotic Assistant for

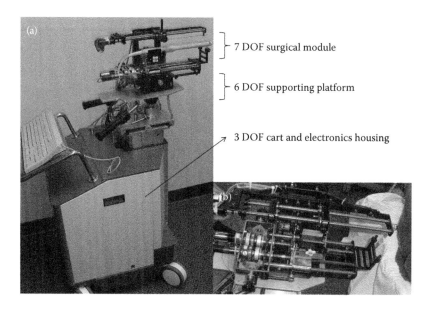

FIGURE 12.2   (a) EUCLIDIAN robotic system. (b) Close-up view inside an OR.

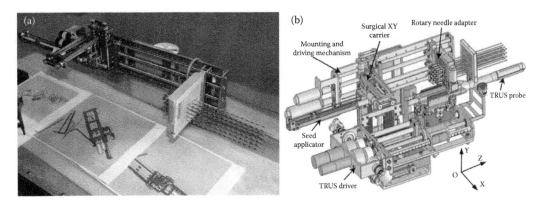

FIGURE 12.3   Multichannel robotic system: (a) prototype, (b) 3D Solidworks™ design.

Brachytherapy (MIRAB) system (Figure 12.3), which can be integrated into the EUCLIDIAN system. MIRAB is a 6 DOF system consisting of five modules

1. The Rotary *Needle Adapter* has rotational capability and can hold 16 needles. Distal ends of the needles are supported and guided by a conventional template. The needles can be simultaneously inserted using one of the servo motors on the Mounting and Driving Module.

2. The 3 DOF *Surgical X–Y Carrier* carries the *Seed Applicator*, which delivers seeds and withdraws the needle.

3. The 2 DOF *mounting and driving mechanism* provides translational motion to the *Needle Adapter* and *Surgical X–Y Carrier* with a maximum of about 240 mm.

4. The motorized *seed applicator* is used to expel the seeds from the seed cartridge automatically according to the dosimetric plan.

5. The 2 DOF *TRUS driver* used in MIRAB can be installed on the EUCLIDIAN by exchanging it with the needle subsystem in the EUCLIDIAN system. This interchangeability allows switching from a single-channel system to a multichannel system. More details description of this system can be found in Podder et al. (2010).

12.2.2.1.4 University of Wisconsin Brachytherapy Robot A team the University of Wisconsin, Madison, developed a prototype 6 DOF brachytherapy robot for automatic and semi-automatic prostate brachytherapy (Meltsner et al., 2005). This robot includes 3 linear motorized slides with a maximum travel of 250 mm, 2 rotary motorized stages, and 1 DOF motorized rotation of the needle. Each linear stage is powered by a servo motor with a theoretical accuracy of 0.25 μm. The theoretical accuracy of the rotary stages is 0.0005°, and the measured accuracy of seed placement in a gel phantom is 0.93 ± 0.48 mm. A 6 DOF magnetic tracking system is used to track the robot in world coordinates, and an attachable magnetic sensor to the TRUS transducer is used for registration of the imaging plane with the robot's workspace. To avoid PAI, the robot can perform angular needle insertions at any position within its workspace at +/− 30° angulation. A force sensor is used to measure the interaction of the needle with tissue, and a needle guide is used to maintain stiffness at the skin surface to reduce possible needle deflection. The robot's insertion slide can be decoupled from the motorized slide via a manual release, allowing positioning of the needle for insertion, and manual insertion of the needle. Thus, the system provides both motorized and manual needle insertion capability.

12.2.2.1.5 Johns Hopkins University: Robot1   A team at the Engineering Research Center and Radiation Oncology of Johns Hopkins University has developed a robotic system that consisted of a TRUS transducer and a spatially co-registered robotic manipulator integrated with an FDA-approved commercial treatment planning system. The system is composed of a small 4 DOF parallel robot attached to the mounting posts of the brachytherapy template. In this way, the robot can replace the template and use the same coordinate system, clinical hardware, workflow, and calibration. The robot consists of two 2 DOF Cartesian X–Y motion stages parallel to each other. The stages can be translated in the X–Y plane that corresponds to the face of the template and relative to the mounting posts. The $\alpha\beta$ stage rides on the XY stage and can provide a maximum needle angulation of about 20° with respect to X- and Y-axis. A pair of carbon fiber fingers on the XY and $\alpha\beta$ stages is used to manually lock them during setup. A passive needle guide sleeve is attached between the fingers using free-moving ball joints. The 4 DOF are encoded allowing tracked angulation of the needle, which the physician manually inserts into the patient. Thus, physicians retain full manual control and natural haptic sensing, while tracking the needle in the live TRUS images. The needle-tip placement errors measured in TRUS are about 1 mm. For more details, refer Fichtinger et al. (2007).

12.2.2.1.6 University of British Columbia 2D TRUS-Guided Brachytherapy Robot   A team at the University of British Columbia, Canada, developed a 4 DOF TRUS-guided robot for prostate brachytherapy. 2 DOF is provided by X–Y translation of the needle guide for alignment of the needle with a target in the prostate. The needle can also be angulated about the *X*- and *Y*-axis with a maximum range of approximately 30°. The robot is compact and mountable on a standard brachytherapy stepper. If the power is turned off, the robot can be manually positioned and the needle oriented over its entire workspace. However, the needle guide must stay in position when the power is off as the robot is not back-drivable. Manual control of each of the motor axes for fine positioning is possible and a quick-release mechanism for gross translation of the needle guide is available. More details can be found in Salcudean et al. (2008).

12.2.2.1.7 Robarts Research Institute 3D TRUS-Guided Brachytherapy Robot   A team at the Robarts Research Institute, Ontario, Canada, developed a 4 DOF robotic system for 3D TRUS-guided prostate brachytherapy (Figure 12.4). The TRUS transducer and the needle guide are mounted on a common shaft acting as a co-axial frame-of-reference. The needle guide is linked to the co-axial shaft by two-hinged parallelograms spaced apart allowing angulation and translation of the needle guide by manipulating these parallelograms. This assembly is mounted onto a stabilizer for use in prostate low-dose rate (LDR) and HDR brachytherapy. The TRUS transducer used with this robot has a side-firing linear array and is coupled to a motorized mover

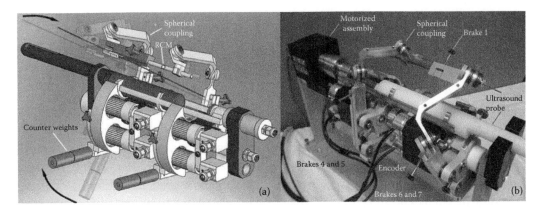

FIGURE 12.4 (a) Schematic of the Robarts Research Institute brachytherapy robot. The device is decoupled through two remote pivot points created from the spherical couplings pinned to each parallelogram. Since the needle guide axis is also aligned with each stationary point, any movement from either parallelogram will result in the needle guide axis to pivot about the stationary point of the opposing spherical linkage, thus resulting in no linear displacement of the intersection point. (b) Photograph of the robot showing the four motors and encoders, which drive the linkage.

assembly that can rotate the transducer about is long axis. To obtain a 3D TRUS image the transducer is rotated over about 160° while 2-dimensional (2D) TRUS images are acquired from the ultrasound machine via a digital frame grabber. These 2D images are reconstructed into a 3D TRUS image as they are acquired so that the 3D image of the prostate is immediately available for dosimetry planning, and dynamic re-planning of needle trajectories to accommodate changes in the plan due to prostate motion or needle deflection. The two-hinged parallelogram mechanism allows coverage of the needle over 6×6 cm^2 area, and needle angulation of about 30° to avoid PAI.

Phantoms testing showed that the geometric error of the 3D TRUS image is less than 0.4 mm. Needle-guidance accuracy tests in agar prostate phantoms showed that the mean error of source placement was 1.6 mm along parallel needle paths, within 1.2 mm of the intended target, and 1° from the preplanned trajectory. At oblique angles of up to 15° relative to the probe axis, sources were placed within 2 mm of the target with an angular error less than 2°. More details can be found in Bax et al. (2011).

12.2.2.1.8 Grenoble University Hospital TRUS-Guided Brachytherapy Robot A team at the Grenoble University Hospital (CHUG), France, has developed a 5 DOF robotic brachytherapy needle-insertion system designed to replace the template used in the manual technique (Hungr et al., 2009). This robot can position and incline a needle within the same workspace as the manual template. To improve needle insertion accuracy, it incorporated provision for needle rotation during insertion into the prostate. The system can be mounted on existing steppers and also easily accommodates existing seed dispensers, such as the Mick® Applicator. The positioning module consists of two pairs of linear translation rails mounted in the form of a parallelogram-like manipulator and allowing for translation and inclination of the insertion module. This system is capable of inserting the needle autonomously; however, the seeds can only be delivered manually. More details are available in Hungr et al. (2009).

12.2.2.2 MR-Based Robotic Prostate Brachytherapy Systems

12.2.2.2.1 University Medical Center Utrecht Robot   A team at University Medical Center Utrecht (UMCU) developed an MR-compatible robotic system, allowing online MRI guidance (van den Bosch et al., 2010). This robot is fixed to a wooden plate that can be slid over the MR table and is placed between the patient's legs. The robot consists of polymers and non-ferromagnetic materials such as brass, copper, titanium, and aluminum to allow use within the magnet's bore. Instead of using MRI compatible motors, the system is pneumatically and hydraulically driven, with needle insertion achieved by a pneumatic tapping device to tap a titanium needle including a stylet stepwise into the patient. Stepwise tapping provides two advantages: (1) high needle insertion speed (momentum of approximately 0.6 Ns) is expected to reduce organ deformation, and (2) needle trajectory can be verified and modified between tapping insertion steps. The maximum insertion depth of the needle per tap is adjustable using a buffer stop and is measured by a potentiometer to an accuracy <1 mm. A controller unit outside the scanning room controls the tapping device and buffer stop, allowing monitoring of the needle insertion on the MR images. This robot has been used clinically to implant fiducial gold markers inside prostates of patients eligible for external beam radiation therapy treatment. In all cases, the robot performs needle tapping but marker deposition is done manually. Prostate HDR brachytherapy is expected to be the next application.

12.2.2.2.2 Johns Hopkins University: MrBot for Prostate Interventions   An MR-compatible 4 DOF robot, called *MrBot*, developed for transperineal prostate intervention was developed by the Urology Department of Johns Hopkins University (Figure 12.5). The entire robot is built with nonmagnetic and dielectric materials and is designed to perform fully automated brachytherapy

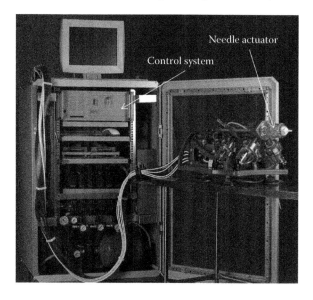

FIGURE 12.5   JHU MrBot—MRI-guided robotic system for prostate brachytherapy. (Image courtesy of Podder et al. 2014. *Med Phys* 41(10):101501.)

source placement within a closed MR scanner. This robot can be used to accommodate various needle drivers for various percutaneous interventions such as biopsy, thermal ablations, and brachytherapy. After initial rough alignment of the robot toward the prostate, the robot aligns the needle to a selected target in the prostate.

The robot uses a specially designed pneumatic step motor (*PneuStep®*) to be compatible for operation inside the bore of an MRI magnet. To achieve MR compatibility, the brachytherapy source placement subsystem uses an MR compatible needle injector end-effector; however, non-MR-compatible components are in the control cabinet. The brachytherapy source dispenser is composed of a jar with a funneled bottom that is shaken using a motor, and source locking, sending, and counting mechanisms. As the jar is shaken, the preloaded sources in the jar are dropped into the funnel leading to the sending system through a tube.

Testing of the robot-source injector mechanism, registration, and image guidance algorithms involved placing the sources at targets specified in the MR image. Results showed that the mean source placement error in agar test phantoms was approximately 1.2 ± 0.4 mm, and the source placement reproducibility was a fraction of a millimeter. Testing of the system's ability to guide a needle to a target in a 3 T MRI scanner showed that the needle tip can be placed within 1 mm of a selected target in the MR image. Experiments with four dogs showed that the median error for MR-guided needle positioning and source positioning was 2.0 mm (range, 0.9–3.2 mm) and 2.5 mm (range, 1.5–10.5 mm), respectively. Details are available in Stoianovici et al. (2007).

12.2.2.2.3 Johns Hopkins University: Robot3 for Prostate Interventions Another MR compatible robot with a remotely actuated manipulator for prostate intervention was developed by the Engineering Research Center and the Department of Mechanical Engineering of the Johns Hopkins University (Krieger et al., 2011). This robot can be used for MRI-guided needle biopsy, fiducial marker placements, and therapy delivery in closed and open MRI scanners.

The robot includes a rectal sheath, which is placed adjacent to the prostate in the patient's rectum, and a needle guide containing a curved needle channel. After placement of the rectal sheath in the rectum, it is held stationary while the needle guide can rotate and translate within the sheath to align the needle toward the target in the prostate. For optimal coverage of targets in the prostate the needle can exit the needle guide through a window in the sheath at a 45° between axis of the guide and the needle. In this way, the robot provides 3 DOF (rotation and translation of the needle guide and insertion of the needle) to guide the needle to targets in the prostate. The manipulator contains two types of MR coils: an imaging coil, which is looped around the window of the sheath, and tracking coils for position encoding. Testing of this robot showed that placement error of fiducial-markers in canines and biopsy procedures with patients were about 2 mm. More details are available in Krieger et al. (2011).

12.2.2.2.4 Johns Hopkins University: Robot4 A different robotic system was developed in collaboration between the Engineering Research Center at the Johns Hopkins University and the Department of Radiology at the Brigham and Women's Hospital. This 6 DOF robot is pneumatically operated under remote control for guiding needles into the prostate through the perineum while the patient is in a 3T closed-bore MRI scanner (Tokuda et al., 2012). Because of the 60 cm diameter MR bore, the patient's legs are spread less than what is used for TRUS-guided brachytherapy, and the knees must be lowered into a semi-lithotomy position. The robot base includes two prismatic motions and two rotational motions upon a manual linear slide. The slide positions the robot in the access tunnel and allows fast removal for reloading brachytherapy needles or collecting biopsy specimens. Additional application-specific motions are also provided and include needle insertion, cannula retraction, needle rotation, and biopsy gun actuation. Details can be found in Tokuda et al. (2012).

12.2.2.3 Robotic Brachytherapy Systems for Other Organs

12.2.2.3.1 University of Western Ontario: MIRA Robot for Lung Brachytherapy A team at the University of Western Ontario (UWO), Canada, developed the Minimally Invasive Robot Assistant (MIRA) for image-guided lung brachytherapy (Trejos et al., 2008). The system is used for accurate radioactive source placement in the lung with commercially available dosimetry planning software. MIRA-V robotic system has dosimetry planning software and a modified version of previously developed software for needle guidance in prostate brachytherapy. Two automated endoscopic system for optimal positioning (AESOP) arms are used to control the video camera and the instrument via voice control and the InterNAV3.0™ interface, respectively. Although these robotic arms are no longer commercially available, they were used in this application for the prototype evaluation and proof-of-concept as they are available for research purposes at the Canadian Surgical Technology and Advanced Robotics (CSTAR). Needle-tip position is monitored using 5 DOF Aurora electromagnetic (EM) tracking sensor (NSI, Waterloo, Ontario, Canada). More details are available in various publications (Lin et al., 2008; Trejos et al., 2008).

12.2.2.3.2 PARA-BRACHYROB Robotic System for Thoracic and Abdominal Applications A 5 DOF parallel robot, called PARA-BRACHYROB was developed at the Technical University of Cluj-Napoca (TUCN), Romania for reaching tumors located in hard to reach areas, such as inside the thoracic and abdominal areas (Galdau et al., 2014; Popescu et al., 2015). The robot has a modular design with two modules: (1) a 3 DOF subsystem with three active joints, and (2) a second subsystem with 3 DOF and two active joints. This robot is suitable for general needle placement techniques for medical procedures, such as brachytherapy. Some tests have been performed in laboratory settings.

12.2.2.3.3 Dynamic Modulated Brachytherapy for Rectal Cancer Webster *et al.* at University of California, San Diego (UCSD) have proposed to utilize the dynamic motion of the applicator itself during rectal brachytherapy treatment delivery using a robotic system (Webster et al., 2013) (Figure 12.6). This system, termed as *dynamic modulated brachytherapy* (DMBT), is composed of two piezoelectric motors along with a choice of materials that make the system MR-compatible. The DMBT system has been conceptualized envisioning the delivering of rectal HDR treatment under MR guidance. It uses a cylindrical shaped tungsten alloy shield with a small window on one side to encapsulate an ^{192}Ir source, to create collimation that results in a highly directional beam profile. Using a robotic system, this shield can be dynamically translated and rotated during treatment to create a volumetric modulated arc therapy-type delivery from inside the rectal

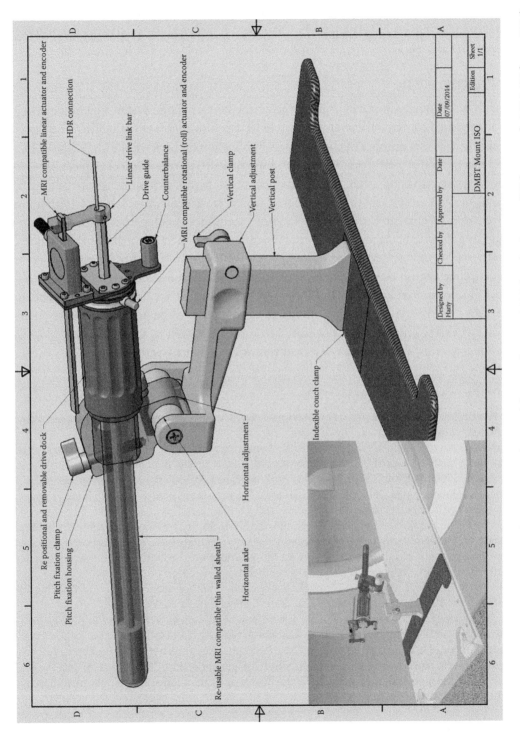

FIGURE 12.6   An MR-compatible DMBT robotic system, designed for rectal cancer HDR brachytherapy (Webster et al., 2013; Webster, 2014).

cavity. Applicators with a shielded design are discussed in Chapter 3. Monte Carlo simulations and planning optimization algorithms have been developed to evaluate the effectiveness of this new approach using some clinical treatments to quantify the potential clinical benefit. Results from simulations of ideal phantom geometries are encouraging and reported in publications (Webster et al., 2013; Webster, 2014).

12.3 CONCLUSION

Clinical brachytherapy would be significantly benefited if robotic systems could be developed or modified to make the systems simple to use with increased reliability and easily accessible to a greater number of clinical sites. Publication of the AAPM TG-192 and GEC-ESTRO has generated significant interest and momentum toward clinical implementation of robotic systems in brachytherapy, especially in prostate cancer treatment with both LDR and HDR brachytherapy and in gynecological cancer treatment with HDR brachytherapy. However, due to current healthcare policy and reimbursement pattern, brachytherapy is struggling despite being one of the most common and effective treatment modalities. Brachytherapy is underrated and shadowed by the EBRT and IMRT/VMAT as well as new technology like proton beam therapy (Popescu et al., 2015). Cost effectiveness of brachytherapy in comparison to other radiotherapeutic treatment modalities is discussed in Chapter 29. It is expected that with anticipated changes in healthcare policies, brachytherapy may get a stronger foothold in cancer care and new as well as advanced technologies, like robotic systems, will claim its deserved position in cancer treatment.

REFERENCES

Bax, J., Smith, D., Bartha, L. et al. 2011. A compact mechatronic system for 3D ultrasound guided prostate interventions. *Med. Phys.* 38:1055–69.

Beaulieu, L., Evans, D. A., Aubin, S. et al. 2007. Bypassing the learning curve in permanent seed implants using state-of-the-art technology. *Int. J. Radiat. Oncol. Biol. Phys.* 67:71–77.

Dhillon, B. S. 2000. *Medical Device Reliability and Associated Areas.* CRC Press, Boca Raton, FL.

Fichtinger, G., Fiene, J. et al. 2007. Robotic assistance for ultrasound guided prostate brachytherapy. *Med. Image Comput. Comput. Assist. Interv.* 10:119–27.

Galdau, B., Plitea, N., Vaida, C. et al. 2014. Design and control system of a parallel robot for brachytherapy. *IEEE International Conference on Automation, Quality and Testing*, Robotics. 1–6.

Hungr, N., Troccaz, J. et al. 2009. Design of an ultrasound-guided robotic brachytherapy needle-insertion system. *Conf. Proc. IEEE Eng. Med. Biol. Soc.* 250–53.

Krieger, A., Iordachita, I. et al. 2011. An MRI-compatible robotic system with hybrid tracking for MRI-guided prostate intervention. *IEEE Trans Biomed Eng.* 11:3049–3060.

Lin, A. W., Trejos, A. L., Mohan, S. et al. 2008. Electromagnetic navigation improves minimally invasive robot-assisted lung brachytherapy. *Comput Aided Surg.* 13:114–123.

Martinez, E., Daidone, A., Gutierrez, C. et al. 2015. Permanent seed brachytherapy for clinically localized prostate cancer: Long-term outcomes in a 700 patient cohort. *Brachytherapy.* 14:166–72.

Meltsner, M., Ferrier N., and B. Thomadsen. 2005. Design and quantitative analysis of a novel brachytherapy robot. *Med. Phys.* 32:1949.

Peschel, R. E. and J. W. Colberg. 2003. Surgery, brachytherapy, and external-beam radiotherapy for early prostate cancer. *Lancet Oncol.* 4:233–41.

Podder, T. K., Beaulieu, L., Caldwell, B. et al. 2014. AAPM and GEC-ESTRO guidelines for image-guided robotic brachytherapy: Report of Task Group 192. *Med. Phys.* 41:101501.

Podder, T. K., Buzurovic, I., K. Huang, K. et al. 2010. Reliability of EUCLIDIAN: An autonomous robotic system for image-guided prostate brachytherapy. *Med. Phys.* 38:96–106.

Podder, T. K., Buzurovic, I. and Y. Yu. 2010. Multichannel robot for image-guided brachytherapy. IEEE International Conference on Bioinformatics and Biomedical Engineering (BIBE), Philadelphia, PA, pp. 209–13.

Popescu, T., Kacsó, A. C., Pisla, D. and Kacsó, G. 2015. Brachytherapy next generation: robotic systems. *J Contemp Brachytherapy* 7:510–514.

Rivard, M. J., Evans, D. A. et al. 2005. A technical evaluation of the Nucletron FIRST system: Conformance of a remote afterloading brachytherapy seed implantation system to manufacturer specifications and AAPM Task Group report recommendations. *J. Appl. Clin. Med. Phys.* 6:22–50.

Salcudean, S. E., Parananta, T. D. et al. 2008. A robotic needle guide for prostate brachytherapy. *International Conference on Robotics and Automation.* Pasadena, CA, USA, May 19–23, pp. 2975–81.

Stoianovici, D., Song, D. et al. 2007. MRI Stealth robot for prostate interventions. *Minim Invasive Ther Allied Technol.* 16:241–48.

Tokuda, J., Song, S. E., Fischer, G. S. et al. 2012. Preclinical evaluation of an MRI-compatible pneumatic robot for angulated needle placement in transperineal prostate interventions. *Int. J. Comput. Assist. Radiol. Surg.* 7:949–957.

Trejos, L., Lin, A. W., Mohan, S. et al. 2008. MIRA V: An integrated system for minimally invasive robot-assisted lung brachytherapy. *2008 IEEE International Conference on Robotics and Automation.* Pasadena, CA, USA, May 19–23.

van den Bosch, M. R., Moerland, M. A., Lagendijk, J. J. et al. 2010. New method to monitor RF safety in MRI-guided interventions based on RF induced image artifacts. *Med. Phys.* 37:814–21.

Webster, M. J. 2014. Dynamic Modulated Brachytherapy (DMBT) and Intensity Modulated Brachytherapy (IMBT) for Treatment of Rectal and Breast Carcinomas. *UC San Diego Electronic Theses and Dissertations.* Local Identifier: b8183065. http://escholarship.org/uc/item/49r7c82w

Webster, M. J., Devic, S., Vuong, T. et al. 2013. Dynamic modulated brachytherapy (DMBT) for rectal cancer. *Med. Phys.* 40:011718.

Wust, P., von Borczyskowski, D. W., Henkel, T. et al. 2004. Clinical and physical determinants for toxicity of 125-I seed prostate brachytherapy. *Radiother Oncol.* 73:39–48.

Yamada, Y., Potters, L., Zaider, M. et al. 2003. Impact of intraoperative edema during transperineal permanent prostate brachytherapy on computer-optimized and preimplant planning techniques. *Am J Clin Oncol.* 26:e130–35.

Yu, Y., Podder, T. K., Zhang, Y. D. et al. 2007. Robotic system for prostate brachytherapy. *Comput. Aided. Surg.* 12:366–70.

Zelefsky, M. J., Yamada, Y., Marion, C. et al. 2003. Improved conformality and decreased toxicity with intraoperative computer-optimized transperineal ultrasound-guided prostate brachytherapy. *Int. J. Radiat. Oncol. Biol. Phys.* 55:956–63.

II

Imaging for Brachytherapy Guidance

CHAPTER 13

Optical Imaging and Navigation Technologies

Robert Weersink

CONTENTS

13.1 INTRODUCTION

White light endoscopy (WLE) has long played a critical role in diagnostics and disease staging of cancer for many intraluminal sites. With the advent of laparoscopic surgery, endoscopy is also critical for minimally invasive surgery. Evolving optical imaging techniques that provide greater sensitivity to disease detection, such as fluorescence and optical coherence tomography (OCT), are becoming part of clinical practice. In parallel, image-guided interventions that employ tracking and navigation tools are now commonplace in surgery and interventional radiology. Neurosurgery in particular has used optical tracking of surgical tools to guide the surgeon between the visualization of the tumor in a three-dimensional (3D) image and the surgical field of view. Together, endoscopy and tool tracking has led to new forms of augmented visualization during interventional procedures.

To date, endoscopy and surgical navigation tools have had only limited roles in brachytherapy; electromagnetic (EM) tracking is under evaluation for catheter segmentation and quality assurance (Chapters 4 and 10, respectively) while endoscopy is principally used to guide applicator insertion for lung and esophagus brachytherapy. This chapter will briefly discuss new techniques in endoscopy, introduce several evolving optical imaging modalities that are proving valuable in intraluminal disease detection, and outline applications of EM navigation in other fields. Readers will be directed to recent surveys where appropriate, while this chapter focuses on how these methods can be applied in brachytherapy. Given the small role that each of these technologies currently has in brachytherapy, this section will be, by necessity, somewhat speculative but it is hoped that this encourages further consideration on how to implement these technologies in brachytherapy.

13.2 OPTICAL IMAGING

13.2.1 Why Optical Imaging?

The standard concern with optical imaging in the context of radiation therapy is that optical imaging is essentially superficial while much of radiation therapy is 3D. Optical photons are highly absorbed and scattered, limiting their mean free paths to about 1–2 mm. While this property limits the depth of penetration into tissue, it also provides its utility, with signal differentiation arising from strong variations in attenuation for different tissue volumes. Optical imaging works as a very sensitive molecular imaging modality, using both endogenous and exogenous markers. Several endogenous markers are used in both white light and fluorescence imaging for the identification of tumor. Likewise, optical contrast agents can be easily conjugated to functional tracers for the imaging of multiple molecular signals. At the bulk tissue scale, optical imaging provides excellent gross functional imaging of features where there are differences in hemoglobin levels and saturation, which can be used for the assessment of therapy response.

13.2.2 WLE Techniques

Several new technologies developed in the last 10–15 years have contributed to significant improvements in WLE image quality (Urquhart et al., 2013). The most significant advancement for flexible endoscopes is the move from fiber-based scopes to high-definition endoscopes in which the detection chip is located at the endoscope tip covered by wide-field optical lenses. End-face detection removes the "pixilation" present in the images due to the collection fiber bundle and eliminates the dark spots that occur as fibers in the bundle break. This move to end-face high-resolution detectors has demonstrated some incremental improvement disease detection, such as in colonic polyp detection.

Spectral imaging, using tissue stains, differential light delivery or detection is used to improve early disease detection. Tissue stains, such as Lugol's iodine or methylene blue, are applied topically or orally, to highlight tissue features associated with the disease state and generate a strong color contrast to the normal tissue. Virtual spectral imaging is based on differential light penetration and absorption across the visible light spectrum due to hemoglobin absorption in the blue and green bands and more diffuse light penetration in the red. Narrow band imaging uses two optical band pass filters at the source that limit illumination to narrow blue and green optical bands that overlap with major oxyhemoglobin

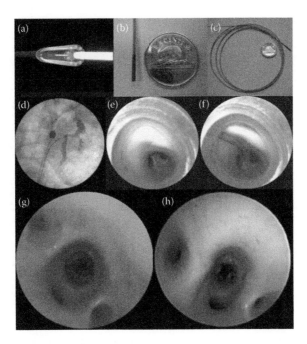

FIGURE 13.1   Novel endoscopic platform using scanning fiber endoscopes (SFEs): (a) Schematic showing the operation of SFE. (b) SFE tip. (c) Flexible SFE fiber. (d) Fluoroscopic image of the SFE in place in the renal artery with a balloon in the descending aorta for proximal flow control. (e) SFE real-time image at the end of the guide catheter, showing a smooth white endothelium and the first bifurcation of the renal artery. (f) SFE image of renal branch selection using a 0.014-inch microwire. (g) Comparison of endoscope images between SFE (left) and conventional type bronchoscopy (right). (h) SFE fiber can reach to the most peripheral bronchus and provide clear images. (Courtesy of P. McVeigh.)

absorption bands (Kara et al., 2006), emphasizing superficial capillaries that strongly absorb blue light and highlighting superficial mucosal details.

Microelectromechanical systems (MEMs) technology has led to the development of several new endoscopic techniques with high magnification and with very small footprint, allowing these techniques to be employed in small lumens (Gu et al., 2014; Louie et al., 2014). At larger fields of view, scanning fiber endoscopy (SFE) offers an excellent high-magnification alternative to chip-based technology within a small flexible probe (Lee et al., 2010). It uses a central source fiber delivering red, green, and blue laser light to the surface surrounded by light collection fibers connected to three wavelength-separated detectors. Source and detection wavelengths can be altered to achieve the same effect as narrow band imaging, or offer fluorescence and Raman imaging. The SFE has so far shown applications in rapid screening of the esophagus, renal artery imaging for stent and microcoil deployments, and bronchoscopy-guided biopsies of deep nodules. Figure 13.1 shows the use of a scanning fiber endoscope.

13.2.3 Fluorescence

Fluorescence imaging is based on either endogenous fluorophores (known as autofluorescence), such as collagen, nicotinamide adenine dinucleotide (NADH), flavin adenine dinucleotide (FAD), and naturally occurring porphyrins (Georgakoudi et al., 2002), or

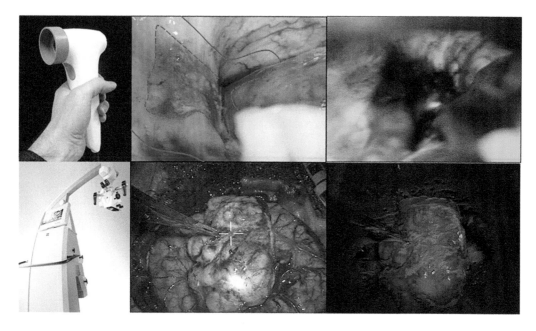

FIGURE 13.2 Tissue fluorescence using endogenous markers (top row) and 5-aminolevulinic acid (ALA) contrast. Top row shows a simple fluorescence device for fluorescence imaging in the oral cavity (left) with WL and green/red fluorescence shown in middle and right images. Lower row shows a surgical microscope (Leica®) for detecting ALA-induced fluorescence in gliomas. Bright signal in the right image indicates the presence of tumor.

exogenous markers that exhibit preferential uptake in malignant tissue. In most clinical autofluorescence systems, blue light excitation (~400–450 nm) is used for tissue excitation with fluorescence collected in separate green and red channels. Tumor appearance is manifested as a reduction of the green fluorescence (via collagen cross-links breakdown) and increase in the red fluorescence (due to an increase of local porphyrins) (Lee et al., 2009). In Figure 13.2, two examples of tissue fluorescence with endogenous markers and 5-aminolevulinic acid (ALA) are demonstrated. Autofluorescence imaging systems have been developed for many intraluminal sites such as oral (Lane et al., 2006; Jayaprakash et al., 2009), laryngeal (Kraft et al., 2011), cervical (Huh et al., 2004), gastric (Hanaoka et al., 2010), and colorectal (Inoue et al., 2010) cancer. While extremely sensitive to early detection of cancer (>90%) (Lane et al., 2006; Kraft et al., 2011), there is still some debate on its specificity (Huh et al., 2004; Lane et al., 2006; Jayaprakash et al., 2009; Kraft et al., 2011).

Fluorescence imaging of exogenous contrast agents uses near-infrared light for deeper tissue penetration, separation from autofluorescence, and simultaneous imaging and over-lay of the fluorescence image on white light images of the surgical field using a second camera and spectrally separating optics. Currently, the only approved fluorescent agents are indocyanine green (ICG), ALA, fluorescein isothiocyanate (FITC), and methylene blue, which has slowed the progress in developing new clinical applications of fluorescence imaging. ICG is used primarily in mapping vascular structures and lymphatic mapping (Troyan et al., 2009; Crane et al., 2011), where, after direction injection into the tumor, real-time fluorescence imaging maps the kinetics of the ICG as it travels through the lymphatic structure.

To date, the only fluorescent agent that directly targets tumor cells is ALA, which induces the synthesis and accumulation of porphyrins in tumors. Fluorescence-guided surgery (FGS) using ALA has shown positive clinical results in the resection of malignant gliomas (Stummer et al., 2006) and bladder cancer (Lerner and Goh, 2015). An example of an ALA-induced fluorescence for detection of gliomas is shown in Figure 13.2. The broader clinical impact of FGS will really only come with the use of active molecular or tumor targeting agents (Achilefu et al., 2000), such as the deployment of tumor-targeted fluorescent agent, folate-FITC, for ovarian cancer patients (van Dam et al., 2011). Many targeted agents have been demonstrated in preclinical studies, but await formal clinical testing and regulatory clearance.

For intraluminal or intracavitary brachytherapy, fluorescence imaging could reduce treatment margins by improving the visualization of disease extent. With targeted molecular markers, it may also offer information not only on the location of disease, but on brachytherapy response based on molecular changes during treatment progression, similar to the hypothesis that positron emission tomography (PET) imaging can be used to inform an adaptive approach to external beam radiotherapy (EBRT).

13.2.4 Optical Coherence Tomography

OCT is often described as the optical analog to ultrasound (US) imaging, using low coherence reflected optical signals rather than US for detection (Huang et al., 1991; Zysk et al., 2007). While the low penetration of light limits the depth of imaging (1–3 mm), the resolution of OCT approaches that of pathology sampling (~10 μm). Several detection schemes have been developed with acquisition speeds improving dramatically since its initial development. OCT has seen rapid clinical acceptance in ophthalmology since it provides exquisite images showing almost histological-level detail of the retina enabling the diagnosis of numerous diseases of the macula. As with US imaging, flow can also be measured, using Doppler techniques (Standish et al., 2008) or forms of speckle imaging (Mariampillai et al., 2008). An interesting example in the context of radiation therapy has shown that in patients with late oral radiation toxicity, Doppler and speckle variance OCT can differentiate normal versus irradiated tissue, based on smaller capillaries and increased flow present in irradiated tissue (Davoudi et al., 2013).

OCT systems have been developed for diagnosis of gastrointestinal diseases using a balloon applicator that can be inserted in the esophagus (Blackshear et al., 2013). Within the balloon, the OCT device radially scans the esophagus wall (~6 cm in 2 min) with resolution of several microns to a depth of 2 mm (Suter et al., 2008). In Figure 13.3, the OCT image of a Barrett's esophagus is shown.

As with fluorescence, OCT has several possible roles in intraluminal/intracavitary brachytherapy, such as improved definition of disease extent or in assessing treatment response. Functional information can be provided by Doppler OCT either in target definition or in response monitoring. For several indications, such as esophagus, vaginal, or lung, the depth imaging provided by OCT may be useful in assigning a patient-specific prescription depth rather than using a standard prescription depth.

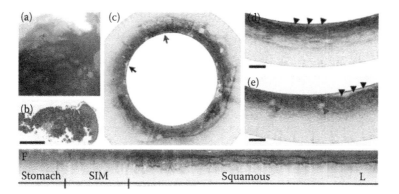

FIGURE 13.3 Optical coherence tomography (OCT) images of Barrett's esophagus. (a) WLE image reveals patchy mucosa. (b) Histopathologic image shows intestinal metaplasia and low-grade dysplasia. (c) Cross-sectional OCT image demonstrating high-grade dysplasia (arrow). (d) Expanded view of C taken from the region denoted by the gray arrow in C. (e) Expanded view of C taken from the region denoted by the black arrow in C. (f) A longitudinal slice highlights the transition from gastric cardia, through a 9-mm segment of specialized intestinal metaplasia and finally into squamous mucosa. Scale bars and tick marks represent 1 mm. (Reprinted with permission from Suter, M. J. et al. *Gastrointestinal Endoscopy* 68(4); 2008: 745–53.)

13.3 TRACKING TECHNOLOGIES

Tool tracking in image-guided procedures typically employ optical tracking techniques using stereo cameras in combination with triangulation to estimate position and orientation based on imaging multiple markers fixed to a tool. While accurate and robust, for typical brachytherapy applications that require the insertion of a device or catheter into the patient, optical tracking is not useful because of size and the obvious obstruction of the line of sight once the device is inserted. Device tracking is still of great interest and several brachytherapy applications using EM tracking are in development (Yaniv et al., 2009; Franz et al., 2014).

EM tracking systems consists of three primary components: a field generator, sensors, and processing unit (Franz et al., 2014). Sensors are typically between 0.2 and 1 mm in diameter and vary in length between 5 and 9 mm, making them ideal for tracking thin, flexible devices such as needles, catheters, and endoscopes. In commercially available systems, field generators are available in several configurations, allowing several options for positioning around a patient. Figure 13.4 shows examples of a sensor and field generators for EM tracking. Optimal accuracy in EM tracking is ~1 mm, but this accuracy reduces the farther the sensor is from the field generator and the closer metal devices are to either the sensor or field generator (Nixon et al., 1998; Poulin and Amiot, 2002). If the metal objects are static within the measurement volume, the positions and orientations can be mapped and corrected. A recent detailed survey of EM tracking discusses the technology, accuracy measurements, and clinical applications (Franz et al., 2014).

EM tracking has been widely used clinically for percutaneous needle insertions in a large cross section of clinical sites (Grand et al., 2011; Xu et al., 2012; Wong et al., 2013). Often,

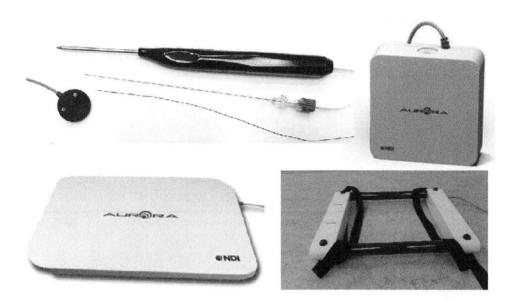

FIGURE 13.4 Technology for EM tracking including sensors (top left) and three types of generators, including mounted (fixed in position with a mounting arm), table top (placed under the patient), and window (to allow simultaneous fluoroscopic imaging).

the percutaneous needle insertion uses a preoperative computer tomography (CT) or magnetic resonance image (MRI) to define the target location with EM-guided needle insertion based solely on these images (Meyer et al., 2008; Penzkofer et al., 2011). Alternatively, an US probe can be simultaneously tracked, and the US image plane registered to prior 3D images and hence to the target location. Needle insertion is performed in the plane of the US probe, with EM tracking used to visualize the needle path. Compared to standard approaches using fluoroscopy, this method provides greater accuracy (~3–6 mm) and much less radiation dose to both the patient and clinical staff.

Navigated endoscopy can be performed by integrating EM tracking with flexible endoscopes, such as in navigated bronchoscopy, as an aid in the biopsy of peripheral pulmonary lesions (PPL). Early CT screening can identify PPLs, which can be removed using a partial lobectomy, leaving the patient cured with minimal impact on quality of life (Aberle et al., 2011). However, many of these small lesions are benign (Rubins et al., 2000), and so the decision to treat depends on biopsy confirmation. While these biopsies often use a transthoracic/transpleural approach, there is great interest in a transbronchial approach to improve the safety of these procedures (Leong et al., 2012).

Several research groups and vendors have developed navigated bronchoscopy systems using EM navigation. They require high-quality CT images for the identification of the PPL location and include planning of the path from the bronchus to the PPL. During the bronchoscopy procedure, the navigation system displays orthogonal views of the CT image at the tip of the bronchoscope. A virtual endoscope view is also displayed based on a surface rendering of the bronchial tree, where the virtual camera pose is set by the closest location of the tracked endoscope to the inside of the bronchial tree. The planned path

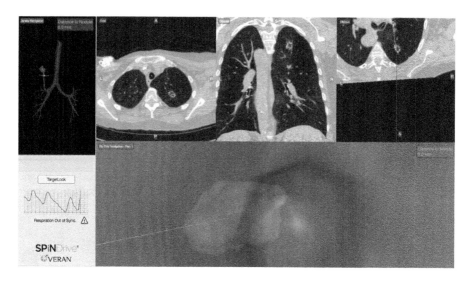

FIGURE 13.5 Navigated bronchoscopy with (top left) planned bronchoscopic path to peripheral lung nodule shown in 3D rendering of bronchus from CT image. (Top right) Orthogonal CT slices shown centered at EM tracking coordinates as the endoscope approaches the target lesion. (Bottom right) Virtual endoscopic view at EM tracking coordinates as the endoscope approaches the target tumor. The planned path, displayed in this virtual view is used to guide the clinician while they simultaneously observe the actual endoscopic view. (Courtesy of Veran Inc.)

is displayed on both the virtual view and CT images. In Figure 13.5, the bronchoscopic path to a peripheral lung nodule is illustrated with an endoscopic navigation approach. Numerous clinical studies have demonstrated the feasibility of EM-guided bronchoscopy and evaluated its utility (Leong et al., 2012). The maturing of this technology into commercially available systems invites the utilization of this technology to other procedures and interventions, such as transbronchial locally ablative therapies.

13.4 POSSIBLE APPLICATIONS OF ENDOSCOPY AND NAVIGATION IN BRACHYTHERAPY

13.4.1 Endoscopic Placement of HDR Catheters in Head and Neck Brachytherapy

There have been limited reports of using endoscopy to guide the placement of high-dose rate (HDR) brachytherapy catheters in head and neck cancers. In one case, catheter placement was performed during the bulk salvage resection of macroscopic disease in patients with recurrent disease following EBRT (Tagliaferri et al., 2015). HDR delivery followed on the third postoperation day. Endoscopic guidance was used to ensure correct positioning of the catheters relative to the site of tumor resection and any remaining tumor. Catheters were then stabilized in position using either molds or sutures, depending on the tumor location. Following the surgery, patients underwent a CT scan for treatment planning, including segmentation of the catheters and any critical structures. The primary benefit of HDR brachytherapy for these patients was the ablation of the disease with negligible toxicities.

(a)  (b) 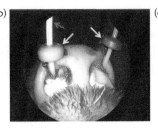 (c)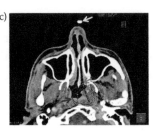

FIGURE 13.6 Endoscopic-guided brachytherapy. (a) The instruments and applicators used in endoscopic-guided interstitial brachytherapy, including rigid endoscope and applicators. (b) The nasal outward view of applicators sewed to the nose wings. (c) CT images of tumor residue, applicators, and isodose line. Orange arrows indicate the path of the applicator, while the red arrow indicated the 100% isodose curve covering the whole gross target volume.

In a large trial of patients with nasopharyngeal cancer, interstitial brachytherapy was compared to intracavitary brachytherapy for patients who had already received EBRT (Wan et al., 2014). Intracavitary brachytherapy following EBRT improves local relapse-free survival in patients with T1-2 nasopharyngeal carcinomas from ~60%–80% without brachytherapy boost to 91%–95% with brachytherapy. However, in more advanced disease, local control is less satisfactory and is associated with a higher risk of complications as the dose is increased. The impact of inserting applicators into deep-seated nasopharyngeal carcinomas on planned dose and patient toxicities was examined. Critical to the insertion was the utilization of a rigid sinus endoscope to guide the needle insertion. Between two and four applicators were inserted, limited by the space needed for the applicators and endoscope. In Figure 13.6, an example of an endoscopic-guided applicator insertion in a nasopharyngeal tumor is presented. CT simulation was used to segment the applicator placement and plan treatments. Applicators were placed such that the maximum distance from the tumor margin to the nearest applicator <1 cm and the maximum distance between applicators <2 cm. Intracavitary applicators were removed between fractions, while interstitial applicators were left in place for all fractions. Although the interstitial approach provided a more homogeneous dose compared to the superficial delivery using intracavitary applicators, clinical results were very similar for both treatment methods, including almost identical overall survival.

Endoscopic guidance was considered critical for the applicator insertions in these cases. Given the insertion planning rules above, the catheter placement precision was deemed not critical.

13.4.2 Applicator Insertion Using Endoscopic and Navigation Technology

HDR brachytherapy for esophagus and lung patients uses an applicator inserted into the lumen using a combination of endoscopy and fluoroscopy. Applicator position and planned target length are determined by using WLE to find the distal and proximal ends of the tumor. Simultaneous fluoroscopic imaging of the endoscope in combination with temporary radio-opaque skin markers is used as guides for the insertion of the brachytherapy applicator. After applicator insertion, catheter mapping of a guide wire in the applicator using fluoroscopy is used to define the treatment dwell positions.

Visualization using this method is only two-dimensional (2D) and because the applicator and tissue surface markers are separated perpendicular to the imaging plane, the fluoroscopy device must remain fixed in position to avoid any positioning errors. We are developing a combined navigated endoscopy and EM applicator mapping procedure that is potentially being more accurate than current practice. Using navigation endoscopy, the endoscope position is recorded during the initial patient exam and locations of the tumor's proximal and distal ends recorded. After applicator insertion, an EM tracking tool records the path of the applicator channel inside the lumen. This path can be used with the recorded endoscopy recording to reconstruct the dwell positions relative to the target volume for treatment planning. Our initial phantom testing has demonstrated the feasibility and accuracy of this applicator insertion technique. Clinical validation of this method is now underway.

This combination of endoscopy and navigated insertion fits well within the current procedure for lung and esophagus applicator insertion, but it could also serve a role in the insertion of some gynecological applicators. An alternative approach would be to fully integrate the endoscopy imaging into the applicator. The small diameter of the SFE is approximately the same size as the transfer tubes used in HDR applicators; hence an applicator with an endoscopic view from the distal end as it is inserted is conceivable. EM tracking of the whole applicator would enable the orientation of the ring with respect to the tandem or for the guidance of needle insertions with respect to the primary applicator.

13.4.3 Contouring and Dose Visualization Using Navigated Endoscopy

For many intraluminal cancers, endoscopy may be invaluable in visualizing the superficial, radiographically occult extent of disease. In many of these clinical sites, however, this information is only used diagnostically to confirm the disease stage and extent. The relation of these clinical findings to the planning environment, however, is either absent altogether or based primarily on the treating clinician's recall of the endoscopically visible disease relative to anatomical landmarks, a process subject to substantial contouring bias, error, and uncertainty in delineating the target. Using technology similar to the navigated bronchoscopy described earlier, we have recently developed a method of quantitatively registering endoscopic information with the radiation treatment planning images (Weersink et al., 2011). Initial interest has been on the treatment of head and neck disease treated with EBRT, where there is significant evidence that coverage of superficial disease is inadequate based on planning that only uses radiological imaging (Daisne and Gregoire, 2006; Thiagarajan et al., 2012).

Here, the EM tracking is used to register the 2D endoscopic images to the 3D radiation treatment planning CT. Using standard surface rendering techniques and by creating a virtual camera in the same position and orientation as the actual endoscope (derived from the tracking device), simultaneous real and virtual views of the lumen can be generated. Superficial disease not present on the CT image can be viewed on the real endoscopic view, where it can be contoured, projected into 3D image space, and imported into the treatment planning software. Figure 13.7 shows an example of the process. For intraluminal brachytherapy applications, such endoscopic contouring can be used to define the extent of superficial disease

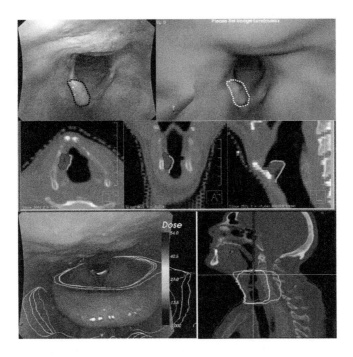

FIGURE 13.7 Demonstration of endoscopic contouring and endoscopic dose display. Top row left: Real endoscopic image showing larynx tumor contoured in green. Top row right: Virtual endoscopic image matching real endoscopic image matching the real view. The contour has been projected onto the virtual view. Middle row: Endoscopic contours mapped onto the treatment planning CT, shown in comparison to the gross tumor volume in dark gray. Lower row: Isodose display of radiation dose overlaid on endoscopic image with image at right showing the dose and relative position of the endoscope.

while 3D imaging can be used to delineate bulk disease, with tracking of the endoscope used to integrate both contouring sets into a single comprehensive treatment plan.

The endoscopic navigation technology can also be used to display radiation dose on the endoscopic image (Qiu et al., 2012). For those cases that use brachytherapy as a boost following external beam radiation therapy, this dose visualization may be useful in combining EBRT and brachytherapy dose calculations, or aiding in catheter placement based on past radiation dose coverage. The dose visualization can also be used during posttreatment evaluation of treatment response relative to delivered dose, or during surgery to project radiation dose onto the surgical field. Rather than visualizing radiation dose, other 3D parametric parameters can be visualized such as PET or MRI signals that may be relevant to an applicator insertion.

13.5 CONCLUSION

This chapter presents two medical fields not normally associated with brachytherapy: (1) endoscopic imaging and (2) image-guided navigation technologies. Both of these fields, especially in combination with each other, offer some new approaches to treatment planning and delivery in brachytherapy. Navigation technologies have become commonplace in surgery and interventional radiology while endoscopic tracking integrated with volumetric imaging presents new "augmented" methods of visualizing clinical information. As applicators become more sophisticated, improved methods of guiding their insertion are needed and navigation technologies are sure to play a role, not only in catheter reconstruction (as described in Chapter 4), but also for placement guidance.

Finally, image guidance in brachytherapy in industrialized countries has become reliant on the latest advances in MRI, CT, and US imaging. However, for developing countries, not-so-widespread implementation of MR and CT for brachytherapy may hinder the wider adaptation of the latest advances in brachytherapy. Here, the integration of optical and navigation technology offers a cost-effective alternative solution for several brachytherapy applications.

REFERENCES

Aberle, D. R., A. M. Adams, C. D. Berg et al. Reduced lung-cancer mortality with low-dose computed tomographic screening. *New England Journal of Medicine* 365(5);2011:395–409.

Achilefu, S., R. B. Dorshow, J. E. Bugaj et al. Novel receptor-targeted fluorescent contrast agents for in vivo tumor imaging. *Investigative Radiology* 35(8);2000:479–85.

Blackshear, L., E. Aranda-Michel, H. Wolfsen et al. Volumetric laser endomicroscopy (VLE): An OFDI case study of Barrett's esophagus with dysplasia. *American Journal of Gastroenterology* 108;2013:S656–S56.

Crane, L. M. A., G. Themelis, R. G. Pleijhuis et al. Intraoperative multispectral fluorescence imaging for the detection of the sentinel lymph node in cervical cancer: A novel concept. *Molecular Imaging and Biology* 13(5);2011:1043–49.

Daisne, J.-F. and V. Gregoire. Multimodalities imaging for target volume definition in radiotherapy. *Bulletin Du Cancer* 93(12);2006:1175–82.

Davoudi, B., M. Morrison, K. Bizheva et al. Optical coherence tomography platform for microvascular imaging and quantification: Initial experience in late oral radiation toxicity patients. *Journal of Biomedical Optics* 18(7);2013:76008.

Franz, A. M., T. Haidegger, W. Birkfellner et al. Electromagnetic tracking in medicine—A review of technology, validation, and applications. *IEEE Transactions on Medical Imaging* 33(8);2014:1702–25.

Georgakoudi, I., B. C. Jacobson, M. G. Muller et al. NAD(P)H and collagen as in vivo quantitative fluorescent biomarkers of epithelial precancerous changes. *Cancer Research* 62(3);2002: 682–87.

Grand, D. J., M. A. Atalay, J. J. Cronan et al. Ct-guided percutaneous lung biopsy: Comparison of conventional Ct fluoroscopy to Ct fluoroscopy with electromagnetic navigation system in 60 consecutive patients. *European Journal of Radiology* 79(2);2011:E133–E36.

Gu, M., H. Bao, and H. Kang. Fibre-optical microendoscopy. *Journal of Microscopy* 254(1);2014:13–18.

Hanaoka, N., N. Uedo, A. Shiotani et al. Autofluorescence imaging for predicting development of metachronous gastric cancer after *Helicobacter pylori* eradication. *Journal of Gastroenterology and Hepatology* 25(12);2010:1844–49.

Huang, D., E. A. Swanson, C. P. Lin et al. Optical coherence tomography. *Science* 254(5035);1991:1178–81.

Huh, W. K., R. M. Cestero, F. A. Garcia et al. Optical detection of high-grade cervical intraepithelial neoplasia in vivo: Results of a 604-patient study. *American Journal of Obstetrics and Gynecology* 190(5);2004:1249–57.

Inoue, K., N. Wakabayashi, Y. Morimoto et al. Evaluation of autofluorescence colonoscopy for diagnosis of superficial colorectal neoplastic lesions. *International Journal of Colorectal Disease* 25(7);2010:811–16.

Jayaprakash, V., M. Sullivan, M. Merzianu et al. Autofluorescence-guided surveillance for oral cancer. *Cancer Prevention Research* 2(11);2009:966–74.

Kara, M. A., M. Ennahachi, P. Fockens et al. Detection and classification of the mucosal and vascular patterns (mucosal morphology) in Barrett's esophagus by using narrow band imaging. *Gastrointestinal Endoscopy* 64,(2);2006:155–66.

Kraft, M., C. S. Betz, A. Leunig et al. Value of fluorescence endoscopy for the early diagnosis of laryngeal cancer and its precursor lesions. *Head and Neck—Journal for the Sciences and Specialties of the Head and Neck* 33(7);2011:941–48.

Lane, P. M., T. Gilhuly, P. Whitehead et al. Simple device for the direct visualization of oral-cavity tissue fluorescence. *Journal of Biomedical Optics* 11(2);2006:024006.

Lee, C. M., C. J. Engelbrecht, T. D. Soper et al. Scanning fiber endoscopy with highly flexible, 1 mm catheterscopes for wide-field, full-color imaging. *Journal of Biophotonics* 3(5-6);2010:385–407.

Lee, P., R. M. van den Berg, S. Lam et al. Color fluorescence ratio for detection of bronchial dysplasia and carcinoma in situ. *Clinical Cancer Research* 15(14);2009:4700–05.

Leong, S., H. Ju, H. Marshall et al. Electromagnetic navigation bronchoscopy: A descriptive analysis. *Journal of Thoracic Disease* 4(2);2012:173–85.

Lerner, S. P. and A. Goh. Novel endoscopic diagnosis for bladder cancer. *Cancer* 121(2);2015:169–78.

Louie, J. S., R. Richards-Kortum, and S. Anandasabapathy. Applications and advancements in the use of high-resolution microendoscopy for detection of gastrointestinal neoplasia. *Clinical Gastroenterology and Hepatology* 12(11);2014:1789–92.

Mariampillai, A., B. A. Standish, E. H. Moriyama et al. Speckle variance detection of microvasculature using swept-source optical coherence tomography. *Optics Letters* 33(13);2008:1530–32.

Meyer, B. C., O. Peter, M. Nagel et al. Electromagnetic field-based navigation for percutaneous punctures on C-Arm Ct: Experimental evaluation and clinical application. *European Radiology* 18(12);2008:2855–64.

Nixon, M. A., B. C. McCallum, W. R. Fright et al. The effects of metals and interfering fields and on electromagnetic trackers. *Presence-Teleoperators and Virtual Environments* 7(2);1998:204–18.

Penzkofer, T., P. Bruners, P. Isfort et al. Free-hand Ct-based electromagnetically guided interventions: Accuracy, efficiency and dose usage. *Minimally Invasive Therapy & Allied Technologies* 20(4);2011:226–33.

Poulin, F., and L. P. Amiot. Interference during the use of an electromagnetic tracking system under OR conditions. *Journal of Biomechanics* 35(6);2002:733–37.

Qiu, J., A. J. Hope, B. C. J. Cho et al. Displaying 3D radiation dose on endoscopic video for therapeutic assessment and surgical guidance. *Physics in Medicine and Biology* 57(20);2012:6601–14.

Rubins, J. B., S. L. Ewing, S. Leroy et al. Temporal trends in survival after surgical resection of localized non-small cell lung cancer. *Lung Cancer* 28(1);2000:21–27.

Standish, B. A., K. K. C. Lee, X. Jin et al. Interstitial Doppler optical coherence tomography as a local tumor necrosis predictor in photodynamic therapy of prostatic carcinoma: An in vivo study. *Cancer Research* 68(23);2008:9987–95.

Stummer, W., U. Pichlmeier, T. Meinel et al. Fluorescence-guided surgery with 5-aminolevulinic acid for resection of malignant glioma: A randomised controlled multicentre phase III trial. *Lancet Oncology* 7(5);2006:392–401.

Suter, M. J., B. J. Vakoc, P. S. Yachimski et al. Comprehensive microscopy of the esophagus in human patients with optical frequency domain imaging. *Gastrointestinal Endoscopy* 68(4);2008:745–53.

Tagliaferri, L., F. Bussu, M. Rigante et al. Endoscopy-guided brachytherapy for sinonasal and nasopharyngeal recurrences. *Brachytherapy* 14(3);2015:419–25.

Thiagarajan, A., N. Caria, H. Schoeder et al. Target volume delineation in oropharyngeal cancer: Impact of PET, MRI, and physical examination. *International Journal of Radiation Oncology Biology Physics* 83(1);2012:220–27.

Troyan, S. L., V. Kianzad, S. L. Gibbs-Strauss et al. The Flare((Tm)) intraoperative near-infrared fluorescence imaging system: A first-in-human clinical trial in breast cancer sentinel lymph node mapping. *Annals of Surgical Oncology* 16(10);2009:2943–52.

Urquhart, P., R. DaCosta, and N. Marcon. Endoscopic mucosal imaging of gastrointestinal neoplasia in 2013. *Current Gastroenterology Reports* 15(7);2013:330–30.

van Dam, G. M., G. Themelis, L. M. A. Crane et al. Intraoperative tumor-specific fluorescence imaging in ovarian cancer by folate receptor-alpha targeting: First in-human results. *Nature Medicine* 17(10);2011:1315–U202.

Wan, X.-B, R. Jiang, F.-Y. Xie et al. Endoscope-guided interstitial intensity-modulated brachytherapy and intracavitary brachytherapy as boost radiation for primary early T stage nasopharyngeal carcinoma. *Plos One* 9(3);2014:e90048.

Weersink, R. A., J. Qiu, A. J. Hope et al. Improving superficial target delineation in radiation therapy with endoscopic tracking and registration. *Medical Physics* 38(12);2011:6458–68.

Wong, S. W., A. U. Niazi, K. J. Chin et al. Real-time ultrasound-guided spinal anesthesia using the SonixGPS® needle tracking system: A case report. *Canadian Journal of Anesthesia—Journal Canadien D Anesthesie* 60(1);2013:50–53.

Xu, H. X., M. D. Lu, L. N. Liu et al. Magnetic navigation in ultrasound-guided interventional radiology procedures. *Clinical Radiology* 67(5);2012:447–54.

Yaniv, Z., E. Wilson, D. Lindisch et al. Electromagnetic tracking in the clinical environment. *Medical Physics* 36(3);2009:876–92.

Zysk, A. M., F. T. Nguyen, A. L. Oldenburg et al. Optical coherence tomography: A review of clinical development from bench to bedside. *Journal of Biomedical Optics* 12(5);2007:051403.

CHAPTER 14

Ultrasound

Maximilian P. Schmid, Luc Beaulieu, Nicole Nesvacil,
Bradley R. Pieters, Arnoud W. Postema, Stefan G. Schalk,
Frank-André Siebert, and Hessel Wijkstra

CONTENTS

14.1 INTRODUCTION AND GENERAL ASPECTS

Clinical ultrasound (US) or sonography is an imaging modality using high frequency sound waves typically between 1 and 20 MHz. The US pulses are created in transducers, often a piezo-electrical crystal, which converts changes of thickness of the crystal into sound waves by alternating applying voltage. The emerging sound waves are refracted and reflected in the tissue. Echoes of these sound waves can be received by the transducer. The transducer changes that deformation into an electrical potential which can be detected. Today, US transducers are mostly used in brightness (B-) mode. That means sound waves are sent out from an array of transducers in different directions and are received a while later. From the detected signals, a two-dimensional (2D) image plane can be computed where each pixel is a representation of the amplitude of the reflected signal from that position.

In brachytherapy, transrectal ultrasound (TRUS) probes are commonly used. They can be used for prostate and anal cancers, as well as, with some restrictions, for gynecological brachytherapy treatments. Mostly transversal and longitudinal viewing planes are in use. A three-dimensional (3D) reconstruction can be performed after scanning a volume by longitudinal movement of the probe or using a fan-like scan while rotating the probe in the patient. Another usage for US in brachytherapy is the abdominal transducers which can be used, for example, for breast and gynecological brachytherapy.

US is extensively used for diagnostic purposes in the field of oncology. The high image resolution and good soft-tissue contrast allow for assessment of different oncologic sites such as abdominal or pelvic tumors, breast cancer, or various lymph node regions. For selected indications (e.g., anal cancer, prostate cancer, gynecologic cancer, cervical and inguinal lymph nodes), US is considered equal or even superior to computed tomography (CT) or magnetic resonance imaging (MRI). Beside the diagnostic possibilities, the easy handling, missing ionizing radiation exposure, and high flexibility qualifies for repetitive imaging for monitoring treatment response in the frame of, for example, currently investigated adaptive radiotherapeutic treatment regimes. In addition to standard morphologic imaging using B-mode US, color Doppler can provide insights in the vascularization of tumors. Further developments such as contrast-enhanced US seem to facilitate visualization of local tumor extension (e.g., liver and prostate; see Section 14.2), and elastography allows for measurement and depiction of the elasticity of tissues (e.g., for liver, breast, or thyroid gland). Thus, US can be also considered a functional imaging modality. US can be applied for various steps within the brachytherapy treatment chain: the tumor/target volume can be assessed directly intraoperatively before insertion of the applicator, which enables different preplanning possibilities, simple tumor/target volume measurements to support x-ray or CT-based contouring/treatment planning (e.g., gynecologic tumors, head and neck cancers), tumor/target volume assessment to facilitate selection of the appropriate applicator system/implant, full 3D treatment planning to predefine and evaluate the exact interstitial needle and dwell point positions (e.g., prostate cancer). However, especially, the attribute of dynamic real time "live" imaging appears attractive for image-guided applications. The insertion of interstitial needles can be visualized intraoperatively to optimize the coverage of the target volume and the geometry of the implant. In Figure 14.1, flexible tubes in buccal mucosa are imaged on US. Interstitial needles generally depict a hyperechogenic signal, but the quality varies according to the used material and the surrounding tissue. Plastic, steel, and titanium needles appear to be suitable, but differ in the extent

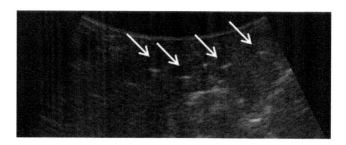

FIGURE 14.1 Verification of the implant geometry in an interstitial brachytherapy of the buccal mucosa. The white arrows indicate four flexible plastic tubes.

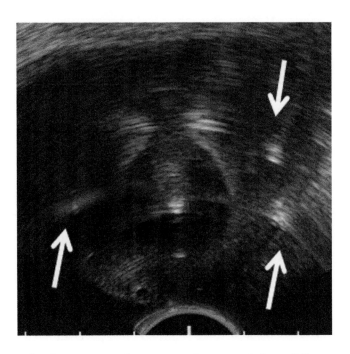

FIGURE 14.2   Example of a Vienna tandem ring applicator with three different interstitial needles (marked by the white arrows: plastic needle left, steel needle upper right, and titanium needle lower right) visualized in a gel phantom.

of artifacts around and behind the needle. This difference in visualization can be seen from a phantom experiment in Figure 14.2. The final implant can be verified by US and be adapted. For certain indications, treatment planning can be realized directly in the US dataset. Such a time saving approach can improve the current logistic challenges of image-guided brachytherapy using CT or MRI, if the respective imaging modalities are not directly integrated into the brachytherapy department.

In principle, all the different brachytherapy indications, which are accessible for US imaging, would be suitable for an US-guided procedure. So far, most experience is available for prostate brachytherapy, where the complete procedure (i.e., needle insertion, applicator reconstruction, and treatment planning) can be performed under US guidance as illustrated in Figure 14.3. Similar techniques have been reported for brachytherapy of the anal canal (Figure 14.4) and are under development for gynecologic malignancies. However, also for breast and head and neck (Figure 14.1), brachytherapy US imaging is of increasing value and interest.

14.2 US FOR PROSTATE BRACHYTHERAPY

Traditionally, prostate brachytherapy was aimed to a rather homogeneous dose to the whole gland. However, with the advent of advanced and improved imaging of the prostate gland, new radiation techniques are developed to target visible tumor areas. Multi-parametric MRI (mpMRI) has a proven value to discriminate so-called intraprostatic lesions (IL), particularly tumor lesions larger than 5 mm or with a Gleason score higher than 6 (Turkbey et al., 2011).

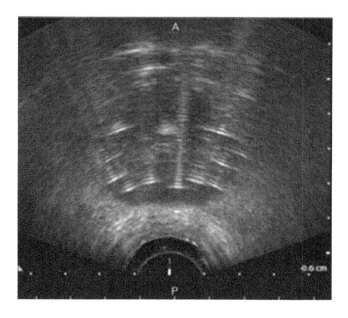

FIGURE 14.3   Transversal US image showing an HDR prostate implant using 11 implant needles.

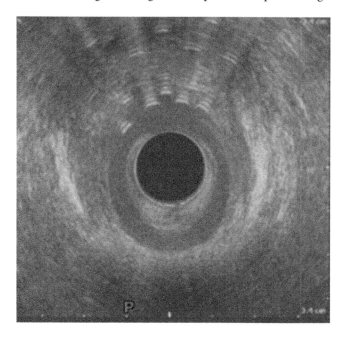

FIGURE 14.4   360° US image of a brachytherapy anal implant with seven needles.

If separate tumor lesions can be identified, two strategies regarding targeted brachytherapy can be followed. The first strategy is called focused brachytherapy. With focused brachytherapy, radiation is given to the whole prostate gland with a boost dose on the IL. Whole gland brachytherapy in this setting can be part of a boost dose in addition to external beam radiotherapy (EBRT) (Kazi et al., 2010; Schick et al., 2011). The focused boost can be considered as the second coned-down boost. Patients with prostate cancer of intermediate or high-risk profile, according to, for example, NCCN risk classification, are considered suitable for EBRT with a brachytherapy boost.

The second strategy is called focal brachytherapy. With focal brachytherapy, only a part of the prostate gland is treated. There is no general definition for the term focal therapy, which can vary from strictly brachytherapy to the visible lesion to a subtotal prostate gland irradiation. With focal brachytherapy, EBRT is omitted and only a part of the prostate gland containing the visible IL is treated (Nguyen et al., 2012; Cosset et al., 2013). Focal therapy is reserved for patients with a tumor of low-risk profile.

For both strategies, visualization of the ILs is essential to accurately direct the position of the brachytherapy sources and consequently the dose distribution. In this perspective, contrast-enhanced ultrasound (CEUS) is a technique that is able to identify ILs and can aid in targeting the lesions with brachytherapy.

The ultrasound contrast agents (UCA) used in CEUS imaging consist of small bubbles of gas (microbubbles) encapsulated in a biocompatible shell, which resonate when excited by US waves at certain frequencies. CEUS imaging exploits the nonlinearity of the back-scattered signal coming from resonating microbubbles to image blood flow and suppress signals coming from the surrounding tissue. Because the size of the microbubbles is comparable to that of red blood cells, they can travel through all vessels including the smallest capillaries allowing assessment of microvascular flow patterns. This concept can be used to reveal the altered blood flow related to angiogenesis, which is a key indicator of prostate cancer presence and aggressiveness (Russo et al., 2012).

During CEUS examination, the inflow and enhancement pattern is recorded for 1–2 minutes after intravenous injection of the UCA. In a visual assessment of CEUS videos, common markers used to detect prostate tumors are rapid UCA wash-in, high peak concentration with respect to surrounding tissue, and asymmetry of the intraprostatic vessels (Sano et al., 2011; Seitz et al., 2011). In addition, several quantitative techniques have been proposed aimed at detecting an increased perfusion as a result of angiogenesis (Hudson et al., 2009; Cosgrove and Lassau, 2010; Barrois et al., 2013). However, the effect of angiogenesis on blood perfusion can be dual. On one hand, flow resistance is reduced by the lack of vasomotor control and increase in arteriovenous shunts. On the other hand, flow resistance is increased by a higher interstitial pressure and by a smaller diameter and higher tortuosity of the microvasculature (Cosgrove, 2003).

Recently, contrast ultrasound dispersion imaging (CUDI) has been proposed as a technique estimating the local dispersion rather than perfusion to locate angiogenic tissue. Initially, time-intensity curves (TICs, i.e., image intensity vs. time) were fitted in each pixel by a model to estimate the dispersion (Kuenen et al., 2011), but later it was shown that the local similarity between TICs is indicative of the presence of prostate cancer (Mischi et al., 2012; Kuenen et al., 2013). In a preliminary validation (Kuenen et al., 2013), parametric maps of the correlation between TICs were made for 12 planes in eight patients in which regions of cancerous and healthy tissue were drawn based on radical prostatectomy. Figure 14.5 shows an example of CUDI by correlation between neighboring TICs. The resulting sensitivity and specificity were 77% and 86%, respectively. A recent trial in 82 patients revealed that the tumor status of individual prostate sectors could be predicted using a

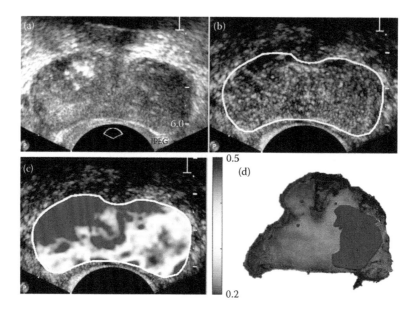

FIGURE 14.5   Example of CUDI by correlation between neighboring TICs: (a) fundamental image, (b) CEUS frame, (c) parametric map of correlation, and (d) corresponding histology slice.

similar technique for parametric analysis of CEUS recordings (Postema et al., 2015). Using parametric maps, 411/651 biopsy locations were predicted to be benign of which 23 (5.6%) revealed >10% biopsy core involvement of Gleason ≥7 disease.

CEUS techniques provide visibility of ILs and can be used to target these lesions with brachytherapy. The advantage of CEUS is that this technique is based on US, which is by far the most straightforward technique for prostate implantations. With an US technique, the complexity of using MR registration can be avoided. At the Academic Medical Center, The Netherlands, a proof-of-principle analysis was performed investigating the consequences on dose distribution with CEUS (Pieters et al., 2012). For eight patients, CEUS images were available. ILs were delineated on CEUS images and rigidly registered on available US images obtained from the implantation procedure as can be seen in Figure 14.6. Two brachytherapy plans were designed. The first plan was a conventional plan without knowledge of the localization of the ILs. The second plan was performed with the knowledge and visibility of the ILs. Dose objectives for whole gland prostate, rectum, and urethra were kept similar. The CEUS-based plans had an extra dose objective to have at least 95% of the ILs volumes to be covered by the 140% isodose. Comparison of both plans showed no difference in V_{100} prostate, D_{90} prostate, D_{2cm3} rectum, D_{max} urethra, and TRAK (Total Reference Air Kerma). However, a 15%–18% increase of D_{90} was observed on the ILs. The $V_{140\%}$ coverage of the ILs increased by 25%–30%. In Figure 14.7, the improvement of IL dose coverage is illustrated if the location of the IL can be taken into account with CEUS. This study showed that with knowledge of the localization of ILs by using CEUS, better targeting of the ILs could be achieved.

Further developments that are needed for daily practice is to implement the CEUS technique in US machines and probes suitable for brachytherapy. The CEUS lesions should be preferentially marked in the treatment planning software for brachytherapy targeting.

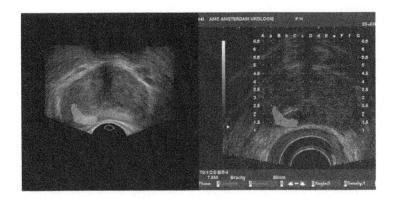

FIGURE 14.6 Intraprostatic lesion delineated on CEUS (left) rigidly registered on TRUS image for brachytherapy (right). (Reproduced with permission from Pieters B R et al. 2012. *J Contemp Brachytherapy* 4, 67–74.)

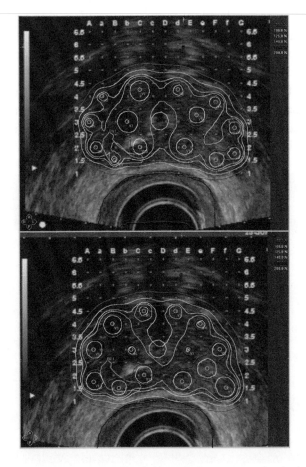

FIGURE 14.7 Intraprostatic lesion partially (left) and almost completely (right) covered by the 140% isodose. (Reproduced with permission from Pieters B R et al. 2012. *J Contemp Brachytherapy* 4, 67–74.)

14.3 US FOR GYNECOLOGICAL BRACHYTHERAPY

Even though US is a relevant and well-established imaging modality in gynecologic oncology, there are only a few reports in the field of gynecologic brachytherapy. However recently, based on the advances in US-guided brachytherapy in prostate cancer, there is also increasing interest in using the distinct advantages of US for complex gynecologic brachytherapy procedures and to implement US into the concept of (MR-) image-guided adaptive brachytherapy in particular for cervical cancer. Transabdominal and transrectal US have been described as suitable for cervix cancer brachytherapy. Transabdominal ultrasound (TAUS) provides a good overview of the pelvic organs, but with restrictions in image quality especially for more deeply located structures such as the cervix uteri or rectum. Due to the direct proximity to the cervix uteri, TRUS allows an excellent image quality in this crucial area but is limited by the relatively small field of view.

14.3.1 US for Image-Guided Applicator Insertion

Several studies support the use of TAUS-guidance for placement of the tandem into the uterine cavity in cervix cancer brachytherapy using tandem ring or tandem ovoid applicators. Such an image-guided approach shortens the time to complete the implantation, facilitates the insertion in patients with challenging anatomy (e.g., retroflected uterus and cervical stenosis) and reduces the rate of uterine perforation from 2%–14% to 0%–1.4% (Rao and Ghosh, 2015). A full bladder is necessary to provide optimal scanning conditions and to improve image quality. Figure 14.8 shows a correctly positioned tandem in the uterine cavity. Beside cervical cancer, TAUS can also be used to guide the insertion of Heyman-capsules for primary brachytherapy in patients with endometrial cancer (Figure 14.9).

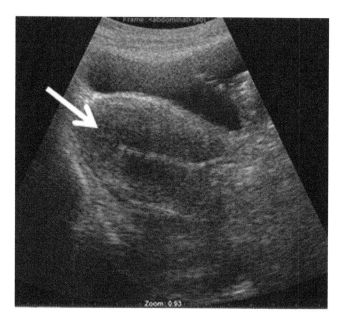

FIGURE 14.8 TAUS in sagittal orientation. The white arrow shows the correctly positioned tandem in the uterine cavity. Anterior: the urinary bladder with a Foley catheter. Posterior: rectum.

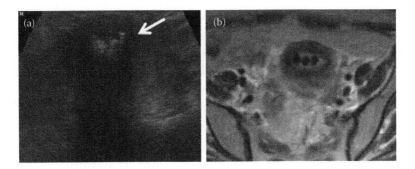

FIGURE 14.9   (a) Transverse TAUS of the uterine fundus with three Heyman capsules and (b) corresponding T2-weighted MRI.

TRUS provides a more detailed image of the uterine cervix, thus qualifying for the depiction of interstitial needles. TRUS is considered a practical and effective imaging device for "preplanning" of gynecologic implants and guiding interstitial needles in patients with gynecologic malignancies such as cervical and vaginal cancer or pelvic wall recurrences. During TRUS-guidance, the interstitial needles can be accurately placed (Stock et al., 1997; Sharma et al., 2010) and the final implant can be verified before submitting the patient for additional imaging for treatment planning purposes. The possibility of intraoperative real-time image guidance substantially enlarges the scope of gynecological brachytherapy and allows the performance of complex interstitial or combined interstitial–intracavitary implants. Figure 14.10 shows an US-guided insertion of interstitial needles in a patient with a squamous cell carcinoma on the right pelvic wall.

14.3.2 US for Target Volume Definition

So far, a concrete target volume concept has not been reported in gynecological malignancies with TAUS or TRUS. However, for cervical cancer, a comparison of several reference points along the uterine surface between TAUS and T2-weighted MRI with applicator in place indicate an excellent correlation, but with limitations at the posterior uterine wall and the parametria (Mahantshetty et al., 2012). Such reference points and distances in

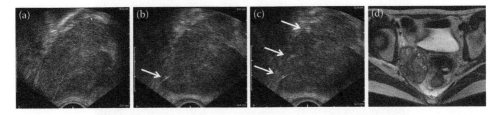

FIGURE 14.10   TRUS-guided needle insertion in a patient with a squamous cell carcinoma of the right pelvic wall. This patient was treated preoperatively with 60 Gy intensity modulated radiotherapy and weekly cisplatin. In addition, a brachytherapy boost in the area of the expected (positive) resection margin at the pelvic wall was delivered. To reach this area, the interstitial needles were inserted through the perineum under TRUS guidance. The images illustrate the step-by-step procedure: (a) preplan TRUS, (b) insertion of the first needle, (c) insertion of third needle, and (d) final implant on T2-weighted MRI. After radiochemotherapy and brachytherapy, the tumor could be completely resected.

relation to the applicator can then also be used for slight individualization of standard 2D (Point A-based) treatment plans (Van Dyk et al., 2009). First clinical results from Australia and Thailand support this method (Narayan et al., 2014; Tharavichitkul et al., 2015)).

In contrast, with 3D-TRUS, high image resolution volumetric imaging with and without applicator in place is feasible for cervical cancer. The presence of the applicator leads to certain artifacts but image quality is overall good to excellent. Depiction of the MRI-derived high risk CTV (CTV_{HR}) is possible and a comparison of CTV_{HR} dimensions between 3D-TRUS and T2-weighted MRI was shown to be non-inferior. Especially, the highly relevant CTV_{HR} width was not statistically significantly different and also standard deviations were in a comparable range. Moreover, 3D-TRUS seems to be superior to CT for CTV_{HR} assessment (Schmid et al., 2013, 2016). Since CT provides an excellent depiction of the applicator including interstitial needles and organs at risk (OARs), combinations of TRUS and CT appear to be an attractive alternative to MRI-based image-guided adaptive brachytherapy (see subsequent sections below). The use of Power Doppler additionally enables the visualization of the (residual) GTV. In fact, a recent prospective multicenter study demonstrated the high diagnostic accuracy of TRUS for early cervical cancer undergoing surgery with comparable to and even superior results to MRI (Epstein et al., 2013).

Beside cervical cancer, TRUS or transvaginal ultrasound (TVUS) can be helpful to assess the thickness of the vaginal wall for vaginal brachytherapy, for example, in case of adjuvant treatment of intermediate risk endometrial cancer. An example of measurement of the vaginal wall thickness with US is illustrated in Figure 14.11.

14.3.3 3D-TRUS Volume Acquisition with Brachytherapy Applicator *in situ*

TRUS imaging for brachytherapy treatment planning of cervical cancer requires both the target volume and the applicator to be identified with high precision in the acquired 3D

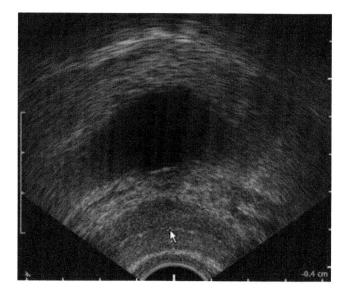

FIGURE 14.11 TRUS in a patient planned for adjuvant brachytherapy of the vagina. The thickness of the vaginal wall can be easily measured and used for dose prescription.

image volume. Volume acquisition can be done with similar stepper devices as used for prostate brachytherapy, although for gynecological applications the TRUS probes should ideally be longer than what is needed for prostate imaging. TRUS image acquisition can be done with transversal or sagittal arrays, by pullback or rotation of the probe. First attempts with an in-house developed system for TRUS-based cervix cancer brachytherapy (Nesvacil et al., 2016) indicated a superiority of volume acquisition by rotating the probe, as pulling it back can cause small motions of the applicator which are then observed as ripples in the resulting image set.

14.3.4 Applicator Reconstruction on TRUS

For image-guided cervix cancer brachytherapy, intracavitary applicators and interstitial needles made of plastic or titanium are typically used. A first series of tests with plastic tandem and ring applicators (Elekta Brachytherapy, Veenendaal, The Netherlands) showed excellent visibility of the tandem in both transabdominal and transrectal images. The ring applicator can usually not be depicted by transabdominal scans, but due to the TRUS probe's vicinity to the ring applicator, the posterior part can be well identified. For applicator reconstruction using 3D library models, it is necessary to define at least three points on the applicator with high precision. On TRUS images, the axis of the tandem and the cranial border of the posterior part of the ring can be identified in most cases, but artifacts of the applicator itself or other anatomical structures lead to a decreased precision compared to MRI-based applicator reconstruction, which is typically less than 1–2 mm. If interstitial needles are present, their depicted positions can be used to adjust the correct rotation of the intracavitary applicator, based on which holes have been occupied in the ring template. In a test series of consecutive patients, however, full applicator reconstruction with clinically acceptable precision could be achieved only for 1/5th of patients (Nesvacil et al., 2016). To overcome this limitation, future technological developments will therefore involve indirect determination of the applicator position, by tracking the applicator from the outside. This could be accomplished by electromagnetic tracking (see Chapter 4) or optical tracking (see Chapter 13) using markers placed on the TRUS probe and the applicator. Such an automated approach would facilitate and accelerate the overall procedure of applicator reconstruction and would further pave the way for online treatment planning in cervical cancer brachytherapy. First tests using optical tracking are currently ongoing at the Medical University of Vienna and indicate the high potential of this method.

14.3.5 TRUS for Treatment Planning

As target depiction with TRUS has been shown to be comparable to MRI, attempts are being made to use TRUS for brachytherapy treatment planning in settings where access to MR machines is limited. As the field of view for TRUS is however relatively small, combinations with other imaging modalities, in particular CT, could be applied to overcome this limitation. A combination of TRUS for target volume definition and CT for OARs definition as well as for treatment planning would allow the taking advantage of both imaging modalities and could serve as an alternative to MRI-based image-guided adaptive brachytherapy.

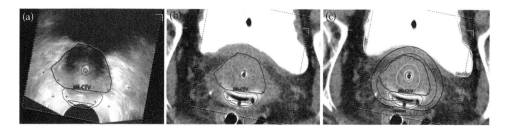

FIGURE 14.12 (a) Transverse 3D TRUS image with CTV_{HR} contoured in red and reconstructed tandem ring applicator, (b) transverse CT image with the transferred TRUS target volume via applicator-based image fusion, and (c) CT-based treatment planning including the target volume defined by 3D TRUS and the OARs defined by CT.

At first, the target volume is contoured on TRUS and the applicator position is assessed and reconstructed in the 3D-TRUS volume. If these (target and applicator) remain fixed to each other, for example, by thorough vaginal packing, the applicator position can be used for image registration with other 3D volumes acquired with the applicator *in situ*. A similar workflow has been presented for combining MRI and CT by Nesvacil et al. (2013) and could be equally applied for TRUS and CT imaging. This combined planning workflow would allow for a full treatment plan optimization to all involved tissue types, and overcome the low soft-tissue contrast provided by CT for target contouring (Figure 14.12). First comparisons of combined TRUS/CT treatment plans with MRI or CT-only treatment plans indicate the dosimetric superiority of the combined TRUS/CT workflow to CT-only based treatment planning as well as comparable results to the MRI-based treatment plan (Nesvacil et al., 2016). This is mainly due to a systematic overestimation of the target volume on CT, which has been reported multiple times in the literature.

14.4 US FOR BREAST BRACHYTHERAPY

The basic characteristics that make US a good real-time image guidance modality, of course, also applies to breast (Sehgal et al., 2006). US is part of the physicians' arsenal when it comes to breast cancer detection in particular for dense breasts, for which the sensitivity of mammography decreases (Kelly et al., 2010; Brem et al., 2015). A recent review indicates that the addition of US to mammography significantly increases cancer detection (Brem et al., 2015). It has been recognized for some time now that US is also an excellent modality for real-time image-guided biopsy procedures, where the needle can be continuously visualized due to the high refresh rate of this imaging technique as is illustrated in Figure 14.13 (Smith et al., 2000; Helbich et al., 2004). Furthermore, it was pointed-out early on that US-guided core breast biopsy was a highly cost-effective approach (Liberman et al., 1998). US-guidance of breast surgery (Harlow et al., 1999; Kaufman et al., 2003; Thompson and Klimberg, 2007; Krekel et al., 2013) as well as interventional cryoablation (Kaufman et al., 2002) and thermal therapy (van Esser et al., 2009) procedures are also commonly found in the literature.

It is interesting to find that the use of US to provide reference information of the lumpectomy cavity for radiation boost planning was investigated over 20 years ago (Leonard et al., 1993), when DeBiose et al. from William Beaumont Hospital demonstrated the use

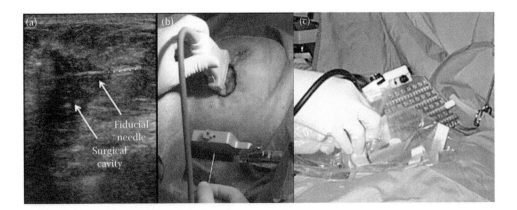

FIGURE 14.13   Examples of use of US real-time imaging capability for needle/catheter insertion guidance, for permanent seed implants ((a) and (b) From Pignol J-P et al. 2006. *Int J Radiat Oncol Biol Phys* 64, 176–81, Online: http://linkinghub.elsevier.com/retrieve/pii/S0360301605011624.), and template-based multi-catheter interstitial HDR breast brachytherapy (c).

of intraoperative US of lumpectomy cavity visualization and needle insertion guidance for interstitial breast brachytherapy (DeBiose et al., 1997). Breast-conserving surgery and accelerated partial breast irradiation (Arthur and Vicini, 2005; Patel et al., 2007; Polgár et al., 2010) have brought different approaches to breast brachytherapy in addition to the multi-catheters interstitial technique, such as breast balloon (intracavitary applicators) brachytherapy and more recently permanent seed implants (Pignol et al., 2006; Leonard and Limbergen, 2012). Freehand 2D ultrasound (2DUS) imaging is commonly used for these procedures to determine the optimal brachytherapy template positioning relative to the surgical cavity prior to an interstitial procedure, as shown in Figure 14.13, or helping in the positioning of the balloon applicators (Zannis et al., 2003; Dickler et al., 2005). In most clinical practice, 2DUS will also be used to guide the insertion of needles (or catheters) during the interstitial brachytherapy procedure. This is particularly true for the deepest row implanted if in close proximity to the chest wall in order to avoid puncturing the lung.

14.4.1 3D US Imaging

Most of the literature mentioned above deal with 2DUS and technologies that have not been designed specifically for radiation therapy or brachytherapy. It is, however, interesting to note that 3D ultrasound (3DUS) systems have slowly made their apparition in radiation oncology at the end of the 1990s and early 2000s (Fenster et al., 2001). In fact, prostate brachytherapy was one of the first radiation therapy applications of 3DUS using a standard TRUS probe mounted on a motorized stepper unit with precise encoders enabling full 3D image reconstruction from the oversampling of overlapping 2D images (Fenster et al., 1996). The resulting dataset can further be displayed in any arbitrary planes without image degradation. Another implementation, namely optical tracking of a regular 2D probe was also implemented and used for external beam radiation therapy (EBRT) guidance, including breast EBRT treatment guidance. A modern review of 2DUS and 3DUS guidance in EBRT has recently been published by Fontanarosa et al. (2015).

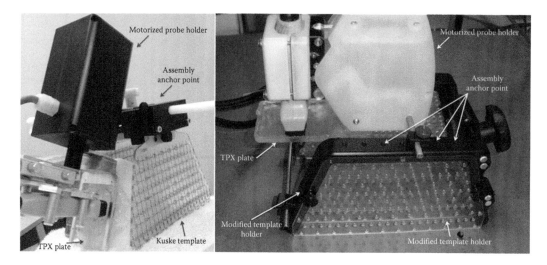

FIGURE 14.14 Left: early prototype of a computer-controlled motorized 3DUS guidance system for breast HDR brachytherapy. (From De Jean et al. 2009. *Med Phys* 36, 5099–5106.) Right: a modified version enabling greater flexibility for positioning the US motorized assembly on the template. (From Poulin E et al. 2015a. *Med Phys* 42, 6830–9.)

3DUS for needle insertion guidance in the breast first appears in relation to breast biopsies where phantom studies showed greater accuracy than freehand 2DUS guidance in terms of the ability for the operator to reach specific targets within the designated volume (Smith et al., 2001, 2002; Surry et al., 2002). The system proposed by Fenster et al. was further developed as an add-on to a mammography system where the biopsy needle was guided in between the device compression plate (Fenster et al., 2004).

Based on the concept behind the breast biopsy system, De Jean et al. designed a computer-controlled motorized 3DUS guidance system for breast high-dose rate (HDR) brachytherapy (De Jean et al., 2009). Figure 14.14 provides an overview of the system, which is based on a standard 2DUS breast probe that is mounted on a motorized probe holder. The particularity of this system is that it was designed as an attachment to a widely used interstitial breast brachytherapy template, the Kuske Template (Elekta Brachytherapy, Veenendaal, The Netherlands). The system was also designed such that the 2DUS probe could be moved to the insertion plane of a needle and to track its insertion in real time without further motion, similar to existing prostate brachytherapy systems (Beaulieu et al., 2007). One important issue in breast brachytherapy is the mobility of this external organ. Moving an US probe on the surface of the breast is likely to induce deformation and displacement. This is avoided in part by the use of an US compatible polymethylpentene (TPX™, Mitsui Chemicals America Inc., Rye Brook, New York, USA) plate. The plate is placed on the breast and the pressure adjusted to minimize the impact at depth. Once setup, the 2DUS probe is moved on this plate instead of the breast itself. This ensures stability of the organ shape and location.

The above-mentioned prototype was used for an in-house imaging-only study (unpublished). Some deficiencies were noted, in particular its limited positioning configuration on the template did not allow for imaging of more anterior surgical beds (Poulin et al., 2015a). A new system was built that integrated a combination of linear and tilt motion of the probe as well as enhanced freedom of position of the motorized probe holder from a redesign of the template attachments (right panel of Figure 14.14) (Poulin et al., 2015a). The probe holder further incorporates encoders that track both the motion of the probe and the angle of the TPX plate. Since the geometry of the template is fixed, the resulted 3DUS image volume is fully registered with the needle insertion axis and template-hole geometry. A full complement of software allows for registration, contouring, and insertion guidance. Poulin et al. further demonstrated various workflow hypothesis and end-to-end procedures (Poulin et al., 2015b), on phantom, using inverse planning that incorporates the optimization of the catheter numbers and their positions (Poulin et al., 2013). This system has yet to be utilized in a clinical trial. However, its design makes for an excellent basis for permanent breast seed implant procedures described by Pignol et al. (2006).

14.4.2 Next Generation US Imaging

The system described in the previous section unfortunately addresses only the interstitial component of breast brachytherapy. For balloon-based techniques, no advanced 3DUS systems are currently available. However, in practice, one could envision the use of existing 3DUS systems already used for breast EBRT (Fontanarosa et al., 2015) except that deformation and organ motion need to be taken into account. However, new technological opportunities might help mitigate this effect. On the one hand, 3DUS probes based on regular 2D probe technology but integrated mechanical motion within the probe are available and one has already been commercialized for prostate EBRT (Figure 14.15, left panel) (Lachaine and Falco, 2013). Another, more forward-looking approach is the new

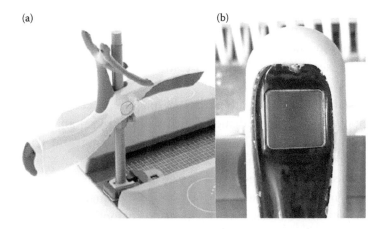

FIGURE 14.15 Example of new technologies that could be beneficial for breast brachytherapy in the form of a 3DUS probe where mechanical translation of the transducers are integrated within the probe assembly directly ((a) From Lachaine M and Falco T. 2013. *Med Phys Int* 1, 72–80) and of a matrix transducer array technology allowing the construction of large 2D transducer plane for direct volumetric imaging ((b) the X7-2t live 3D array composed of 2500 elements from Philips Medical Systems, USA).

2D matrix technology which packs an array of transducers for which beamforming and other parameters are fully controlled by software (Figure 14.15, right panel) (Wygant et al., 2008; Powers and Kremkau, 2011; Lachaine and Falco, 2013). With such technology, probe motion is not necessary to acquire 3D volumetric image sets and usually the repetition rate can be one or more sets per second.

14.5 CONCLUSION

US offers obvious advantages: high soft-tissue contrast, dynamic imaging, low cost, and high flexibility clearly qualify it for use in brachytherapy. For prostate brachytherapy, complete systems for image-guided applications and treatment planning are available and are used worldwide on a routine basis. The absence of such dedicated systems for other indications seems to currently hamper the use in daily clinical practice. However, increasing evidence indicates the high potential of US for, for example, gynecologic, breast, or anal brachytherapy and respective US systems, workflows, and solutions for brachytherapy are currently under development. Especially the possibility of intraoperative image-guidance is at the center of attention as it enables the precise performance of complex implantations. This outstanding attribute is still not fully exploited for brachytherapy and can clearly increase the scope of brachytherapy. Besides morphologic imaging, US provides a variety of functional imaging options. While those are more immediately associated with nuclear medicine, MRI, or even CT, there is a resurgence of interest for functional imaging with US. In prostate cancer, for example, much emphasis is placed nowadays on mpMRI to better identify and guide biopsy and therapy of suspected lesions (Mendez et al., 2015; Diaz de Leon et al., 2016). It turns out that mpUS may be around the corner with the promise of real-time multidimensional image sets in the form of Doppler and Power Doppler imaging, image texture and raw single analysis, CEUS, and shear wave elastography (SWE) (Postema et al., 2015b; Grey and Ahmed, 2016). Already, raw single analysis has given rise to a commercial product named HistoScanning™ (Advanced Medical Diagnostics s.a., Waterloo, Belgium) (Simmons et al., 2012; Hamann et al., 2013) and Doppler and Power Doppler are readily accessible on many US platforms, while CEUS and SWE are being actively investigated (Postema et al., 2015a).

It is further quite interesting to note that US imaging of cell death (Czarnota et al., 1997), namely apoptosis, has been demonstrated using spectral analysis of the radiofrequency signal (Kolios et al., 2002). The same group further demonstrated that such prowess could be achieved in the conventional clinical frequency range (5–12 MHz) (Sadeghi-Naini et al., 2013a). This technique has been used to study the effect of radiation therapy *in vitro* (Vlad et al., 2008) and more recently for the noninvasive quantitative assessment of the response of locally advanced breast cancer patients to chemotherapy (Sadeghi-Naini et al., 2013b). As such, *US based-spectroscopy* might make a worthy addition to mpUS (Willmann et al., 2008; Kolios and Czarnota 2009).

REFERENCES

Arthur D W and Vicini F A. 2005. Accelerated partial breast irradiation as a part of breast conservation therapy. *J Clin Oncol* 23, 1726–35.

Barrois G, Coron A, Patyen T, Dizeux A and Bridal L. 2013. A multiplicative model for improving microvascular flow estimation in dynamic contrast-enhanced ultrasound (DCE-US): Theory and experimental validation. *IEEE Trans Ultrason Ferroelectr Freq Control* 60, 2284–94.

Beaulieu L, Evans D, Aubin S, Angyalfi S, Husain S, Kay I, Martin A, Varfalvy N, Vigneault E and Dunscombe P. 2007. Bypassing the learning curve in permanent seed implants using state-of-the-art technology. *Int J Radiat Oncol Biol Phy* 67, 71–7.

Brem R F, Lenihan M J, Lieberman J and Torrente J. 2015. Screening breast ultrasound: Past, present, and future. *Am J Roentgenol* 204, 234–40.

Cosgrove D. 2003. Angiogenisis imaging-ultrasound. *Br J Radiol* 76(1), S43–9.

Cosgrove D and Lassau N. 2010. Imaging perfusion using ultrasound. *Eur J Nucl Med Mol Imaging* 37(Suppl 1), S65–85.

Cosset JM, Cathelineau X, Wakil G, Pierrat N, Quenzer O, Prapotnich D, Barret E, Rozet F, Galiano M and Vallancien G. 2013. Focal brachytherapy for selected low-risk prostate cancers: A pilot study. *Brachytherapy* 12, 331–7.

Czarnota G J, Kolios M C, Vaziri H, Benchimol S, Ottensmeyer F P, Sherar M D and Hunt J W. 1997. Ultrasonic biomicroscopy of viable, dead and apoptotic cells. *Ultrasound Med Biol* 23, 961–5.

De Jean P, Beaulieu L and Fenster A. 2009. Three-dimensional ultrasound system for guided breast brachytherapy. *Med Phys* 36, 5099–106.

DeBiose D A, Horwitz E M, Martinez A A, Edmundson G K, Chen P Y, Gustafson G S, Madrazo B, Wimbish K, Mele E and Vicini F A. 1997. The use of ultrasonography in the localization of the lumpectomy cavity for interstitial brachytherapy of the breast. *Int J Radiat Oncol Biol Phys* 38, 755–9.

Diaz de Leon A, Costa D and Pedrosa I. 2016. Role of multiparametric MR imaging in malignancies of the urogenital tract. *Magn Reson Imaging Clin N Am* 24, 187–204.

Dickler A, Kirk M C, Chu J and Nguyen C. 2005. The MammoSite breast brachytherapy applicator: A review of technique and outcomes. *Brachytherapy* 4, 130–6.

Epstein E, Testa A, Gaurilcikas A et al. 2013. Early-stage cervical cancer: Tumor delineation by magnetic resonance imaging and ultrasound—A European multicenter trial. *Gynecol Oncol* 128(3), 449–53.

Fenster A, Downey D B and Cardinal H N. 2001. Three-dimensional ultrasound imaging. *Phys Med Biol* 46, R67–99.

Fenster A, Surry K J M, Mills G R and Downey D B. 2004. 3D ultrasound guided breast biopsy system. *Ultrasonics* 42, 769–74.

Fontanarosa D, Van Der Meer S, Bamber J, Harris E, O'Shea T and Verhaegen F. 2015. Review of ultrasound image guidance in external beam radiotherapy: I. Treatment planning and interfraction motion management. *Phys Med Biol* 60, R77–114.

Grey A and Ahmed H U. 2016. Multiparametric ultrasound in the diagnosis of prostate cancer. *Curr Opin Urol* 26, 114–9.

Hamann M F, Hamann C, Schenk E, Al-Najar A, Naumann C M and Jünemann K-P. 2013. Computer-aided (HistoScanning) biopsies versus conventional transrectal ultrasound-guided prostate biopsies: Do targeted biopsy schemes improve the cancer detection rate? *Urology* 81, 370–5.

Harlow S P, Krag D N, Ames S E and Weaver D L. 1999. Intraoperative ultrasound localization to guide surgical excision of nonpalpable breast carcinoma. *J Am Coll Surg* 189, 241–6.

Helbich T H, Matzek W and Fuchsjäger M H. 2004. Stereotactic and ultrasound-guided breast biopsy. *Eur Radiol* 14, 383–93.

Hudson J M, Karshafian R and Burns P N. 2009. Quantification of flow using ultrasound and microbubbles: A disruption replenishment model based on physical principles. *Ultrasound Med Biol* 35, 2007–20.

Kaufman C S, Bachman B, Littrup P J, White M, Carolin K A, Freman-Gibb L, Francescatti D et al. 2002. Office-based ultrasound-guided cryoablation of breast fibroadenomas. *Am J Surg* 184, 394–400.

Kaufman C S, Jacobson L, Bachman B and Kaufman L B. 2003. Intraoperative ultrasonography guidance is accurate and efficient according to results in 100 breast cancer patients. *Am J Surg* 186, 378–82.

Kazi A, Godwin G, Simpson J and Sasso G 2010. MRS-guided HDR brachytherapy boost to the dominant intraprostatic lesion in high risk localised prostate cancer. *BMC Cancer* 10, 472.

Kelly K M, Dean J, Comulada W S and Lee S-J. 2010. Breast cancer detection using automated whole breast ultrasound and mammography in radiographically dense breasts. *Eur Radiol* 20, 734–42.

Kolios M C and Czarnota G J. 2009. Potential use of ultrasound for the detection of cell changes in cancer treatment. *Future Oncol* 5, 1527–32.

Kolios M C, Czarnota G J, Lee M and Hunt J W. 2002. Ultrasonic spectral parameter characterization of apoptosis. *Ultrasound Med Biol* 28, 589–97.

Krekel N, Haloua M H, Cardozo A and de Wit R H. 2013. Intraoperative ultrasound guidance for palpable breast cancer excision (COBALT trial): A multicentre, randomised controlled trial. *Lancet Oncol* 14, 48–54.

Kuenen M P J, Mischi M and Wijkstra H. 2011. Contrast-ultrasound diffusion imaging for localization of prostate cancer. *IEEE Trans Med Imaging* 30, 1493–502.

Kuenen M P J, Saidov T A, Wijkstra H, de la Rosette J J M C H and Mischi M. 2013. Spatiotemporal correlation of ultrasound contrast agent dilution curves for angiogenesis localization by dispersion imaging. *IEEE Trans Ultrason Ferroelectr Freq Control* 60, 2665–9.

Lachaine M and Falco T. 2013. Intrafractional prostate motion management with the Clarity Autoscan system. *Med Phys Int* 1, 72–80.

Leonard C, Harlow C L, Coffin C and Dross J. 1993. Use of ultrasound to guide radiation boost planning following lumpectomy for carcinoma of the breast. *Int J Radiat Oncol Biol Phy* 27, 1193–7.

Leonard K L and Limbergen E V. 2012. Brachytherapy for breast cancer. In: Venselaar, J, Meigooni, A S, Baltas, D and Hoskin, P J (eds.) *Comprehensive Brachytherapy: Physical and Clinical Aspects*, Chapter 22, Taylor & Francis, London, pp. 319–31.

Liberman L, Feng T L, Dershaw D D, Morris E A and Abramson A F. 1998. US-guided core breast biopsy: Use and cost-effectiveness *Radiology* 208, 717–23.

Mahantshetty U, Khanna N, Swamidas J et al. 2012. Trans-abdominal ultrasound (US) and magnetic resonance imaging (MRI) correlation for conformal intracavitary brachytherapy in carcinoma of the uterine cervix. *Radiother Oncol* 102(1), 130–4.

Mendez M H, Joh D Y, Gupta R and Polascik T J. 2015. Current trends and new frontiers in focal therapy for localized prostate cancer. *Curr Urol Rep* 16, 35.

Mischi M, Kuenen M P J and Wijkstra. H 2012. Angiogenesis imaging by spatio-temporal analysis of ultrasound-contrast dispersion kinetics. *IEEE Trans Ultrason Ferroelectr Freq Control* 59, 621–9.

Narayan K, van Dyk S, Bernshaw D et al. 2014. Ultrasound guided conformal brachytherapy of cervix cancer: Survival, patterns of failure, and late complications. *J Gynecol Oncol* 25, 206–13.

Nesvacil N, Pötter R, Sturdza A, Hegazy N, Federico M and Kirisits C. 2013. Adaptive image guided brachytherapy for cervical cancer: A combined MRI-/CT-planning technique with MRI only at first fraction. *Radiother Oncol* 107, 75–81.

Nesvacil N, Schmid M P, Pötter R et al. 2016. Combining transrectal ultrasound and CT for image-guided adaptive brachytherapy of cervical cancer: Proof of concept. *Brachytherapy*. 15(16), 30538–4. doi: 10.1016/j.brachy.2016.08.009.

Nguyen P L, Chen M H, Zhang Y, Tempany C M, Cormack R A, Beard C J, Hurwitz M D, Suh W W and D'Amico A V. 2012. Updated results of magnetic resonance imaging guided partial prostate brachytherapy for favorable risk prostate cancer: Implications for focal therapy. *J Urol* 188, 1151–6.

Patel R R, Becker S J, Das R K and Mackie T R. 2007. A dosimetric comparison of accelerated partial breast irradiation techniques: Multicatheter interstitial brachytherapy, three-dimensional conformal radiotherapy, and supine versus prone helical tomotherapy. *Int J Radiat Oncol Biol Phys* 68, 935–42.

Pieters B R, Wijkstra H, van Herk M, Kuipers R, Kaljouw E, de la Rosette J and Koning C. 2012. Contrast-enhanced ultrasound as support for prostate brachytherapy treatment planning. *J Contemp Brachytherapy* 4, 67–74.

Pignol J-P, Keller B, Rakovitch E, Sankreacha R, Easton H and Que W. 2006. First report of a permanent breast 103Pd seed implant as adjuvant radiation treatment for early-stage breast cancer. *Int J Radiat Oncol Biol Phys* 64, 176–81, Online: http://linkinghub.elsevier.com/retrieve/pii/S0360301605011624.

Polgár C, van Limbergen E, Pötter R, Kovács G, Polo A, Lyczek J, Hildebrandt G, Niehoff P, Guinot J L et al. and Strnad V GEC-ESTRO Breast Cancer Working Group. 2010. Patient selection for accelerated partial-breast irradiation (APBI) after breast-conserving surgery: Recommendations of the Groupe Européen de Curiethérapie-European Society for Therapeutic Radiology and Oncology (GEC-ESTRO) breast cancer working group based on clinical evidence (2009). *Radiother Oncol* 94, 264–73.

Postema A W, Frinking P J, Smeenge M, de Reijke T M and de la Rosette J J C M. 2015. Dynamic contrast-enhanced ultrasound parametric imaging for the detection of prostate cancer. *BJU Int* Doi: 10.1111/bju.13116.

Postema A, Idzenga T, Mischi M, Frinking P, la Rosette de J and Wijkstra H. 2015a. Ultrasound modalities and quantification: Developments of multiparametric ultrasonography, a new modality to detect, localize and target prostatic tumors. *Curr Opin Urol* 25, 191–7.

Postema A, Mischi M, la Rosette de J and Wijkstra H. 2015b. Multiparametric ultrasound in the detection of prostate cancer: A systematic review. *World J Urol* 33, 1651–9.

Poulin E, Fekete C-A C, Létourneau M, Fenster A, Pouliot J and Beaulieu L. 2013. Adaptation of the CVT algorithm for catheter optimization in high dose rate brachytherapy. *Med Phys* 40, 111724.

Poulin E, Gardi L, Barker K, Montreuil J, Fenster A and Beaulieu L. 2015a. Validation of a novel robot-assisted 3DUS system for real-time planning and guidance of breast interstitial HDR brachytherapy. *Med Phys* 42, 6830–9.

Poulin E, Gardi L, Fenster A, Pouliot J and Beaulieu L. 2015b. Towards real-time 3D ultrasound planning and personalized 3D printing for breast HDR brachytherapy treatment. *Radiother Oncol* 114, 335–8.

Powers J and Kremkau F. 2011. Medical ultrasound systems. *Interface Focus* 1, 477–89.

Rao P B and Ghosh S. 2015. Routine use of ultrasound guided tandem placement in intracavitary brachytherapy for the treatment of cervical cancer—A South Indian institutional experience. *J Contemp Brachytherapy* 7, 352–6.

Russo G, Mischi M, Scheepens W, de la Rosette J J M C H and Wijkstra H. 2012. Angiogenesis in prostate cancer: Onset, progression and imaging. *BJU Int* 110, E794–808.

Sadeghi-Naini A, Papanicolau N and Falou O. 2013a. Low-frequency quantitative ultrasound imaging of cell death in vivo. *Med Phys* 40(8), 082901.

Sadeghi-Naini A, Papanicolau N, Falou O, Zubovits J, Dent R, Verma S, Trudeau M et al. 2013b. Quantitative ultrasound evaluation of tumor cell death response in locally advanced breast cancer patients receiving chemotherapy. *Clin Cancer Res* 19, 2163–74.

Sano F, Terao H, Kawahara T, Miyoshi Y, Sasaki T, Noguchi K, Kubota Y and Uemura H. 2011. Contrast-enhanced ultrasonography of the prostate: Various imaging findings that indicate prostate cancer. *BJU Int* 107, 1404–10.

Schick U, Popowski Y, Nouet P, Bieri S, Rouzaud M, Khan H, Weber D C and Miralbell R. 2011. High-dose-rate brachytherapy boost to the dominant intra-prostatic tumor region: Hemi-irradiation of prostate cancer. *Prostate* 71, 1309–16.

Schmid M P, Nesvacil N, Pötter R et al. 2016. Transrectal ultrasound for image-guided adaptive brachytherapy in cervix cancer—An alternative to MRI for target definition? *Radiother Oncol* 120(3), 467–72.

Schmid M P, Pötter R, Brader P et al. 2013. Feasibility of transrectal ultrasonography in the assessment of locally advanced cervical cancer in the course of primary radiochemotherapy. *Strahlenther Onkol* 189, 123–8.

Sehgal C M, Weinstein S P, Arger P H and Conant E F. 2006. A review of breast ultrasound. *J Mammary Gland Biol Neoplasia* 11, 113–23.

Seitz M, Gratzke C, Schlenker B, Buchner A, Karl A, Roosen A, Singer B B et al. 2011. Contrast-enhanced transrectal ultrasound (CE-TRUS) with cadence-contrast pulse sequence (CPS) technology for the identification of prostate cancer. *Urol Oncol* 29, 295–301.

Sharma D N, Rath G K, Thulkar S et al. 2010. Use of transrecal ultrasound for high dose rate interstitial brachytherapy for patients of carcinoma of uterine cervix. *J Gynecol Oncol* 21, 12–17.

Simmons L A M, Autier P, Záťura F, Braeckman J, Peltier A, Romic I, Stenzl A et al. 2012. Detection, localisation and characterisation of prostate cancer by prostate HistoScanning.™ *BJU Int* 110, 28–35.

Smith L F, Rubio I T, Henry-Tillman R, Korourian S and Klimberg V S. 2000. Intraoperative ultrasound guided breast biopsy. *Am J Surg* 180, 419–23.

Smith W L, Surry K J, Kumar A, McCurdy L, Downey D B and Fenster A. 2002. Comparison of core needle breast biopsy techniques: Freehand versus three-dimensional US guidance. *Acad Radiol* 9, 541–50.

Smith W L, Surry K J, Mills G R, Downey D B and Fenster A. 2001. Three-dimensional ultrasound-guided core needle breast biopsy. *Ultrasound Med Biol* 27, 1025–34.

Stock R G, Chan K, Terk M et al. 1997. A new technique for performing Syed–Neblett template interstitial implants for gynecologic malignancies using transrectal-ultrasound guidance. *Int J Radiat Oncol Biol Phys* 37, 819–25.

Surry K J, Smith W L, Campbell L J, Mills G R, Downey D B and Fenster A. 2002. The development and evaluation of a three-dimensional ultrasound-guided breast biopsy apparatus. *Med Image Anal* 6, 301–12.

Tharavichitkul E, Tippanya D, Jayavasti R et al. 2015. Two-years of transabdominal ultrasound guided brachytherapy for cervical cancer. *Brachytherapy* 2, 238–44.

Thompson M and Klimberg V S. 2007. Use of ultrasound in breast surgery. *Surg Clin North Am* 87, 469–84–x.

Tong S, Downey D B, Cardinal H N and Fenster A. 1996. A three-dimensional ultrasound prostate imaging system. *Ultrasound Med Biol* 22, 735–46.

Turkbey B, Mani H, Shah V, Rastinehad A R, Bernardo M, Pohida T, Pang Y et al. 2011. Multiparametric 3 T prostate magnetic resonance imaging to detect cancer: Histopathological correlation using prostatectomy specimens processed in customized magnetic resonance imaging based molds. *J Urol* 186, 1818–24.

Van Dyk S, Narayan K, Fisher R and Bernshaw D. 2009. Conformal brachytherapy planning for cervical cancer using transabdominal ultrasound. *Int J Radiat Oncol Biol Phys* 75, 64–70.

van Esser S, Stapper G, van Diest P J, van den Bosch M A A J, Klaessens J H G M, Mali W P T M, Borel Rinkes I H M and van Hillegersberg R. 2009. Ultrasound-guided laser-induced thermal therapy for small palpable invasive breast carcinomas: A feasibility study. *Ann Surg Oncol* 16, 2259–63.

Vlad R M, Alajez N M, Giles A, Kolios M C and Czarnota G J. 2008. Quantitative ultrasound characterization of cancer radiotherapy effects in vitro. *Int J Radiat Oncol Biol Phys* 72, 1236–43.

Willmann J K, van Bruggen N, Dinkelborg L M and Gambhir S S. 2008. Molecular imaging in drug development. *Nat Rev Drug Discov* 7, 591–607.

Wygant I O, Zhuang X, Yeh D T, Oralkan O, Ergun A S, Karaman M and Khuri-Yakub B T. 2008. Integration of 2D CMUT arrays with front-end electronics for volumetric ultrasound imaging. *IEEE Trans Ultrason Ferroelectr Freq Control* 55, 327–42.

Zannis V J, Walker L C, Barclay-White B and Quiet C A. 2003. Postoperative ultrasound-guided percutaneous placement of a new breast brachytherapy balloon catheter. *Am J Surg* 186, 383–5.

CHAPTER 15

X-Ray and Computed Tomography

Martin T. King and Michael J. Zelefsky

CONTENTS

15.1 INTRODUCTION

Treatment planning in brachytherapy requires accurate imaging of both the implanted materials and the surrounding anatomy. Ultrasound (US) and magnetic resonance imaging (MRI), which can provide excellent soft-tissue contrast for visualizing anatomy, can be limited in depicting radioactive seeds or interstitial needles. X-ray and computed tomography (CT), however, can provide excellent visualization of most implanted materials and electron density information (Carey, 2013). The purpose of this chapter is to provide an overview of the use of x-ray and CT-based imaging for brachytherapy.

15.2 PROSTATE

Transrectal ultrasound (TRUS) guidance is the main imaging modality for modern prostate brachytherapy. TRUS allows for excellent anatomical visualization and real-time intraoperative feedback. Needle insertion, radioactive seed (^{125}I or ^{103}Pd) deposition, and treatment planning can be performed completely under TRUS guidance. However, both x-ray and CT continue to play important roles for intraoperative planning and postimplant dosimetry.

15.2.1 Intraoperative Planning for Permanent Prostate Seed Implants

Since the 1990s, intraoperative treatment planning has emerged as an increasingly viable alternative to preplanning. During intraoperative planning, prostate imaging, treatment planning, and radioactive seed implantation are performed within a single session (Zelefsky et al., 2003). TRUS has remained the primary modality for intraoperative imaging. However, artifacts from hyperechogeneic blood, prostate calcifications, and surrounding needles may affect the accurate localization of individual needles on TRUS. Furthermore, TRUS has marked limitations in identifying deposited seeds. An observer study showed that only 74% of implanted seeds on TRUS were correctly identified (Han et al., 2003). Furthermore, at least 95% of seeds need to be accurately identified for reliable dosimetric analysis (Su et al., 2005).

Over the past couple of decades, brachytherapists have been interested in real-time planning techniques, which can account for intraoperative changes in prostate dosimetry due to seed migration, prostate edema, needle positioning error, and hemorrhage. A critical component of real-time planning, termed dynamic dose calculation by the American Brachytherapy Society (ABS), involves knowledge of the actual locations of deposited seeds for real-time feedback (Nag et al., 2001; Polo et al., 2010). Although real-time planning can be performed under TRUS guidance with careful identification of deposited seeds from a Mick® applicator (Meijer et al., 2006), advanced techniques utilizing x-ray and CT may provide more robust dosimetric assessments.

The basic premise of x-ray fluoroscopy for intraoperative dose assessment is that the three-dimensional (3D) source distribution of radioactive seeds can be reliably reconstructed from multiple fluoroscopic images obtained at different angles (Narayanan et al., 2002; Lee and Zaider, 2003). Isodose lines from the fluoroscopy images can be visualized, and additional seeds can be added if regions of the prostate appear underdosed. In a series of 26 patients, intraoperative fluoroscopy alone was shown to improve prostate dosimetry (Reed et al., 2005). Methods for intraoperative dosimetry utilizing TRUS–fluoroscopy fusion have also been developed (Todor et al., 2003; French et al., 2005). TRUS–fluoroscopy fusion allows for isodose lines generated from the fluoroscopic images to be overlaid onto the TRUS images. This technique was shown to improve prostate dosimetry in a series of 16 patients (Orio et al., 2007).

Within the past several years, investigators have also utilized CT-based planning for intraoperative dosimetry. Intraoperative CT-based planning is most often performed with kilovoltage cone-beam CT (CBCT) scanners (Jaffray et al., 2002). Although the image quality of CBCT is inferior to that of CT largely due to increased scatter, CBCT is more readily available than dedicated CT-on-rail systems within the operating room. The basic workflow is similar to that for fluoroscopic-based intraoperative dosimetry. Isodose lines generated from deposited seeds within the CBCT image are overlaid onto the TRUS image. The physician then adds seeds to underdosed areas of the prostate. In one report, a C-arm system was used to acquire an intraoperative CBCT scan with the patient in the supine position. Underdosage was identified in 9 of 20 patients. Four additional seeds were implanted on average, and dosimetry improved (Westendorp et al., 2007). In another report, the intraoperative CBCT scan was obtained with an O-arm system. Since the O-arm system

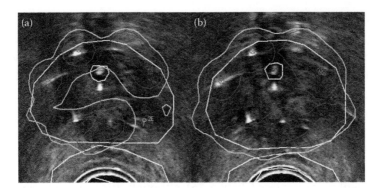

FIGURE 15.1   Real-time intraoperative CBCT dosimetry superimposed onto TRUS images for a permanent prostate implant with the patient in the dorsal lithotomy position. (a) TRUS image showing area of decreased coverage within the left posterior midgland-apex peripheral zone. (b) Subsequent TRUS image showing improved coverage of left posterior midgland-apex peripheral zone after implantation of additional seeds. Isodose lines (percent prescription): light gray—150%; dark gray—120%; white (thick)—100%.

has a large bore (~96 cm), the CBCT scans could be acquired with the patient in the dorsal lithotomy position with the TRUS probe in place. As a result, the prostate deformation from the TRUS probe could be accounted for in the isodose lines generated from the CBCT scan (Zelefsky et al., 2010). Figure 15.1 shows an example of real-time intraoperative dosimetry generated from TRUS–CBCT fusion images.

15.2.2 Postimplant Dosimetry for Permanent Prostate Seed Implants

Postimplant dosimetry is an essential component for evaluating the quality of permanent prostate seed implants, given the correlation between dosimetric outcomes and important biochemical endpoints (Stock et al., 1998; Zelefsky et al., 2012). CT scans have been uniformly utilized for postimplant dosimetry. Deposited seeds can be easily identified, often with computerized algorithms. Furthermore, surrounding critical structures including the rectum and bladder can be delineated (Davis et al., 2012). However, a key disadvantage of postimplant CT is that the prostate capsule is not easily discernible, leading to interobserver variability in contouring, volume estimation (with a tendency for overestimation), and dosimetric indices (De Brabandere et al., 2012). Although MRI scans can provide more reliable visualization of the prostate capsule, especially at the base and apex (Smith et al., 2007), precise seed localization has remained a challenge due to signal voids (Thomas et al., 2009). CT–MRI fusion with accurate image registration can provide both excellent source localization and prostate capsule delineation. Studies have demonstrated that CT–MRI fusion can lead to improved postimplant dosimetry, and allow for accurate assessment of intraprostatic edema after implantation (Polo et al., 2004; Crook et al., 2004).

15.2.3 High-Dose Rate Brachytherapy

In high-dose rate (HDR) brachytherapy, the prostate is irradiated by a radioactive source (typically ^{192}Ir) through interstitial catheters placed under TRUS guidance within the prostate gland. In early reports, HDR treatment planning was performed with intra-operative US (Martinez et al., 2001) and orthogonal x-ray films (Demanes et al., 2000). Subsequently, treatment planning with CT gained traction due to improved 3D visualization of the needles (Hsu et al., 2005). Recently, CBCT has been utilized for determining the magnitude of catheter displacement between CT planning and treatment (Holly et al., 2011). Another group incorporated CBCT with TRUS in the intraoperative setting for treatment planning. CBCT eliminated the need to transport the patient for a separate CT scan, and also reduced the uncertainty in identifying needle tips from TRUS images alone (Even et al., 2014). MRI has also been utilized for guiding the insertion of HDR catheters and treatment planning (see Chapter 16), although this technique can be more time-consuming than US and CT-based planning methods (Ménard et al., 2004).

15.3 GYNECOLOGY

15.3.1 Intracavitary Brachytherapy

For cervical brachytherapy treatment planning after applicator insertion, x-ray imaging is limited due to its inability to accurately image either the cervix or the surrounding organs. CT imaging has fundamentally changed brachytherapy treatment planning, since the cervix, bladder, and rectum can be delineated for each individual patient. Dose volume histogram (DVH) analysis has shown that CT-based planning allows for better target coverage and conformality than prescription to Point A (Shin et al., 2006). Furthermore, International Commission on Radiation Units and Measurements (ICRU) bladder and rectal points may underestimate maximal bladder and rectal doses (Pelloski et al., 2005). Recently, a prospective multicenter study demonstrated that patients who underwent 3D-based dosimetry, predominantly with CT, had improved local control and decreased toxicity compared to those who were treated with two-dimensional (2D) techniques (Charra-Brunaud et al., 2012). An added benefit is that CT-based imaging allows for the detection of uterine perforation, which is a serious complication often not easily appreciated on conventional x-ray (Barnes et al., 2007).

The primary limitation of CT is that the boundaries of the cervix are not clearly discernible, since the cervix is often isointense to both the parametria as well as the uterus. MRI, on the other hand, can provide the soft-tissue contrast necessary for accurately outlining the cervix (Haie-Meder et al., 2005). A prospective study demonstrated that, although the rectum and bladder contours were similar between CT and MRI, the cervix was overcontoured on CT, especially in the lateral extent (Viswanathan et al., 2007). Furthermore, in a contouring study involving 23 experts in gynecologic radiation oncology, MRI volumes were smaller than CT volumes, especially for cervical cancers with parametrial extension that demonstrated a radiographic response to treatment. However, this study also showed improved interobserver agreement for CT-based rather than MRI-based contours (Viswanathan et al., 2014). Despite the limitations in delineating cervical target volumes on CT, the majority of cervical brachytherapy procedures are performed using this modality, largely due to its availability in most radiation oncology departments (Viswanathan and Erickson, 2010). To overcome the problem of overcontouring on CT and the limited availability of MRI, methods for fusing CT with US images are being developed. In Chapter 14, the developments in US are discussed in-depth.

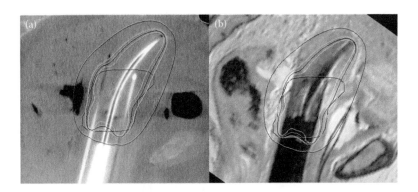

FIGURE 15.2 CT–MRI fusion for interstitial brachytherapy treatment planning of a cervical lesion after external beam radiation therapy. (a) Sagittal CT image showing clear delineation of the interstitial needles and tandem for dosimetric purposes. (b) Sagittal MRI with soft-tissue delineation for delineating the cervical target volume (red line). Isodose lines (percent prescription): yellow—150%; blue—120%; green (thick)—100%; magenta—50%.

15.3.2 Interstitial Brachytherapy

Interstitial brachytherapy is another option for treating gynecologic cancers, especially bulky cervical or vaginal cancers (Viswanathan and Thomadsen, 2012; Beriwal et al., 2012). Interstitial needles are inserted into the tumor through a perineal template. A tandem can also be utilized in order to provide dosimetric coverage to the upper cervical or uterine cavity if necessary. X-ray and CT can be utilized for intraoperative guidance. X-ray fluoroscopy can ensure that needles are sufficiently parallel in order to aid in radiation treatment planning (Nag et al., 1998). Intraoperative CT allows for serial adjustments in needle depth in order to ensure adequate tumor coverage while minimizing the risk of penetration into the surrounding bladder or rectum (Lee et al., 2013). US (Stock et al., 1997) and MRI (Viswanathan et al., 2013) have also been incorporated into the intraoperative setting.

After needle insertion, treatment planning is typically performed with CT, which facilitates the delineation of surrounding organs at risk (OAR), including the bladder and rectum, the identification of individual needles, as well as the generation of an optimal 3D-based treatment plan (Beriwal et al., 2012). MRI has also been incorporated into the treatment planning process, due to the excellent soft-tissue visualization provided by this modality (Viswanathan et al., 2013). At our institution, treatment planning is performed with CT–MRI fusion, which can provide excellent visualization of both the soft tissues and interstitial needles. Figure 15.2 depicts an example of CT–MRI fusion for interstitial brachytherapy treatment planning of a cervical lesion.

15.4 OTHER SITES

X-ray and CT have played a critical role in brachytherapy for other disease sites. For head-and-neck cancers such as the oral tongue or floor-of-mouth, interstitial needles are inserted through the submandibular space and into the tumor, often under fluoroscopic

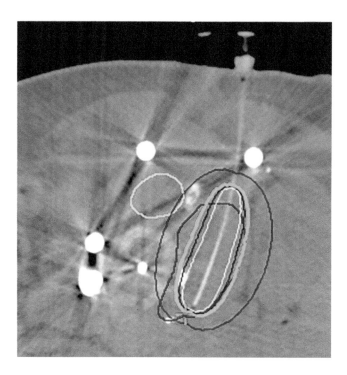

FIGURE 15.3   Axial CBCT scan acquired for HDR brachytherapy treatment planning for a right paravertebral metastasis at L3. The patient had undergone prior decompressive surgery and two prior courses of radiation therapy to the lumbar region. Contours: red—clinical target volume; light blue—spinal cord. Isodose lines (percent prescription): yellow—150%; blue—120%; green (thick)—100%; magenta—50%.

x-ray guidance (Nag et al., 2001). In accelerated partial breast irradiation for breast cancer, interstitial catheters or intracavitary applicators can be inserted into the seroma cavity after the lumpectomy is performed (Shah et al., 2013). For sarcoma, interstitial catheters can be positioned in the intraoperative bed after tumor resection (Holloway et al., 2013). For all these disease sites, a CT scan is then acquired after applicator insertion. Treatment planning is then performed prior to irradiation with a HDR source.

X-ray and CT have also been incorporated into joint brachytherapy procedures between radiation oncology and interventional radiology. For liver malignancies, an interventional radiologist can insert multiple catheters into the tumor under x-ray and CT guidance. Radiation treatment planning is then performed on CT or MRI prior to irradiation with an 192Ir source (Ricke and Wust, 2011). At our institution, HDR brachytherapy is a treatment option for patients with recurrent spinal malignancies that have been previously irradiated. An interventional radiologist can insert needles into the spinal tumors under x-ray and CT-fluoroscopic guidance with the patient in the prone position. A CBCT is then acquired for radiation treatment planning, such as that shown in Figure 15.3 (Folkert et al., 2013). Both types of procedures can yield highly conformal treatment plans with minimal irradiation of critical organs.

15.5 CONCLUSION

X-ray and CT imaging continue to play an integral role in brachytherapy. These modalities provide excellent visualization of implanted sources/catheters needed for accurate treatment planning, and can aid in the intraoperative placement of brachytherapy applicators. Although MRI does provide the best soft tissue contrast for tumor delineation, CT remains the most widely available modality for treatment planning. As such, advances in CT imaging for brachytherapy can yield profound benefits for the field. Further advances in multimodality image registration and the incorporation of advanced navigation systems will expand the usability of x-ray and CT for brachytherapy.

REFERENCES

Barnes E a., Thomas G, Ackerman I et al. Prospective comparison of clinical and computed tomography assessment in detecting uterine perforation with intracavitary brachytherapy for carcinoma of the cervix. *Int J Gynecol Cancer*. 2007;17(4):821–826.

Beriwal S, Demanes DJ, Erickson B et al. American Brachytherapy Society consensus guidelines for interstitial brachytherapy for vaginal cancer. *Brachytherapy*. 2012;11(1):68–75.

Carey BM. Imaging for post-implant dosimetry. In: Kovács G, Hoskin P, eds. *Interstitial Prostate Brachytherapy*. Berlin, Heidelberg: Springer, 2013:119–140.

Charra-Brunaud C, Harter V, Delannes M et al. Impact of 3D image-based PDR brachytherapy on outcome of patients treated for cervix carcinoma in France: Results of the French STIC prospective study. *Radiother Oncol*. 2012;103(3):305–313.

Crook J, McLean M, Yeung I, Williams T, Lockwood G. MRI-CT fusion to assess postbrachytherapy prostate volume and the effects of prolonged edema on dosimetry following transperineal interstitial permanent prostate brachytherapy. *Brachytherapy*. 2004;3(2):55–60.

Davis BJ, Horwitz EM, Lee WR et al. American Brachytherapy Society consensus guidelines for transrectal ultrasound-guided permanent prostate brachytherapy. *Brachytherapy*. 2012;11(1):6–19.

De Brabandere M, Hoskin P, Haustermans K, Van den Heuvel F, Siebert F-A. Prostate post-implant dosimetry: Interobserver variability in seed localisation, contouring and fusion. *Radiother Oncol*. 2012;104(2):192–198.

Demanes DJ, Rodriguez RR, Altieri GA. High dose rate prostate brachytherapy: The California Endocurietherapy (CET) method. *Radiother Oncol*. 2000;57(3):289–296.

Even AJG, Nuver TT, Westendorp H, Hoekstra CJ, Slump CH, Minken AW. High-dose-rate prostate brachytherapy based on registered transrectal ultrasound and in-room cone-beam CT images. *Brachytherapy*. 2014;13(2):128–136.

Folkert MR, Bilsky MH, Cohen GN et al. Intraoperative and percutaneous iridium-192 high-dose-rate brachytherapy for previously irradiated lesions of the spine. *Brachytherapy*. 2013;12(5):449–456.

French D, Morris J, Keyes M, Goksel O, Salcudean S. Computing intraoperative dosimetry for prostate brachytherapy using TRUS and fluoroscopy1. *Acad Radiol*. 2005;12(10):1262–1272.

Haie-Meder C, Pötter R, Van Limbergen E et al. Recommendations from Gynaecological (GYN) GEC-ESTRO Working Group★ (I): Concepts and terms in 3D image based 3D treatment planning in cervix cancer brachytherapy with emphasis on MRI assessment of GTV and CTV. *Radiother Oncol*. 2005;74(3):235–245.

Han BH, Wallner K, Merrick G, Butler W, Sutlief S, Sylvester J. Prostate brachytherapy seed identification on post-implant TRUS images. *Med Phys*. 2003;30(5):898–900.

Holloway CL, DeLaney TF, Alektiar KM, Devlin PM, O'Farrell DA, Demanes DJ. American Brachytherapy Society (ABS) consensus statement for sarcoma brachytherapy. *Brachytherapy*. 2013;12(3):179–190.

Holly R, Morton GC, Sankreacha R et al. Use of cone-beam imaging to correct for catheter displacement in high dose-rate prostate brachytherapy. *Brachytherapy*. 2011;10(4):299–305.

Hsu I-CJ, Cabrera AR, Weinberg V et al. Combined modality treatment with high-dose-rate brachytherapy boost for locally advanced prostate cancer. *Brachytherapy*. 2005;4(3):202–206.

Jaffray DA, Siewerdsen JH, Wong JW, Martinez AA. Flat-panel cone-beam computed tomography for image-guided radiation therapy. *Int J Radiat Oncol*. 2002;53(5):1337–1349.

Lee EK, Zaider M. Intraoperative dynamic dose optimization in permanent prostate implants. *Int J Radiat Oncol*. 2003;56(3):854–861.

Lee LJ, Damato AL, Viswanathan AN. Clinical outcomes of high-dose-rate interstitial gynecologic brachytherapy using real-time CT guidance. *Brachytherapy*. 2013;12(4):303–310.

Martinez AA, Pataki I, Edmundson G, Sebastian E, Brabbins D, Gustafson G. Phase II prospective study of the use of conformal high-dose-rate brachytherapy as monotherapy for the treatment of favorable stage prostate cancer: A feasibility report. *Int J Radiat Oncol*. 2001;49(1):61–69.

Meijer GJ, van den Berg HA, Hurkmans CW, Stijns PE, Weterings JH. Dosimetric comparison of interactive planned and dynamic dose calculated prostate seed brachytherapy. *Radiother Oncol*. 2006;80(3):378–384.

Ménard C, Susil RC, Choyke P et al. MRI-guided HDR prostate brachytherapy in standard 1.5 T scanner. *Int J Radiat Oncol*. 2004;59(5):1414–1423.

Nag S, Cano ER, Demanes DJ, Puthawala AA, Vikram B. The American Brachytherapy Society recommendations for high-dose-rate brachytherapy for head-and-neck carcinoma. *Int J Radiat Oncol*. 2001;50(5):1190–1198.

Nag S, Ciezki JP, Cormack R et al. Intraoperative planning and evaluation of permanent prostate brachytherapy: Report of the American Brachytherapy Society. *Int J Radiat Oncol*. 2001;51(5):1422–1430.

Nag S, Martínez-Monge R, Ellis R et al. The use of fluoroscopy to guide needle placement in interstitial gynecological brachytherapy. *Int J Radiat Oncol*. 1998;40(2):415–420.

Narayanan S, Cho PS, Marks II RJ. Fast cross-projection algorithm for reconstruction of seeds in prostate brachytherapy. *Med Phys*. 2002;29(7):1572–1579.

Orio III PF, Tutar IB, Narayanan S et al. Intraoperative ultrasound-fluoroscopy fusion can enhance prostate brachytherapy quality. *Int J Radiat Oncol*. 2007;69(1):302–307.

Pelloski CE, Palmer M, Chronowski GM, Jhingran A, Horton J, Eifel PJ. Comparison between CT-based volumetric calculations and ICRU reference-point estimates of radiation doses delivered to bladder and rectum during intracavitary radiotherapy for cervical cancer. *Int J Radiat Oncol*. 2005;62(1):131–137.

Polo A, Cattani F, Vavassori A et al. MR and CT image fusion for postimplant analysis in permanent prostate seed implants. *Int J Radiat Oncol*. 2004;60(5):1572–1579.

Polo A, Salembier C, Venselaar J, Hoskin P. Review of intraoperative imaging and planning techniques in permanent seed prostate brachytherapy. *Radiother Oncol*. 2010;94(1):12–23.

Reed DR, Wallner KE, Narayanan S, Sutlief SG, Ford EC, Cho PS. Intraoperative fluoroscopic dose assessment in prostate brachytherapy patients. *Int J Radiat Oncol*. 2005;63(1):301–307.

Ricke J, Wust P. Computed tomography–guided brachytherapy for liver cancer. *Semin Radiat Oncol*. 2011;21(4):287–293.

Shah C, Vicini F, Wazer DE, Arthur D, Patel RR. The American Brachytherapy Society consensus statement for accelerated partial breast irradiation. *Brachytherapy*. 2013;12(4):267–277.

Shin KH, Kim TH, Cho JK et al. CT-guided intracavitary radiotherapy for cervical cancer: Comparison of conventional point A plan with clinical target volume-based three-dimensional plan using dose–volume parameters. *Int J Radiat Oncol*. 2006;64(1):197–204.

Smith WL, Lewis C, Bauman G et al. Prostate volume contouring: A 3D analysis of segmentation using 3DTRUS, CT, and MR. *Int J Radiat Oncol*. 2007;67(4):1238–1247.

Stock RG, Chan K, Terk M, Dewyngaert JK, Stone NN, Dottino P. A new technique for performing Syed-Neblett template interstitial implants for gynecologic malignancies using transrectal-ultrasound guidance. *Int J Radiat Oncol.* 1997;37(4):819–825.

Stock RG, Stone NN, Tabert A, Iannuzzi C, DeWyngaert JK. A dose–response study for I-125 prostate implants. *Int J Radiat Oncol.* 1998;41(1):101–108.

Su Y, Davis BJ, Herman MG, Manduca A, Robb RA. Examination of dosimetry accuracy as a function of seed detection rate in permanent prostate brachytherapy. *Med Phys.* 2005;32(9):3049–3056.

Thomas SD, Wachowicz K, Fallone BG. MRI of prostate brachytherapy seeds at high field: A study in phantom. *Med Phys.* 2009;36(11):5228–5234.

Todor DA, Zaider M, Cohen GN, Worman MF, Zelefsky MJ. Intraoperative dynamic dosimetry for prostate implants. *Phys Med Biol.* 2003;48(9):1153.

Viswanathan AN, Dimopoulos J, Kirisits C, Berger D, Pötter R. Computed tomography versus magnetic resonance imaging-based contouring in cervical cancer brachytherapy: Results of a prospective trial and preliminary guidelines for standardized contours. *Int J Radiat Oncol.* 2007;68(2):491–498.

Viswanathan AN, Erickson BA. Three-dimensional imaging in gynecologic brachytherapy: A survey of the American Brachytherapy Society. *Int J Radiat Oncol.* 2010;76(1):104–109.

Viswanathan AN, Erickson B, Gaffney DK et al. Comparison and consensus guidelines for delineation of clinical target volume for CT- and MR-based brachytherapy in locally advanced cervical cancer. *Int J Radiat Oncol Biol Phys.* 2014;90(2):320–328.

Viswanathan AN, Szymonifka J, Tempany-Afdhal CM, O'Farrell DA, Cormack RA. A prospective trial of real-time magnetic resonance–guided catheter placement in interstitial gynecologic brachytherapy. *Brachytherapy.* 2013;12(3):240–247.

Viswanathan AN, Thomadsen B. American Brachytherapy Society consensus guidelines for locally advanced carcinoma of the cervix. Part I: General principles. *Brachytherapy.* 2012;11(1):33–46.

Westendorp H, Hoekstra CJ, van't Riet A, Minken AW, Immerzeel JJ. Intraoperative adaptive brachytherapy of iodine-125 prostate implants guided by C-arm cone-beam computed tomography-based dosimetry. *Brachytherapy.* 2007;6(4):231–237.

Zelefsky MJ, Chou JF, Pei X et al. Predicting biochemical tumor control after brachytherapy for clinically localized prostate cancer: The Memorial Sloan-Kettering Cancer Center experience. *Brachytherapy.* 2012;11(4):245–249.

Zelefsky MJ, Worman M, Cohen GN et al. Real-time intraoperative computed tomography assessment of quality of permanent interstitial seed implantation for prostate cancer. *Urology.* 2010;76(5):1138–1142.

Zelefsky MJ, Yamada Y, Marion C et al. Improved conformality and decreased toxicity with intraoperative computer-optimized transperineal ultrasound-guided prostate brachytherapy. *Int J Radiat Oncol Biol Phys.* 2003;55(4):956–963.

CHAPTER 16

Magnetic Resonance Imaging

Cynthia Ménard, Uulke A. van der Heide,
Maroie Barkati, and Eirik Malinen

CONTENTS

16.1 GENERAL PRINCIPLES OF MRI ACQUISITION FOR BRACHYTHERAPY PLANNING

16.1.1 Defining Implanted Soft-Tissue Tumor Targets

The treatment targets for brachytherapy universally consist of small volumes of soft tissue, whereby accurately distinguishing tumor boundaries and uninvolved tissues and organs at risk (OARs) of injury is paramount. In virtually all brachytherapy applications, tumor targets are best visualized, characterized, and defined using magnetic resonance imaging (MRI).

As a general rule, high-resolution T2-weighted fast-spin-echo (FSE) images are considered anatomic reference, with or without complementary use of contrast-enhanced T1-weighted or diffusion images. The imaging volume should be individually tailored and limited to the target region and implanted devices. Imaging of surface or distant regional anatomy is not required, in contrast to external beam radiotherapy. Similarly, deriving maps of electron density to calculate tissue attenuation is not relevant when using conventional brachytherapy planning tools.

Invariably, signal-to-noise ratio (SNR) issues encountered when imaging small volumes with high resolution must be recovered using increased averages and scan times compared with standard diagnostic practice. It must also be emphasized that applicator displacements relative to the treatment target between image acquisition and radiation delivery must be carefully minimized and monitored during patient transfers.

Sub-millimetric voxel resolution in all three planes can be achieved using two different approaches. First, two 2D FSE acquisitions with high in-plane resolution (but 3–5 mm slice thickness) are sequentially acquired in the axial and sagittal planes. This approach mirrors standard radiology practice, and maintains a more conventional T2 contrast profile. A key disadvantage however is the potential for motion between two acquisitions, and the inability of some commercial planning software tools to support nonaxial acquisitions. A good alternative is to acquire a 3D T2-weighted SE image volume using a variable flip-angle technique resulting in isotropic sub-millimetric voxels. To reduce scan times, images are often acquired in the sagittal plane, and axial reconstructions can be imported in any treatment planning systems without difficulty. This approach does however somewhat alter T2 contrast compared to standard 2D techniques, and may degrade in-plane resolution.

16.1.2 Defining Applicators and Seeds

Requirements for accurate target delineation must be balanced with an equally high requirement for resolving applicator and seed signatures on treatment planning MR images. Meeting both objectives in a single acquisition can be challenging. If devices are poorly resolved in T2-weighted images, three strategies can be considered. In the first, reducing echo-time (TE) and increasing proton-density (PD) weighting until an acceptable balance is reached that will compromise T2 contrast. Alternatively, a separate image is acquired. Options include PD FSE images as shown in Figure 16.1, acquired separately or in a dual-echo sequence. This latter technique specifically highlights signal void signatures from devices or air, but can incur relatively long acquisition times. Faster 3D gradient-echo pulse sequences that accentuate susceptibility artifact from metal or air are often considered (Damato and Viswanathan, 2015). Positive contrast markers can also be inserted to better resolve applicator or catheter lumens (Lim et al., 2014). Finally, models of fixed applicator geometries can be registered using readily identified landmarks (Berger et al., 2009).

16.1.3 Mitigating Geometric Distortions

While MRI provides excellent soft-tissue contrast, the geometrical accuracy of the images requires alertness. The spatial encoding of MRI is based on variations in the magnetic field in space and time. To this end, magnetic field gradients are switched on and off in an MRI sequence. As a consequence, distortions of the magnetic field result in misinterpretation of the location of the signal, presenting as a deformation of the image. The static magnetic field of an MRI (typically 1.5 or 3.0 T) is highly homogeneous in the center of

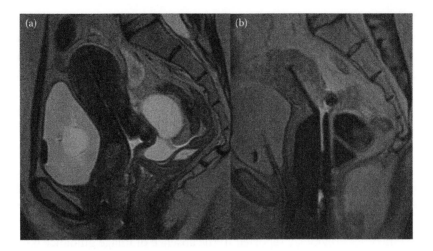

FIGURE 16.1 Sagittal T2w FSE (a) and PD (b) in the presence of applicators for cervix cancer. Applicators are well resolved on the PD acquisition (b).

the scanner where it is most relevant for brachytherapy acquisitions, but starts to deviate toward the outside. The same holds for the magnetic field gradients. As these inhomogene-ities and nonlinearities are static properties of a scanner, they can be measured and used for image correction. All modern MRI scanners have correction algorithms for this pur-pose. A caveat is however that for 2D sequences, such as a standard T2-weighted spin-echo sequence, often a 2D correction is applied. This means that for off-center slices, the image plane itself can be warped. A full 3D correction provides the optimal method to remove static system imperfections (Doran et al., 2005; Reinsberg et al., 2005; Paulson, et al., 2015; Schmidt and Payne, 2015).

In brachytherapy, a more important source of distortions is the patient. The magnetic susceptibility causes variations in magnetic field, particularly at the interface between air (or metal) and tissue. As a consequence, spatial fidelity can be compromised. Susceptibility distortions can be reduced by increasing the read-out bandwidth (which corresponds to reducing the water-fat shift) of a sequence. A drawback is however the loss of SNR. For this reason, in the clinical practice of a diagnostic department, the bandwidth is typically mini-mized. For use in radiotherapy, geometrical fidelity is more critical, requiring dedicated sequences where a different trade-off is made between geometrical accuracy and SNR.

16.1.4 MRI QA for Brachytherapy

When using MRI for brachytherapy planning, some additional issues may play a role. A brachytherapy applicator has to be MRI compatible, and free of metal if possible. However, an MRI-compatible applicator can also cause susceptibility artifacts, especially in the pres-ence of metal (Tanderup et al., 2014). Thus, it is important to maximize read-out bandwidth when metal is present and images are acquired at high field strength. Further, optimal shimming techniques should be applied to create a homogeneous magnetic field in the area of the tumor and applicator.

Brachytherapy seeds, used in prostate cancer, contain nonferrous metal. As with metal fiducial markers, they will present as a signal void on images. However, as the susceptibility of the metal will distort the magnetic field, the shape of the signal void depends on the particular MRI sequence applied. For some sequences, the void may be asymmetric, with a center that does not coincide with that of the seed. Careful validation of sequences and a regular quality assurance (QA) is therefore warranted, when using MRI to guide brachytherapy treatments (Jonsson et al., 2012).

16.2 PROSTATE CANCER

Transrectal ultrasound (TRUS) remains the gold-standard imaging modality for guidance of prostate brachytherapy (Morton, 2014). Its introduction in the 1980s enabled the rebirth of permanent seed low-dose rate (LDR) brachytherapy, whereby the Seattle technique eclipsed local control outcomes previously reported in the absence of image guidance (Grimm et al., 2001). The introduction of MRI to augment or replace the TRUS workflow has logically progressed over the past decades.

The dosimetric quality of TRUS-guided LDR implants, later evaluated using postplanning computed tomography (CT), is highly correlated with prostate-specific antigen (PSA) outcomes (Stock and Stone, 2002). These outcomes can be predicted more specifically within the prostate gland, whereby subregions of underdosage and/or dense tumor burden at the time of brachytherapy seem to correspond to sites of local failure (Crehange et al., 2013). In this regard, MRI is considered state of the art for local tumor staging and visualization. A diagnostic acquisition protocol that includes high-resolution T2-weighted FSE and diffusion-weighted imaging (DWI) with or without dynamic imaging during IV contrast injection can accurately identify regions of gross tumor burden and the presence of gross extracapsular extension or seminal vesicle invasion (Stage T3) (Weinreb et al., 2016). As one would expect, extracapsular invasion on MRI has been shown to predict outcomes after LDR brachytherapy (Riaz et al., 2012).

16.2.1 MRI Postplanning in LDR

The first application of MRI in prostate brachytherapy has been to augment or replace CT in postplanning dosimetric evaluation of LDR implants. Its use was first reported in 1997 (Moerland, et al., 1997) and multiple publications have since emphasized its value (Frank et al., 2008). A more accurate segmentation of the prostate boundary reveals that dosimetric quality can be falsely inflated in the absence of MRI, with much lower D90 and V100 measures compared with TRUS or CT estimates (Brown et al., 2013). This impact is regional, with higher-than-intended dose at the apex, and lower-than-intended doses at the base of the gland (Takiar et al., 2014). High dose to the external urinary sphincter segmented on MRI has also been correlated with adverse urinary outcomes (Register et al., 2013).

16.2.2 MRI Planning in HDR

Prior to the recent development of a more efficient TRUS-only workflow, high-dose rate (HDR) planning was routinely performed using CT. By replacing CT with MRI, depiction of the prostate boundary and OARs is improved (Menard et al., 2004). In the absence

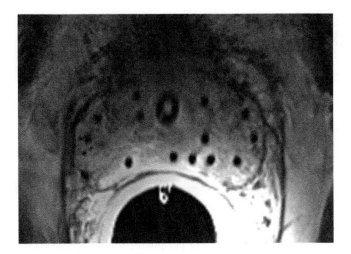

FIGURE 16.2 T2w FSE after catheter placement for prostate HDR brachytherapy boost. Catheters are visualized as signal void. Anatomic boundaries are highly resolved.

of commercial MRI markers, catheter signatures can be accentuated as signal voids in a high-resolution FSE image using an intermediate TE as is illustrated in Figure 16.2. Although this approach presented an improvement in accuracy of delineating the prostate gland, blurring of the apical boundary is evident due to acute needle trauma and bleeding. Depiction of gross tumor is also degraded by edema and bleeding compared with MRI prior to catheter insertion (Murgic et al., 2016).

16.2.3 MRI-Directed TRUS-Guided Implants

Practice-changing impact of MRI in prostate brachytherapy will only be achieved through a paradigm shift in therapeutic approach. The gross target volume (GTV) should be considered in planning and executing brachytherapy for prostate cancer, and in this regard, MRI prior to implantation is paramount. As a first step, the MRI is acquired prior to brachytherapy, the appropriateness of the treatment confirmed, and images cognitively "fused" or considered during the implant to avoid marginal miss of gross tumor. This approach results in a change in treatment plan in a substantial proportion of patients, through the addition of hormonal therapy, the addition of external beam radiotherapy, and/or modification of the implant itself by including sites of extraprostatic extension and/or seminal vesicle invasion (Murgic et al., 2016). Sites of tumor burden can also be considered when trading off target coverage and dose to adjacent OARs, such that undercoverage is permitted only in regions that do not harbor gross tumor.

The next step has been to differentially dose escalate visible tumors, and potentially deescalate dose to microscopically involved prostate gland tissues distant to the GTV. In Figure 16.3, an example of how a focused boost can be given by identifying a tumor lesion within the prostate. A number of publications, predominantly in HDR applications, have demonstrated the ease of escalating dose to tumors without incurring higher dose to OARs (Bauman et al., 2013). We await results of prospective trials to better ascertain the relative gain in effectiveness with this approach. The fact remains that the success or failure of

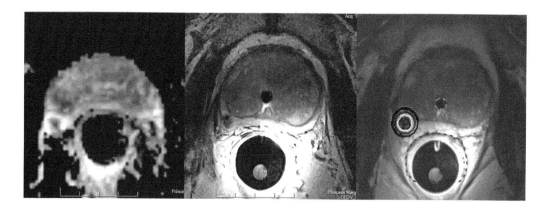

FIGURE 16.3 Example of HDR brachytherapy boost to external beam radiotherapy for dose escalation to the GTV. A 7-mm tumor in the right peripheral zone is identified on diffusion imaging (left) and T2w imaging (center). A single catheter at the center of the tumor enables highly conformal dose escalation.

tumor boost and/or focal-only therapies hinges on highly accurate techniques. Sources of error and uncertainty introduced with MRI-TRUS registration remain to be addressed. Rigid and deformable image registration as image processing possibilities in brachytherapy is discussed in Chapter 7.

16.2.4 MRI-Only Prostate Brachytherapy Workflow

An MRI-only workflow permits MRI to be acquired prior to catheter insertion to aid in implant guidance, and MRI to be acquired after catheter insertion for MRI-based treatment planning. In this manner, registration errors are largely circumvented. The requirement for a separate diagnostic MRI prior to brachytherapy is also removed. Some centers have installed MRI scanners immediately in the HDR delivery suite, removing the need for patient transfer between treatment-planning MRI and delivery of HDR brachytherapy dose. Errors due to motion or swelling are thereby further mitigated, and imaging immediately after (or during) delivery can confirm delivered (in contrast to planned) dose. In Section III (Chapters 19 through 26), several models of brachytherapy suites worldwide and workflow examples are described.

Real-time guidance, where images are acquired during needle insertion, remains challenging by nature of restricted access and inherent limitations of fast MRI. Stereotactic insertions demonstrate substantial needle deflection through tissues. For this reason, new technologies are under development to actively track needles during insertion, and improve the speed, safety, and reliability of the catheter trajectory reconstruction (Wang et al., 2015). New technologies and developments concerning robotics and tracking are discussed in Chapters 4, 10, 12, and 13.

16.3 CERVICAL CANCER

Progress has been made in brachytherapy for gynecological cancers, from 2D LDR Manchester-based planning to HDR 3D image-guided treatments. With the improvement

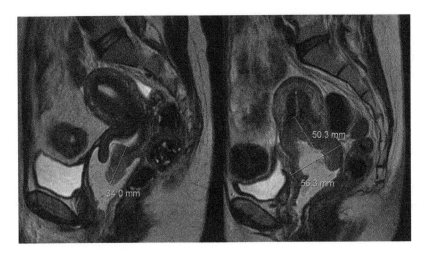

FIGURE 16.4   Sagittal T2w FSE in two patients demonstrating superior visualization of cervical tumor in the absence of applicators and packing.

in hardware and software for brachytherapy planning and the advent of applicators compatible with MRI (Vargo and Beriwal, 2014), the use of MRI-guided brachytherapy for cervix cancer is now considered the standard of care in leading radiation oncology facilities internationally.

T2-weighted FSE MRI using pelvic surface coil is the preferred imaging method (Nag, 2006). The value of MRI in imaging gynecological malignancies lies in its superior contrast resolution for soft tissue, which enables visualization of cervical tumor size and volume, distinction of tumor from normal uterus and cervix, and definition of parametrial infiltration of disease (Figure 16.4).

Consensus recommendations by the American Image-Guided Brachytherapy Working Group and the European Gynecological Brachytherapy Working Group have been published (Nag et al., 2004; Potter et al., 2006). The use of image-guided techniques supported by the recommendations is primarily based on repetitive MRI scans performed before and during treatment. The largest experience with MRI-based planning comes from the Vienna group. By incorporating MRI image-guided planning at each brachytherapy treatment and interstitial needle insertion when needed, they were able to achieve excellent tumor coverage with a mean D_{90} (minimal dose to 90% of the target volume) of 93 ± 13 Gy for the high-risk clinical target volume (CTV_{HR}). The mean D_{2cm3} (minimal dose to the most irradiated 2 cm^3 of normal tissue) for the rectum, sigmoid, and bladder was 65 ± 9, 64 ± 9, and 86 ± 17 Gy, respectively. Complete remission was achieved in 97% of patients and 3-year local control was 95% (Potter et al., 2011).

Following the publication of GEC-ESTRO (Groupe Européen de Curiethérapie-European Society for Therapeutic Radiology and Oncology) recommendations, a multicenter prospective observational study on MRI-guided brachytherapy in locally advanced cervical cancer (EMBRACE) was undertaken, with the most important aim being to implement image-guided brachytherapy for cervix cancer worldwide under high-quality standards (Potter et al., 2008). Based on the huge success of the EMBRACE study (more than 1350 patients), a consecutive EMBRACE II study is under way.

MRI is not universally available in radiation oncology departments. Furthermore, the need for serial volumetric imaging to account for tumor regression and OARs movements is a logistical and financial obstacle that has limited the universal applicability of MRI-based brachytherapy planning. Many radiation oncology departments have overcome these impediments by replacing MRI with CT simulators or have used a hybrid approach using MRI for the first fraction and CT for subsequent treatments (Vargo and Beriwal, 2014). Coregistration of the first fraction MRI and subsequent CT scans allows for better target delineation, while the OARs can be easily contoured on CT scan alone. Alternatively, a hybrid approach with transabdominal ultrasound and MRI has been described by investigators from Melbourne (Narayan et al., 2010).

16.4 MRI-GUIDED DOSE PAINTING

16.4.1 Objectives of Dose Painting

Local disease control is the primary aim of curative radiotherapy. Still, many patients suffer from local relapse and/or experience severe side effects. Because of the latter, increasing the total radiation dose to the patient is not acceptable and other approaches must be considered.

Tumor radioresistance may be due to several underlying intrinsic cellular properties, but also due to factors in the tumor microenvironment. In terms of classical radiation biology, clonogen density, repopulation rate, cellular repair capacity, and hypoxia (oxygen deficiency) are factors that may modify the tumor radiation response. For instance, hypoxia is a known cause of treatment resistance. Thus, if tumor hypoxia can be mapped noninvasively, hypoxic regions may potentially be eradicated by high radiation doses. In dose painting, the aim is to "paint" the tumor dose according to an imaging profile of a given patient to deliver more dose where this is needed (van der Heide et al., 2012). The hypothesis is that this in turn may result in better clinical outcomes.

16.4.2 Magnetic Resonance Images of Relevance for Dose Painting

Medical imaging may provide mapping of the tumor and potentially aggressive subregions. Positron emission tomography (PET), due to its high biological specificity, may be seen as the optimal imaging modality for dose painting. Still, PET examinations are limited to one positron-labeled tracer at a time and give rather low spatial resolution. The possibilities of PET for brachytherapy are discussed in Chapter 17. MRI, on the other hand, has better spatial resolution and may depict several tissue features following one examination. Although these features mostly reflect underlying magnetostructural properties (and not directly cellular or microenvironmental factors), they can be subject to interpretation in terms of tumor physiology.

The most straightforward example of MR-based imaging of relevance for dose painting is perhaps diffusion-weighted (DW) imaging. Typically, echo-planar sequences with variable diffusion-sensitive gradients are used to study the Brownian motion of water protons. To simplify, restricted water motion will be reflected as an increased MR intensity in the DW images. By systematically varying the strength of (or time between) the diffusion-sensitive gradients, the apparent diffusion coefficient (ADC) in a given tissue voxel can be estimated. A low ADC, that is, low water diffusion, points to high diffusion hindrance, which again may be due to high cellularity. In fact, several studies have shown that ADC is negatively correlated with cell density for many tumor types (Chen et al., 2013). Thus, tumor regions with low ADC appear as ideal targets for dose painting. As noted in Section 16.1.4, diffusion-weighted sequences could be prone to geometric artifacts, especially at high magnetic field strengths, which may compromise brachytherapy dose planning and delivery. Also, such distortions may worsen in the presence of metal applicators or needles.

In dynamic contrast-enhanced (DCE) MRI, the tissue distribution of a paramagnetic contrast agent is monitored as a function of time after intravenous bolus injection. The contrast molecules may leak from the blood vessels into the extravascular–extracellular space (EES) where they may reside before again entering the blood stream when renal clearance becomes prominent. Thus, in a given tissue element, a fast contrast "wash-in" may be observed followed by a slower "wash-out." Furthermore, the leakier the vessels and/or the higher the tissue perfusion, the higher the wash-in. Pharmacokinetic models may be fitted to time–intensity curves (preferably converted to time–concentration curves), providing estimates of physiologically relevant parameters. Such parameters may include tissue perfusion and vessel permeability. Low enhancement or low perfusion has been shown to be associated with hypoxia in, among others, head and neck cancers and cancers of the uterine cervix (Horsman et al., 2012). Depending on the application, a high temporal resolution may be warranted in DCE-MRI, which technically can be achieved at the cost of lower SNR and/or spatial resolution. Fast spin-echo or gradient recalled sequences are often employed, giving a temporal resolution in the order of 0.1 s^{-1}. These sequences are less prone to geometrical distortions compared to echo-planar imaging, which, incidentally, can also be used for DCE-MRI.

Varying "chemical shifts" of different paramagnetic molecules are exploited in magnetic resonance spectroscopy (MRS). The method can estimate the amount of metabolites in tissue, which normally are present in much lower concentration than water. Following water suppressing pulses, a spectroscopic, localized pulse may be achieved, for example, by a combination of three orthogonal slice selective pulses. The resulting nuclear magnetic resonance spectrum reflects the most abundant metabolites; the most relevant being lactate, choline, and creatinine in tumor tissues (Nguyen et al., 2014). However, MRS is inherently limited by low SNR, and is also prone to susceptibility artifacts. This may explain the current slight lack of interest in the method, although MRS studies are still conducted.

16.4.3 A Case Example

For illustrating brachytherapy dose painting, a patient with locally advanced cervical cancer was selected. The DCE-MRI-derived "A_{Brix}" parameter has been shown to reflect hypoxia in locally advanced cervical cancer (Halle et al., 2012). Thus, increasing the dose

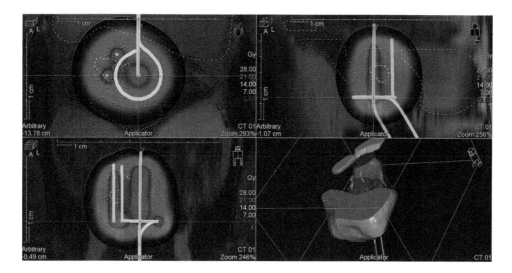

FIGURE 16.5   Brachytherapy dose painting of locally advanced cervical cancer based on hypoxia imaging with DCE-MRI. The images show the Vienna ring applicator and interstitial needles (cyan) with dwell positions given as red points. The resulting dose levels (in %) are shown overlaid on the MR image basis for the axial (top left), sagittal (top right), and coronal (bottom) plane. The gross tumor volume is in red, the hypoxic target volume in green, the bladder in blue, and the rectum in purple. (Courtesy of Anna Li and Taran Paulsen Hellebust, Oslo University Hospital.)

to tumor subregions with low "A_{Brix}" values could lead to increased likelihood of local control. T2-weighted and DCE ("A_{Brix}") images of a selected patient were imported into the treatment planning system, defining the gross tumor volume and the OARs (T2) and the hypoxic target volume (DCE). For dose painting, the hypoxic target volume was boosted to the highest dose possible without violating dose constraints for the OARs. Also, the minimum dose to the rest of tumor was kept equal to that for a conventional dosage.

Figure 16.5 shows the brachytherapy dose painting plan where a Vienna ring applicator and two needles were used. Relative to a conventional brachytherapy dose plan, a 30% dose increase could be achieved in the hypoxic region. However, further dose escalation was limited by the dose constraints imposed by the rectum. The simulations show that brachytherapy dose painting of locally advanced cervical cancer is technically feasible, and that such a strategy may be clinically attractive.

16.5 CONCLUSION

As access to MRI improves in brachytherapy departments, there is a pressing need to develop simplified workflows and tools for a broader adoption of its use worldwide. The integration of MRI-based planning provides an opportunity to develop interventional MRI strategies that minimize patient transfer and improve quality of both implants and dose plans. It also serves as an ideal tool for imaging–pathology correlations, and ultimately biologically guided and highly personalized dose-painted brachytherapy.

REFERENCES

Bauman, G.H.M., van de Heide, U., Menard, C., Boosting of dominant prostate tumors: A systematic review. *Radiother Oncol*, 2013;107(3):274–81.

Berger, D. et al., Direct reconstruction of the Vienna applicator on MR images. *Radiother Oncol*, 2009;93(2):347–51.

Brown, A.P. et al., Improving prostate brachytherapy quality assurance with MRI-CT fusion-based sector analysis in a phase II prospective trial of men with intermediate-risk prostate cancer. *Brachytherapy*, 2013;12(5):401–7.

Chen, L. et al., The correlation between apparent diffusion coefficient and tumor cellularity in patients: A meta-analysis. *PLoS One*, 2013;8(11):e79008.

Crehange, G. et al., Cold spot mapping inferred from MRI at time of failure predicts biopsy-proven local failure after permanent seed brachytherapy in prostate cancer patients: Implications for focal salvage brachytherapy. *Radiother Oncol*, 2013;109(2):246–50.

Damato, A.L. and A.N. Viswanathan, Magnetic resonance-guided gynecologic brachytherapy. *Magn Reson Imaging Clin N Am*, 2015;23(4):633–42.

Doran, S.J. et al., A complete distortion correction for MR images: I. Gradient warp correction. *Phys Med Biol*, 2005;50(7):1343–61.

Frank, S.J. et al., A novel MRI marker for prostate brachytherapy. *Int J Radiat Oncol Biol Phys*, 2008;71(1):5–8.

Grimm, P.D. et al., 10-year biochemical (prostate-specific antigen) control of prostate cancer with (125)I brachytherapy. *Int J Radiat Oncol Biol Phys*, 2001;51(1):31–40.

Halle, C. et al., Hypoxia-induced gene expression in chemoradioresistant cervical cancer revealed by dynamic contrast-enhanced MRI. *Cancer Res*, 2012;72(20):5285–95.

Horsman, M.R. et al., Imaging hypoxia to improve radiotherapy outcome. *Nat Rev Clin Oncol*, 2012;9(12):674–87.

Jonsson, J.H. et al., Internal fiducial markers and susceptibility effects in MRI-simulation and measurement of spatial accuracy. *Int J Radiat Oncol Biol Phys*, 2012;82(5):1612–8.

Lim, T.Y. et al., MRI characterization of cobalt dichloride-N-acetyl cysteine (C4) contrast agent marker for prostate brachytherapy. *Phys Med Biol*, 2014;59(10):2505–16.

Menard, C. et al., MRI-guided HDR prostate brachytherapy in standard 1.5 T scanner. *Int J Radiat Oncol Biol Phys*, 2004;59(5):1414–23.

Moerland, M.A. et al., Evaluation of permanent I-125 prostate implants using radiography and magnetic resonance imaging. *Int J Radiat Oncol Biol Phys*, 1997;37(4):927–33.

Morton, G.C., High-dose-rate brachytherapy boost for prostate cancer: Rationale and technique. *J Contemp Brachytherapy*, 2014;6(3):323–30.

Murgic, J. et al., Lessons learned using an MRI-only workflow during high-dose-rate brachytherapy for prostate cancer. *Brachytherapy*, 2016;15(2):147–55.

Nag, S., Controversies and new developments in gynecologic brachytherapy: Image-based intracavitary brachytherapy for cervical carcinoma. *Semin Radiat Oncol*, 2006;16(3):164–7.

Nag, S. et al., Proposed guidelines for image-based intracavitary brachytherapy for cervical carcinoma: Report from Image-Guided Brachytherapy Working Group. *Int J Radiat Oncol Biol Phys*, 2004;60(4):1160–72.

Narayan, K. et al., Image-guided brachytherapy for cervix cancer: From Manchester to Melbourne. *Expert Rev Anticancer Ther*, 2010;10(1):41–6.

Nguyen, M.L. et al., The potential role of magnetic resonance spectroscopy in image-guided radiotherapy. *Front Oncol*, 2014;4:91.

Paulson, E.S. et al., Comprehensive MRI simulation methodology using a dedicated MRI scanner in radiation oncology for external beam radiation treatment planning. *Med Phys*, 2015;42(1):28–39.

Potter, R. et al., Recommendations from gynaecological (GYN) GEC ESTRO working group (II): Concepts and terms in 3D image-based treatment planning in cervix cancer brachytherapy-3D dose volume parameters and aspects of 3D image-based anatomy, radiation physics, radiobiology. *Radiother Oncol*, 2006;78(1):67–77.

Potter, R. et al., Present status and future of high-precision image guided adaptive brachytherapy for cervix carcinoma. *Acta Oncol*, 2008;47(7):1325–36.

Potter, R. et al., Clinical outcome of protocol based image (MRI) guided adaptive brachytherapy combined with 3D conformal radiotherapy with or without chemotherapy in patients with locally advanced cervical cancer. *Radiother Oncol*, 2011;100(1):116–23.

Register, S.P. et al., An MRI-based dose–reponse analysis of urinary sphincter dose and urinary morbidity after brachytherapy for prostate cancer in a phase II prospective trial. *Brachytherapy*, 2013;12(3):210–6.

Reinsberg, S.A. et al., A complete distortion correction for MR images: II. Rectification of static-field inhomogeneities by similarity-based profile mapping. *Phys Med Biol*, 2005;50(11):2651–61.

Riaz, N. et al., Pretreatment endorectal coil magnetic resonance imaging findings predict biochemical tumor control in prostate cancer patients treated with combination brachytherapy and external-beam radiotherapy. *Int J Radiat Oncol Biol Phys*, 2012;84(3):707–11.

Schmidt, M.A. and G.S. Payne, Radiotherapy planning using MRI. *Phys Med Biol*, 2015;60(22):R323–61.

Stock, R.G. and N.N. Stone, Importance of post-implant dosimetry in permanent prostate brachytherapy. *Eur Urol*, 2002;41(4):434–9.

Takiar, V. et al., MRI-based sector analysis enhances prostate palladium-103 brachytherapy quality assurance in a phase II prospective trial of men with intermediate-risk localized prostate cancer. *Brachytherapy*, 2014;13(1):68–74.

Tanderup, K. et al., Magnetic resonance image guided brachytherapy. *Semin Radiat Oncol*, 2014;24(3):181–91.

van der Heide, U.A. et al., Functional MRI for radiotherapy dose painting. *Magn Reson Imaging*, 2012;30(9):1216–23.

Vargo, J.A. and S. Beriwal, Image-based brachytherapy for cervical cancer. *World J Clin Oncol*, 2014;5(5):921–30.

Wang, W. et al., Evaluation of an active magnetic resonance tracking system for interstitial brachytherapy. *Med Phys*, 2015;42(12):7114–21.

Weinreb, J.C. et al., PI-RADS Prostate Imaging—Reporting and Data System: 2015, Version 2. *Eur Urol*, 2016;69(1):16–40.

CHAPTER 17

Positron Emission Tomography

Alejandro Berlin and Rachel Glicksman

CONTENTS

17.1 INTRODUCTION

Over the past two decades, positron emission tomography (PET) and particularly positron emission tomography/computed tomography (PET/CT) have become part of standard of care in various areas of cancer medicine. Most commonly, PET/CT utilizes a glucose analog (^{18}F-fluorodeoxyglucose [FDG]) tracer to exploit biochemical and physiological differences between tumor and normal cells. This translates into increased sensitivity to allow for better assessment of tumor burden and spatial localization (e.g., staging and restaging), to plan management accordingly, and in some specific clinical scenarios to monitor or predict treatment results (Farwell et al. 2014). Despite the increasing uptake of PET studies across many different oncology disciplines, the field of brachytherapy has lagged behind. Herein, some of the main work and uses of PET in cervical and prostate cancer are described. We highlight current and future applications of molecular imaging, and describe how brachytherapy can uniquely exploit PET studies for improved cancer treatment.

17.2 TUMOR CHARACTERIZATION: CERVICAL CANCER EXPERIENCE

FDG-PET/CT studies in cervical cancer have particularly focused on its now clearer role in better regional and distant staging and treatment outcomes prognostication (Kidd et al. 2010). The International Federation of Gynecology and Obstetrics (FIGO) staging system for cervical cancer is largely based on physical examination, with limited radiological evaluation for local disease staging. However, magnetic resonance imaging (MRI) has been increasingly used for better determination of primary tumor extension (Haie-Meder et al. 2005), and constitutes the imaging modality for state-of-the-art high-risk clinical target volume (CTV_{HR}) definition for precise brachytherapy planning. Interestingly, few groups have explored the role of primary tumor characterization by PET for brachytherapy treatment planning (to visualize gross tumor volume [GTV]) or response evaluation.

The Washington University group pioneered studies investigating the role of FDG-PET in identifying patients at increased risk of local failure following initial treatment. In the first study, 11 patients underwent baseline and post-tandem/ovoid applicator insertion FDG-PET imaging to visualize and guide three-dimensional (3D) treatment planning, and to determine the feasibility and accuracy of this method when compared to conventional two-dimensional (2D) planning (Mutic et al. 2001; Malyapa et al. 2002). A follow-up study of 24 patients aimed to define the role of sequential FDG-PET studies for characterizing metabolic and dosimetric changes during treatment (Lin et al. 2005). The baseline GTV (described as any area of abnormal FDG uptake; mean 37 cm³) progressively decreased to 17 and 10 cm³ at mid- and end-treatment, respectively, translating in increases in conventionally 2D-planned target isodose coverage from 68% to 76% and 79%, respectively. In a subset of 11 patients (31 intracavitary treatments) with slightly larger initial tumors (mean 56 cm³; range 7–137 cm³) FDG-PET-based optimization increased the percentage of coverage of the target isodose to the metabolically active disease by 5% in the first implant (68 vs. 73%; $p = 0.21$), and 13% in the mid/final implants (70 vs. 83%; $p = 0.02$) while the D_{2cm3} and D_{5cm3} of both bladder and rectum were not significantly increased (Lin et al. 2007).

In parallel to the potential use of PET for treatment planning, the same researchers acknowledged the prognostic value of metabolically determined tumor volume, and the need for standardizing tumor characterization. Lin et al. (2006) described an average reduction of 50% of primary tumor volume within 3 weeks following combined modality treatment, when delineated by threshold method, that is, 40% or higher of the peak tumor intensity. In the most recent continuation study, three metabolic parameters (semiquantitative maximum standardized uptake value [SUVmax], variability of intratumoral FDG-uptake [FDGhetero], and metabolically active tumor volume [MTV]) were studied during treatment (weeks 2 and 4) and compared to baseline and 3-month follow-up FDG-PET/CT to characterize their changes over treatment and determine a potential correlation with disease outcome (Kidd et al. 2013). FDGhetero and MTV at baseline, and FDGhetero and SUVmax at 4 weeks appeared to be the most useful prognostic features for new or persistent disease after concurrent chemoradiation. This study suggests pretreatment and 4-week FDG-PET/CT as valuable time points for outcome prediction, for early response assessment, and potentially for guiding patient selection for treatment intensification strategies.

A corollary of these observational and dosimetric studies is the portrayal of different metabolic response patterns correlating with treatment outcomes, and the potential role for PET-based dose/treatment adaptation. Nonetheless, these findings have not been translated in a prospective study in order to determine if PET-based biomarkers are indeed "actionable," and if its negative prognostic features could be outweighed by treatment escalation and/or dose adaptation. Additionally, the comparison groups in these studies were based on 2D planning, and therefore it is hard to determine if the observed advantages were attributable to PET imaging or 3D planning approach in its inception. Considering recent results reporting local control rates >90% with MRI-based planning (Pötter et al. 2011), the conclusions of the above-mentioned studies warrant contemporary validation in the era of hybrid MRI/PET, FDG-based studies, and novel tracers.

17.3 DEFINING PRECISION TARGETS AND SITE OF RECURRENCE AFTER TREATMENT: PROSTATE CANCER EXPERIENCE

Radiotherapy represents a mainstay curative-intent treatment for localized prostate cancer. Despite improvements in the delivery of radiotherapy, 20–65% of radically treated patients experience relapse or locally persistent/recurrent disease (Jalloh et al. 2015). This in turn has been associated with metastatic progression (Zelefsky et al. 2008), mortality (presented at AUA 2014, from MRC RT-01 trial group (Dearnaley et al. 2014)), and local morbidity (Kaplan et al. 1992). Pioneering studies (Hoskin et al. 2012) have supported the rationale for treatment escalation to the whole gland (Spratt et al. 2014). Recently, in the communicated ASCENDE-RT trial, 398 men with intermediate/high-risk prostate cancer treated with pelvic external beam radiotherapy (EBRT) (46 Gy in 32 fractions) and 8 months of neo-adjuvant and concurrent androgen deprivation therapy were randomized (1:1) to subsequent prostate-EBRT 32 Gy in 16 fractions (total dose, 78 Gy/39 fractions) or low-dose rate (LDR) brachytherapy boost (115 Gy; 125-iodine). The brachytherapy arm had better progression-free survival on multivariate analysis (HR 0.49; 95%CI 0.3–0.8, p = 0.004), with a 5–6% increase in grade 3 genitourinary toxicity prevalence at 5 years (Morris et al. 2015). Although final results and longer follow-up are warranted, brachytherapy boost (either LDR or high-dose rate [HDR]) is likely to become a widely adopted method for dose escalation to improve outcomes for men with nonindolent prostate cancer. However, in high-risk prostate cancer, the main clinical conundrum is identifying *a priori* those men who fail radical treatments due to the presence of occult metastases, and therefore need systemic intensification as local dose escalation alone is likely futile (Bristow et al. 2014). However, the most-studied ^{11}C/^{18}F-choline-PET for staging of untreated prostate cancer seems to have limited sensitivity performance, and remains investigational (Umbehr et al. 2013). Therefore, novel molecular imaging in this setting could allow for more judicious patient selection for brachytherapy boost.

In cases where local recurrences present despite precision radiotherapy, contemporary studies of post-EBRT radical prostatectomy support the rationale of salvage local therapies with favorable biochemical control in a third to half of patients (Chade et al. 2011). However, surgery in this setting is associated with high rates of morbidities, and therefore a pressing need for less invasive salvage methods exists. In our own experience to date, 15 patients have been treated with salvage MRI-guided focal HDR (26 Gy in two separate implants) for biopsy-proven local recurrence. In the 12 patients with at least 1-year follow-up, 10 remain without evidence of biochemical or radiological progression. Likewise, more accurate identification of patients with confined local recurrence is becoming critical for guiding such tailored decisions. For example, Figure 17.1 illustrates a case of a prostate cancer tumor depicted on ^{18}F-fluorocholine PET that was undetected by multiparametric MRI. As

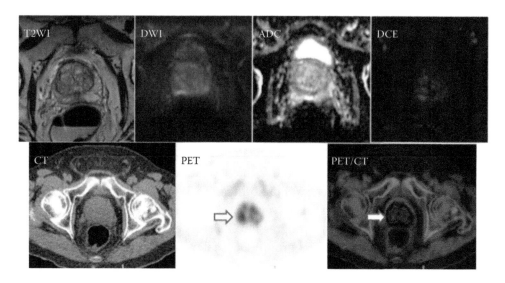

FIGURE 17.1 MRI and ^{18}F-fluorocholine PET/CT studies in high-risk prostate cancer. Eighty-one-year-old man with elevated PSA (9.4 ng/mL). Systematic TRUS-guided biopsies (12) revealed adenocarcinoma Gleason score 4+4 only in cores (3) from the right lateral peripheral zone. Top: Multiparametric MRI showed the absence of definite tumor. Bottom: ^{18}F-fluorocholine PET/CT revealed a focus of increased uptake (arrows) in the right peripheral zone (biopsy confirmed). Moderate heterogeneous uptake in the remaining prostate gland correlated with chronic inflammation seen in original biopsy specimens. (Unpublished, courtesy of Dr. Ur Metser.)

opposed to the primary staging setting, recent meta-analyses have suggested a clearer role for choline-based PET in the restaging of biochemical recurrences after curative-intent treatments, with pooled sensitivity of 85–89% and specificity of 87–88% (Umbehr et al. 2013; Fanti et al. 2016).

More recently, small peptide probes directed against prostate-specific membrane antigen (PSMA) have consistently shown superiority to choline-based PET/CT in characterizing local and distant disease (Afshar-Oromieh et al. 2014; Morigi et al. 2015). Importantly, in one study, ^{68}Ga-PSMA-ligand PET/CT uptake foci were true positives in all nine histological confirmations (Morigi et al. 2015). Moreover, a second-generation PSMA-targeted probe, ^{18}F-DCFPyL, has been developed and has shown increased sensitivity and better image definition (e.g., higher tumor-to-background-signal ratio) compared to ^{68}Ga-PSMA (Szabo et al. 2015). This evidence represents the forthcoming changing landscape of prostate cancer across risk groups and scenarios where brachytherapy has a crucial and growing role. Molecular imaging, PSMA-directed and other novel probes, and newer hybrid PET/MRI technologies will unveil unique research opportunities and treatment approaches for brachytherapy in the management of prostate cancer.

17.4 NOVEL TECHNOLOGIES AND TRACERS: MYRIAD OF OPPORTUNITIES

While PET/CT has been widely adopted due to its unique attributes, it also has pitfalls, especially with respect to accurate local staging, which is essential for robust brachytherapy planning. Further local evaluation with MRI is usually warranted due to its superior soft-tissue contrast and functional imaging capacities. Therefore, newer integrated PET/MRI hybrid systems hold the promise of improved diagnostic, staging, and subsequent restaging investigations across different tumor sites (Rosenkrantz et al. 2015). Nonetheless, evidence demonstrating the benefits of PET/MRI to support its routine clinical adoption is still needed, particularly to justify its added costs and complexities.

Some specific technical considerations of this new imaging technology are worth mentioning. First, given the relative novelty of PET/MRI, to date, there is no standardized acquisition protocols, and those currently used vary widely through the literature and different centers (Fraum et al. 2016). Imaging protocols that maximize morphological, functional, and metabolic data within a reasonable examination time frame need to be developed. Second, considering the lack of CT-derived electron density information, the attenuation correction of PET/MRI integrated systems has to rely on innovative algorithms and segmentation methods (Hofmann et al. 2011), which are still susceptible to motion artifacts, and inhomogeneity-introduced errors. Third, additional devices such as positioning and immobilization apparatuses used for radiotherapy can artifactually lower SUVs derived from MR-based attenuation correction methods (Ferguson et al. 2014). Therefore, the use of PET/MRI for brachytherapy treatment planning warrants specific studies to determine and account for the impact different applicators may have on quantitative PET/MRI. Last, the use of this technology will demand addressing new challenges from a space, workflow, personnel training, and data volume management perspective (Delso et al. 2015).

Increased FDG uptake reflects the increased aerobic glycolysis of many cancer cells (Warburg effect) relative to most normal tissues and benign tumors. Over the last decades, many other PET tracers have been developed and evaluated in preclinical and clinical studies, which account for other distinctive biological tumor hallmarks, such as increased proliferation, low oxygen tension (e.g., hypoxia), elevated amino acid transport, apoptosis, altered receptor expression, induced angiogenesis, and others (Hanahan and Weinberg 2011). Table 17.1 summarizes the different PET tracers, their biological targets, and primary disease sites of clinical use at present.

Our group has a long-standing interest in hypoxia and its impact on disease progression, metastases formation, and treatment response in several solid tumors, including cervical and prostate cancer (Dhani et al. 2015). Targeted PET probes (e.g., ^{18}F-FMISO, ^{18}F-FAZA) allow hypoxia quantification and spatial localization noninvasively. Figure 17.2 illustrates the characterization of tumor hypoxia on PET/CT. Considering hypoxia represents a well-established feature conferring radiation resistance, its identification and quantification by PET studies, and bespoke treatment approaches (including dose intensification by means of brachytherapy) seem well substantiated for prospective clinical evaluation.

In the near future, it is expected that novel hybrid agents (combined PET tracer plus MRI contrast agent) will further exploit the synergy between PET and MRI, allowing for a more accurate signal quantification, measuring both relaxivity and concentration under the same physiological conditions and exact colocalization of the agent in simultaneous PET/MRI (Kiani et al. 2015).

TABLE 17.1   Molecular Imaging Targets with Potential Use for Brachytherapy-Treated Malignancies

Target's Physiological Function	Imaging Probe	Disease Site(s)
Glucose metabolism	^{18}F-FDG, ^{11}C-2-deoxyglucose	Multiple
Amino acids	^{11}C-methionine, ^{18}F-FMT, ^{18}F-FET, ^{18}F-DOPA, ^{18}F-FAMT anti-^{18}F-FACBC, ^{90}Y-MAA, ^{76}Br-BAMP	Multiple
Lipids	^{11}C-choline, ^{18}F-fluorocholine	Prostate cancer
Gene expression	^{18}F-FLT, ^{18}F-FMAU, ^{18}F-fluoroethylspiperone	Multiple
Angiogenesis and vasculature	^{18}F-galacto-RGD, ^{18}F- and ^{64}Cu-labeled tracers for integrin $\alpha_V\beta_3$ (CD61) ^{15}O-water, ^{11}C-carbon monoxide	Melanoma, head and neck cancer, hepatic, lung, brain, and renal cancer Brain tumors
VEGF	^{123}I-VEGF$_{165}$, ^{123}I-VEGF$_{121}$, ^{111}In- and ^{89}Zr-labeled VEGF	Multiple
Hypoxia	^{15}O$_2$, ^{18}F-FMISO, ^{18}F-FAZA, ^{18}F-FETNIM, ^{18}F-EF5, ^{64}Cu-ATSM	Multiple
Apoptosis	^{99m}Tc-N2S2-rh annexin, ^{99m}Tc-HYNIC-annexin V	Multiple
EGFR receptor	^{11}C-gefitinib, ^{11}C-erlotinib	Lung
Estrogen receptor	^{18}F-FES, ^{18}F-FMOX	Breast and endometrial cancer
Androgen receptor	^{18}F-fluorocholine, ^{18}F-FDHT, ^{111}In-capromab pendetide, ^{68}Ga-PSMA-HBED-CC, ^{18}F-DCFPyL PSMA ligand	Prostate cancer

Note: EGFR: epidermal growth factor receptor; VEGF: vascular endothelial growth factor; FDG: fluorode-oxyglucose; FMT: fluoro-methyl-tyrosine; FET: fluoro-ethyl-tyrosine; DOPA: dihydroxy-phenylala-nine; FAMT: fluoro-alpha-methyl-tyrosine; FACBC: fluoro-cyclobutyl-1-carboxylic acid; MAA: human macroaggregated albumin; BAMP: bromo-α-methyl-phenylalanine; FLT: fluorothymidine; FMAU: fluoro-methyl-arabinofuranosyl-uracil; RGD: arginyl-glycyl-aspartic acid; FMISO: fluoromisonida-zole; FAZA: fluoroazomycin-arabinoside; FETNIM: fluoroerythronitroimidazole; EF5: nitro-hydro-gen-imidazol-pentafluoropropyl-acetamide; ATSM: diacetyl-bis-methylthiosemicarbazone; HYNIC: hydrazino nicotinate; FES: fluoroestradiol; FMOX: fluoromoxestrol; FDHT: fluoro-5 alpha-dihydrotes-tosterone; PSMA: prostate-specific membrane antigen; DCFPyL: pyridine-3-carbonyl-amino-pentyl-ureido-pentanedioic acid.

17.5 FUTURE AND CHALLENGES OF PET FOR BRACHYTHERAPY TREATMENT APPROACHES

In the context of the rapidly changing field of radiation oncology and novel precision radiotherapy technologies (Jaffray 2012), there is a pressing need to revitalize brachyther-apy approaches in view of its demonstrated benefits across different disease sites (Petereit et al. 2015). Incorporating molecular imaging studies, particularly novel PET/MRI and tracers, could unveil unique possibilities to exploit brachytherapy's distinct characteristics. Additionally, tissue samples can be collected during brachytherapy procedures. This can allow us to better characterize the performance of novel probes and molecular imaging modalities, and to expand the opportunity to engage in the growing field of multimodal biomarkers, radiomics, and tumor genomics research.

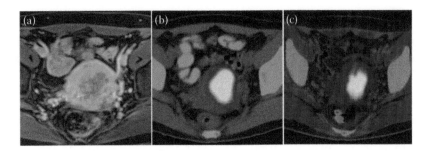

FIGURE 17.2   FAZA PET/CT for tumor hypoxia characterization. Forty-one-year-old female with FIGO stage IIIB cervical squamous cell carcinoma with bilateral pelvic and para-aortic lymphade-nopathy. (a) Baseline MRI sequence T1 VIBE fat saturated post-gadolinium image of the primary tumor. (b) Fused FDG-PET/CT image showing increased tracer uptake in a significant portion of the lesion. (c) Fused FAZA-PET/CT at corresponding level showing only a central subregion with increased tracer uptake, suggesting a hypoxic core within the primary tumor. (Unpublished, cour-tesy of Drs. Anthony Fyles, Kathy Han, and Michael Milosevic.)

PET studies could be foreseeably used in different clinical scenarios where brachyther-apy remains a paramount treatment. In cervical cancer, the use of PET could define regions of increased tumor burden or harboring adverse features (e.g., hypoxia), and subsequently consider the role of upfront increased dosing for better tumor control probabilities, and entertain dose reductions to the remaining metabolically "subclinical" disease, aiming to further improve the treatment's therapeutic index. Additionally, the use of PET for guiding early interventions should be determined. Timely FGD-PET readouts could set grounds for investigational personalized escalation (e.g., systemic chemo- or targeted therapies) or de-escalation approaches in patients with cervical cancer and persistent or rapid metabolic response, respectively. Likewise, intratreatment metabolic changes (PET) could be com-bined with morphological and functional changes (MRI) for target- and treatment-plan adaptation, aiming to maintain and/or increase subvolume dosages while better sparing surrounding normal structures.

Similarly, the use of PET for improved tumor delineation could be particularly relevant in prostate cancer, where the whole gland remains a flawed target surrogate, consider-ing cancer is neither defined nor confined by the prostate boundaries. Advanced imaging modalities during brachytherapy workflow translate into implant and planning modifi-cations to account for optimal (sub)regional dosing (Murgic et al. 2016). Disease-specific tracers (e.g., PSMA-targeted probes) combined with multiparametric MRI allow for unique opportunities to determine the role of combined or focal-only brachytherapy approaches for localized prostate cancer.

In summary, brachytherapy seems uniquely positioned to exploit morphological and functional imaging modalities. Conveying their diagnostic accuracy into the ther-apeutic realm holds the promise of improving treatment outcomes, and could allow for individualized treatment strategies within the personalized radiation treatment *par excellence.*

REFERENCES

Afshar-Oromieh A, Zechmann CM, Malcher A et al. 2014. Comparison of PET imaging with a (68)Ga-labelled PSMA ligand and (18)F-choline-based PET/CT for the diagnosis of recurrent prostate cancer. *Eur J Nucl Med Mol Imaging* 41(1):11–20.

Bristow RG, Berlin A, Dal Pra A. 2014. An arranged marriage for precision medicine: Hypoxia and genomic assays in localized prostate cancer radiotherapy. *Br J Radiol* 87(1035):20130753.

Chade DC, Shariat SF, Cronin AM et al. 2011. Salvage radical prostatectomy for radiation-recurrent prostate cancer: A multi-institutional collaboration. *Eur Urol* 60(2):205–10.

Dearnaley DP, Jovic G, Syndikus I et al. 2014. Escalated-dose versus control-dose conformal radiotherapy for prostate cancer: Long-term results from the MRC RT01 randomised controlled trial. *Lancet Oncol* 15(4):464–73.

Delso G, ter Voert E, de Galiza Barbosa F et al. 2015. Pitfalls and limitations in simultaneous PET/MRI. *Semin Nucl Med* 45(6):552–9.

Dhani N, Fyles A, Hedley D et al. 2015. The clinical significance of hypoxia in human cancers. *Semin Nucl Med* 45(2):110–21.

Fanti S, Minozzi S, Castellucci P et al. 2016. PET/CT with (11)C-choline for evaluation of prostate cancer patients with biochemical recurrence: Meta-analysis and critical review of available data. *Eur J Nucl Med Mol Imaging* 43(1):55–69.

Farwell MD, Pryma DA, Mankoff DA. 2014. PET/CT imaging in cancer: Current applications and future directions. *Cancer* 120(22):3433–45.

Ferguson A, McConathy J, Su Y et al. 2014. Attenuation effects of MR headphones during brain PET/MR studies. *J Nucl Med Technol* 42(2):93–100.

Fraum TJ, Fowler KJ, McConathy J. 2016. PET/MRI: Emerging clinical applications in oncology. *Acad Radiol* 23(2):220–36.

Haie-Meder C, Pötter R, Van Limbergen E et al. 2005. Recommendations from Gynaecological (GYN) GEC-ESTRO Working Group (I): Concepts and terms in 3D image based 3D treatment planning in cervix cancer brachytherapy with emphasis on MRI assessment of GTV and CTV. *Radiother Oncol* 74(3):235–45.

Hanahan D, Weinberg RA. 2011. Hallmarks of cancer: The next generation. *Cell* 144(5):646–74.

Hofmann M, Bezrukov I, Mantlik F et al. 2011. MRI-based attenuation correction for whole-body PET/MRI: Quantitative evaluation of segmentation- and atlas-based methods. *J Nucl Med* 52(9):1392–9.

Hoskin PJ, Rojas AM, Bownes PJ et al. 2012. Randomised trial of external beam radiotherapy alone or combined with high-dose-rate brachytherapy boost for localised prostate cancer. *Radiother Oncol* 103(2):217–22.

Jaffray DA. 2012. Image-guided radiotherapy: From current concept to future perspectives. *Nat Rev Clin Oncol* 9(12):688–99.

Jalloh M, Leapman MS, Cowan JE et al. 2015. Patterns of local failure following radiation therapy for prostate cancer. *J Urol* 194(4):977–82.

Kaplan ID, Prestidge BR, Bagshaw MA et al. 1992. The importance of local control in the treatment of prostatic cancer. *J Urol* 147(3 Pt 2):917–21.

Kiani A, Esquevin A, Lepareur N et al. 2015. Main applications of hybrid PET-MRI contrast agents: A review. *Contrast Media Mol Imaging* 11(2):92–8.

Kidd EA, Siegel BA, Dehdashti F et al. 2010. Lymph node staging by positron emission tomography in cervical cancer: Relationship to prognosis. *J Clin Oncol* 28(12):2108–13.

Kidd EA, Thomas M, Siegel BA et al. 2013. Changes in cervical cancer FDG uptake during chemoradiation and association with response. *Int J Radiat Oncol Biol Phys* 85(1):116–22.

Lin LL, Mutic S, Low DA et al. 2007. Adaptive brachytherapy treatment planning for cervical cancer using FDG-PET. *Int J Radiat Oncol Biol Phys* 67(1):91–6.

Lin LL, Mutic S, Malyapa RS et al. 2005. Sequential FDG-PET brachytherapy treatment planning in carcinoma of the cervix. *Int J Radiat Oncol Biol Phys* 63(5):1494–1501.

Lin LL, Yang Z, Mutic S et al. 2006. FDG-PET imaging for the assessment of physiologic volume response during radiotherapy in cervix cancer. *Int J Radiat Oncol Biol Phys* 65(1):177–81.

Malyapa RS, Mutic S, Low DA et al. 2002. Physiologic FDG-PET three-dimensional brachytherapy treatment planning for cervical cancer. *Int J Radiat Oncol Biol Phys* 54(4):1140–46.

Morigi JJ, Stricker PD, van Leeuwen PJ et al. 2015. Prospective comparison of 18F-fluoromethylcholine versus 68Ga-PSMA PET/CT in prostate cancer patients who have rising PSA after curative treatment and are being considered for targeted therapy. *J Nucl Med* 56(8):1185–90.

Morris JW, Tyldesley S, Rodda S et al. 2015. ASCENDE-RT: A multicenter, randomized trial of dose-escalated external beam radiation therapy (EBRT-B) versus low-dose-rate brachytherapy (LDR-B) for men with unfavorable-risk localized prostate cancer. *J Clin Oncol* 33(Suppl 7):abstr 3.

Murgic J, Chung P, Berlin A et al. 2016. Lessons learned using an MRI-only workflow during high-dose-rate brachytherapy for prostate cancer. *Brachytherapy* 15(2):147–55.

Mutic S, Dempsey JF, Bosch WR et al. 2001. Multimodality image registration quality assurance for conformal three-dimensional treatment planning. *Int J Radiat Oncol Biol Phys* 51(1):255–60.

Petereit DG, Frank SJ, Viswanathan AN et al. 2015. Brachytherapy: Where has it gone? *J Clin Oncol* 33(9):980–82.

Pötter R, Georg P, Dimopoulos JC et al. 2011. Clinical outcome of protocol based image (MRI) guided adaptive brachytherapy combined with 3D conformal radiotherapy with or without chemotherapy in patients with locally advanced cervical cancer. *Radiother Oncol* 100(1):116–23.

Rosenkrantz AB, Koesters T, Vahle AK et al. 2015. Quantitative graphical analysis of simultaneous dynamic PET/MRI for assessment of prostate cancer. *Clin Nucl Med* 40(4):e236–40.

Spratt DE, Zumsteg ZS, Ghadjar P et al. 2014. Comparison of high-dose (86.4 Gy) IMRT vs combined brachytherapy plus IMRT for intermediate-risk prostate cancer. *BJU Int* 114(3):360–7.

Szabo Z, Mena E, Rowe SP et al. 2015. Initial evaluation of [(18)F]DCFPyL for prostate-specific membrane antigen (PSMA)-targeted PET imaging of prostate cancer. *Mol Imaging Biol* 17(4):565–74.

Umbehr MH, Müntener M, Hany T et al. 2013. The role of 11C-choline and 18F-fluorocholine positron emission tomography (PET) and PET/CT in prostate cancer: A systematic review and meta-analysis. *Eur Urol* 64(1):106–17.

Zelefsky MJ, Reuter VE, Fuks Z et al. 2008. Influence of local tumor control on distant metastases and cancer related mortality after external beam radiotherapy for prostate cancer. *J Urol* 179(4):1368–73.

CHAPTER 18

Imaging for Treatment Verification

Taran Paulsen Hellebust, Harald Keller, and Nicole Nesvacil

CONTENTS

18.1 INTRODUCTION

For the last 20 years, radiotherapy has evolved from a two-dimensional (2D) approach using radiographic films to an approach where the treatment is based on three-dimensional (3D) imaging showing the target volumes and the normal tissues. This development is seen for both external beam radiotherapy (EBRT) and brachytherapy. However, the approach for using 3D imaging is different for these two modalities. In EBRT, the treatment is based on pretreatment imaging acquired typically a few days or up to a week before treatment starts, while in brachytherapy, a dedicated imaging procedure is usually performed for each application (or for each treatment), typically a few hours prior to the treatment. To be able to control and guide the EBRT on a daily basis, treatment machines with integrated imaging devices have been developed. Thus, patients are imaged at the treatment couch and the patient's actual anatomical shape and position at the time of treatment is compared to the initial image series. Such in-room imaging systems do not exist in most brachytherapy centers. Usually, the patient is moved from the brachytherapy suite to the imaging device and then to the treatment room. Additionally, an individual treatment plan is made for each fraction and this is a laborious procedure. Anderson et al. (2013) analyzed the movements of organs at risk (OARs) in 36 high-dose rate (HDR) fractions in cervical cancer brachytherapy by performing a second magnetic resonance imaging (MRI) immediately prior to treatment. They found that the D_{2cm3} (dose to the most exposed 2 cm^3) for the rectum and bowel changed by more than 10% in more than half of the fractions (Anderson et al. 2013). In this study, the time between the two MRI scans ranged from 3.2 to 9.9 hours with an average of 4.75 hours. In a study of process efficiency for computed tomography (CT)-guided treatment, Mayadev et al. (2014) reported a bit shorter time interval between imaging and treatment with an average of 3 hours and range of 1.5–8.6 hours. Anyhow, this time gap is still rather long. Also, considering that the patient is moved, there is a high probability that the anatomy can change and/or the applicators can be shifted, which in turn will hamper the accuracy and precision of the dose delivery. In many centers, some rudimentary verification procedures are made to account for such changes, usually by 2D imaging. These procedures can be greatly improved by designing a dedicated brachytherapy suite with an in-room 3D imaging device (e.g., many described in Chapters 19 through 26). With such systems, the movement of the patients could be avoided and the final image verification just prior to treatment can be performed with 3D imaging. In the following sections, in-room imaging with ultrasound (US), CT, MRI, fluoroscopy, and combinations of these are described.

18.2 IN-ROOM IMAGING: US

In-room imaging with US has been used for several decades, especially for prostate brachytherapy where transrectal ultrasound (TRUS) is used. Modern software technology facilitates intraoperative procedures enabling interactive treatment planning techniques (Hoskin et al. 2013; Nath et al. 2009). US is an excellent image modality for delineating the prostate gland. However, it is challenging to identify the tip of the needle for applicator reconstruction during HDR implants or to localize the seeds in permanent implants solely using US imaging. Some new imaging modalities to meet the latter challenge have been described in Chapter 4, as well as ideas to use global positioning system (GPS) technology to identify the tip of the needles is coming up (Tielens et al. 2014).

Since US is widely available, offers fast image acquisition, and is a rather inexpensive modality, it is interesting to use US for body sites other than the prostate. Poulin et al. (2015) have developed a robot-assisted 3D US system for interstitial HDR breast brachytherapy. They validated the system by comparing volume definitions using US, MRI, and CT. They found that the MRI and CT volumes were not significantly different from the 3DUS volumes and that the volumes were a significantly correlated. Furthermore, they compared the 3DUS catheter localization to the one obtained from CT images. They found that the maximum trajectory separation from 3DUS and CT on average was 0.51 ± 0.19 mm for 15 implanted needles and they expected this effect to have a minimal impact on the dose distribution (Poulin et al. 2015). Thus, the 3DUS is a promising imaging modality for breast brachytherapy.

In cervical cancer brachytherapy, US is primarily used to ensure a safe applicator placement, while MRI is the recommended modality for treatment planning (see Section 18.4). At the Peter MacCallum Cancer Center, Australia, 192 cervical cancer brachytherapy patients were imaged with both MRI and transabdominal US between 2007 and 2012, and they concluded that the differences found between these two imaging modalities were within clinically acceptable limits (van Dyk et al. 2014). Mahantshetty et al. (2012) found, on the other hand, that the correlation between transabdominal US and MRI were lowest in the highly relevant areas of the parametrial space and in the posterior uterus surface. Schmid et al. (2013) therefore explored the use of TRUS for cervical cancer brachytherapy and demonstrated that it was feasible to use such technique to assess the thickness of the tumor.

18.3 IN-ROOM IMAGING: CT

Several options for in-room CT imaging have been suggested. Reniers and Verhaegen (2011) described an approach using a cone beam CT (CBCT) solution. They used a simulator with CBCT dedicated for brachytherapy and presented a method to minimize imaging artifacts and optimized the image quality to allow clinical use during brachytherapy treatment planning. Even if the CBCT image quality is inferior to helical CT images, they concluded that such device is a good alternative as a verification tool immediately prior to treatment.

Orcutt et al. (2014) describe an alternative approach for in-room CT imaging. They have installed a CT-on-rails in the brachytherapy suite. The room includes a custom-made operating room-style procedure table with a CT-compatible insert allowing the whole procedure to be performed on that table. During imaging, the CT gantry is moved over the patients, in contrast to conventional CT where the couch is translated through the gantry. The suite has been used for both gynecological and breast implants and Orcutt et al. conclude that such facility allows an efficient brachytherapy procedure with a rapid workflow.

At the Oslo University Hospital, Norway, an in-room CT has been used since 2001 (Hellebust et al. 2007). An in-house made facility allows the whole brachytherapy procedure to be performed while the patient is on the CT couch. During a gynecological implantation, the patient is positioned inside the CT gantry with "feet first" orientation. The head of the patient and the anesthetic cart for general anesthesia is located on one side of the gantry, while the pelvis of the patient, the legs, and the leg stirrups are located on the other side of the gantry (Figure 18.1). The leg stirrups are anchored to the floor (Figure 18.1c). After the application, a table extension is allowing the patient's legs to be put down (Figure 18.1e and f) making it possible to perform an imaging acquisition. The CT-based treatment planning is performed while the patient is still on the CT couch and the treatment is delivered immediately after the planning is finalized. More than 400 gynecological treatments have been delivered yearly since 2001 using this facility and it is considered to be a time-efficient procedure. Since MRI is superior to CT for gynecological cancers, the procedure was adapted in 2009. Since then, an MRI acquisition is performed at least at the first fraction, additional to the CT imaging. The in-room CT could then be used for treatment verification just prior to the treatment.

18.4 IN-ROOM IMAGING: MRI

MRI provides excellent depiction of target and OARs, with better soft-tissue contrast than CT imaging. At the same time, MRI allows to identify the relative position between MRI-compatible brachytherapy applicators and the surrounding anatomy, while keeping radiation exposure of the patient nonexistent. The overall use of MRI for 3D image-guided

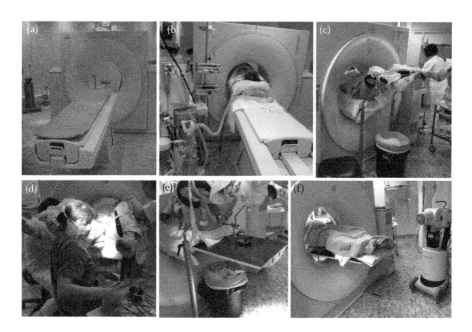

FIGURE 18.1   Illustration of a gynecological brachytherapy procedure with an in-room CT facility at the Oslo University Hospital, Norway. (a) The CT before the preparation of the patient. (b) and (c) The patient is prepared for the procedure with the body through the gantry of the CT and the legs in leg stirrups. (d) The application is performed. (e) A table extension is applied and the applicator is fixed with a suction cup. (f) The leg stirrups are removed, the legs are positioned slightly flexed on the table extension, CT imaging is performed, after treatment planning the treatment is delivered.

brachytherapy treatment planning has increased over the last decade. Some specialized institutions have developed workflows that allow for MRI-guided applicator insertion, MRI for treatment planning, and MRI prior to irradiation, to minimize dosimetric uncertainties due to suboptimal or varying implant geometry, and anatomical variations in relation to the applicator.

At the Medical University of Vienna (MUV), Austria, an MRI suite with an open low-field scanner, located inside the brachytherapy department, has been used for patients with gynecological cancer for over 10 years. This allows an optimized workflow that includes obtaining MR images prior to each treatment fraction, and timely irradiation thereafter. Analyses of the differences between the MRI obtained for planning and the first brachytherapy fraction with images taken the next day before delivery of the second fraction (Lang et al. 2013) have revealed small mean dosimetric differences (a few percent of the planned fraction dose) for target and OAR doses, but the reported standard deviations indicated that for individual patients, clinically significant OAR dose variations could take place. Pooling similar data from a total of six different institutions revealed a random uncertainty due to intra- or interfraction variation of OARs, such as bladder, rectum, and sigmoid, of 10–30% of the planned fraction dose (Nesvacil et al. 2013). For individual patients, whose treatment plans were optimized by pushing the organ doses close to the clinically acceptable limits, in order to allow for optimal target coverage, such high random variations could lead to an unexpected increase of clinical side effects, if they are not detected. Such patients would benefit most from closer monitoring of the actually delivered dose by reimaging shortly before each treatment. A timely detection of large variations would enable clinicians to adapt the treatment plan.

Therefore, optimized workflows to minimize times between MRI and treatment delivery are being implemented in specialized centers where complex tumor geometries are treated with highly conformal dose distributions optimized for high target and OAR doses. The MUV has recently updated their new open MRI facility (0.35 T, Siemens Magnetom C!™) with a patient trolley system that allows to move brachytherapy patients smoothly from a custom-made mobile MRI-compatible operating table that can be used next to the scanner in the MRI suite, to the scanning table and back on the table, which is afterward moved to the treatment room. Patient movement is minimized by keeping the patient placed on the same table top at all times, and small adjustments for arm and leg support can be made during different parts of the procedure to improve patient comfort. By using a rail system to slide the table top from the implant table to the MRI scanner, the risk of applicator movement by manually lifting and repositioning the patient is also reduced. After another verification scan shortly before treatment, the patient can be moved directly to the after-loader room, next to the MRI suite, for prompt irradiation. A flexible patient positioning and transfer device like this also allows for quick implant adjustments, such as needle repositioning, inside the MRI room.

Other solutions for bringing the MRI scanner closer to the brachytherapy delivery machine have been implemented, for example, at the Princess Margaret Cancer Center, Canada. As described by Jaffray et al. (2014), this setup has a rail-mounted mobile 1.5 T MRI scanner (Siemens), which can approach the brachytherapy unit to scan the patient on an operating table. Before treatment, the MRI scanner has to be retracted, as the afterloader is not fully MRI-compatible. In this institution, the mobile MRI scanner is likewise used for on-board patient scanning at the EBRT treatment unit. To further shorten the time between MRI scanning and brachytherapy delivery, a novel development at the University Medical Center Utrecht (UMCU), The Netherlands, allows to bring the afterloader into the MRI suite, just outside the 5 Gauss line, and to treat patients positioned on the MRI table (1.5 T Achieva™, Philips). This requires securing the afterloader by a chain, to prevent it from being attracted by the magnetic field, so as to ensure patient safety. In addition, the MRI room requires to be shielded similar to an HDR afterloader suite. The first results of repetitive imaging prior to the treatment for patients with cervical cancer have been reported by Nomden et al. (2014). For some patients, registration of planning and pretreatment scans and with corresponding dose volume histogram (DVH) verification indicated large intrafraction changes that lead to an adaptation of the plan of the following HDR fractions, to keep the total delivered treatment dose below the clinical dose constraints. Most often, a change in rectum filling, but also of bladder filling, was found to be the reason for large dosimetric variation. If detected prior to treatment, appropriate interventions can be performed to optimize organ volumes to match with the original planning images.

By using such adaptive imaging protocols, based on in-room MRI, an estimated reduction of interfraction variations from 30% down to 10% should become feasible, that is, the level of remaining uncertainties due to interobserver contouring variations (Nesvacil et al. 2013).

A group at the UMCU has explored the possibility to guide HDR prostate brachytherapy using the in-room MRI facility described above. They aim to develop an MRI-based intraprocedural feedback on the exact location of the source with respect to tumor and OARs. de Leeuw et al. (2013) described a new dual-plane MRI technique that offers both sufficient temporal and spatial resolution in order to track the source during treatment inside the MRI system. In a phantom, they tracked the source in 3D with about millimeter accuracy with a temporal resolution of approximately 4 s (de Leeuw et al. 2013).

18.5 IN-ROOM IMAGING: FLUOROSCOPY

In-room imaging using x-ray imaging and fluoroscopy offer the unique potential possibility of verifying the treatment setup and treatment delivery in real time as the HDR source can be directly imaged. Currently, the main application of x-ray imaging and fluoroscopy is for verifying catheter positioning in HDR prostate brachytherapy but can also be used for catheter guidance in endobronchial and endoesophageal brachytherapy treatments (see Chapter 15 for in-depth discussion).

In HDR prostate brachytherapy, the magnitude of catheter displacement can be substantial when the time between catheter insertion and treatment delivery is long, the patient is awake during the time of planning, or the patient has to be transported between different rooms. Whitaker et al. (2011) acquired an anterior–posterior (AP) pelvic x-ray with the marker wires inserted in the catheters to compare their position to a CT scout image at the time of insertion. Displacement was measured along the longitudinal axis. Patients in this study were generally not anesthetized for treatment delivery. They found that in HDR prostate brachytherapy, catheter displacement of 5 mm or more occurred in 67% of implants between CT planning and treatment delivery and recommended that departments performing HDR prostate brachytherapy verify internal catheter positions immediately prior to any treatment delivery.

Keller et al. (2014) described a similar procedure to verify the location of the catheters immediately before the treatment in the HDR suite using the dummy sources of the afterloader, including semiautomated image registration. For each catheter, a lateral x-ray image was acquired on a mobile C-arm while the dummy source was at the first dwell position. Visual comparison of the C-arm images to the reconstruction of the first dwell positions, the Foley balloon, and fiducial markers in the treatment plan resulted in the identification of gross differences, which were corrected for by adjustment of the first dwell position in the treatment plan. In addition, image registration was performed to quantify catheter position represented by the first dwell position using the "CathTrack" software (Figure 18.2). This software first reads the C-arm images, one for each catheter. The distortion of the C-arm images due to the image intensifier is then corrected for. The user then manually selects the AP extension of the Foley balloon in one of the C-arm images and the first dwell position for each catheter. The DICOM structure set, previously exported from the planning system containing the contours of the Foley balloon and fiducial markers, is imported into the software. The structures are projected onto the x-ray image obeying the projection

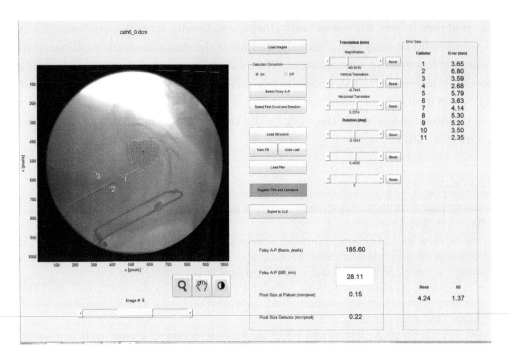

FIGURE 18.2 A screenshot of the CathTrack software.

geometry of the mobile C-arm. The coordinate systems of the planning images and the C-arm device are usually not registered, as there is no isocentric setup of the patient in the C-arm. Therefore, the magnification is determined by the measured and known AP extension of the Foley balloon. The other 5 degrees of freedom (2 translations and 3 rotations) can then be manually adjusted to register the structures to the C-arm image. Alternatively, the position of the fiducial markers can be marked on the x-ray image and automatic registration performed. After registration, the source positions are imported and the distance between the first dwell position in the C-arm image and the planned first dwell position are computed, enabling the first dwell position to be reprogrammed.

A rather similar approach was used to develop a system for localizing seeds in order to perform dynamic dose calculation intraoperatively during permanent seed implantation (Kuo et al. 2014). They used fiducial markers and TRUS in combination with C-arm fluoroscopy. Algorithms were used to locate the fiducial markers and the seeds segmentation, to perform seed matching with the reconstruction, and to register TRUS and fluoroscopy images. The system was evaluated and tested on 10 phantom cases and 37 patients with a fluoroscopy-to-TRUS error of 1.3 mm.

18.6 CONCLUSION

Jaffray (2012) concludes that future radiotherapy will be highly personalized and adaptive. Brachytherapy is a radiotherapy modality where usually new treatment planning is performed at every fraction based on 3D imaging, that is, modern brachytherapy is already highly personalized and adaptive today. However, to further improve the accuracy and precision of these plans, dedicated brachytherapy suite with integrated in-room 3D imaging device should be designed. This will streamline the procedures and facilitate verification imaging just prior to the treatment. This is already a reality in some few institutions, but should be implemented more broadly worldwide. Furthermore, the ultimate aim would be to develop methods for 3D image verification during treatment in order to verify that correct treatment is delivered. The work by the Utrecht group is promising in this respect (de Leeuw et al. 2013).

REFERENCES

Anderson C, Low G, Wills R et al. Critical structure movement in cervix brachytherapy. *Radiother Oncol* 107;2013:39–45.

de Leeuw H, Moerland MA, van Vulpen M, Seevinck PR, Bakker CJG. A dual-plane co-RASOR technique for accurate and rapid tracking and position verification of an Ir-192 source for single fraction HDR brachytherapy. *Phys Med Biol* 58;2013:7829–7839.

Hoskin P, Colombo A, Henry A, Niehoff P, Hellebust TP, Siebert FA, Kovacs G. GEC-ESTRO recommendations on high dose rate afterloading brachytherapy for localised prostate cancer: An update. *Radiother Oncol* 107;2013:325–332.

Hellebust TP, Tanderup K, Bergstrand ES et al. Reconstruction of a ring applicator using CT imaging: Impact of the reconstruction method and applicator orientation. *Phys Med Biol* 52;2007:4893–4904.

Jaffray DA, Carlone MC, Milosevic MF et al. A facility for magnetic resonance-guided radiation therapy. *Semin Radiat Oncol.* 24;2014:193–195.

Keller H, Moseley J, Abed J, Menard C, Rink A. Pre-treatment x-ray verification of catheter positioning for HDR prostate brachytherapy. *Brachytherapy* 13(Supp 1);2014:S56–S57.

Kuo N, Dehghan E, Deguet A et al. An image-guided system for dynamic dose calculation in prostate brachytherapy using ultrasound and fluoroscopy. *Med Phys* 41;2014:91712-1-13.

Lang S, Nesvacil N, Kirisits C et al. Uncertainty analysis for 3D image-based cervix cancer brachytherapy by repetitive MR imaging: Assessment of DVH-variations between two HDR fractions within one applicator insertion and their clinical relevance. *Radiother Oncol* 107;2013:26–31.

Mahantshetty U, Khanna N, Swamidas J et al. Trans-abdominal ultrasound (US) and magnetic resonance imaging (MRI) correlation for conformal intracavitary brachytherapy in carcinoma of the uterine cervix. *Radiother Oncol* 102;2012:130–134.

Mayadev J, Qi L, Lentz S et al. Implant time and process efficiency for CT-guided high-dose-rate brachytherapy for cervical cancer. *Brachytherapy* 13;2014:233–239.

Nath R, Bice W, Butler W et al. AAPM recommendations on dose prescription and reporting methods for permanent interstitial brachytherapy for prostate cancer. *Med Phys* 36;2009:5310–5322.

Nesvacil N, Tanderup K, Hellebust TP et al. A multicentre comparison of the dosimetric impact of inter- and intra-fractional anatomical variations in fractionated cervix cancer brachytherapy. *Radiother Oncol* 107;2013:20–25.

Nomden CN, de Leeuw AA, Roesink JM, Tersteeg RJ, Westerveld H, Jürgenliemk-Schulz IM. Intra-fraction uncertainties of MRI guided brachytherapy in patients with cervical cancer. *Radiother Oncol* 112;2014:217–220.

Orcutt KP, Libby B, Handsfield LL, Moyer G and Showalter TN. CT-on-rails-guided HDR brachytherapy: Single-room, rapid-workflow treatment delivery with integrated image guidance. *Future Oncol.* 10;2014:569–575.

Poulin E, Gardi L, Barker K, Montreuil J, Fenster A, Beaulieu. Validation of a novel robot-assisted 3DUS system for real-time planning and guidance of breast interstitial HDR brachytherapy. *Med Phys* 42;2015:6830–6839.

Reniers B and Verhaegen F. Cone beam CT imaging for 3D imaging guided brachytherapy for gynecological HDR brachytherapy. *Med Phys* 38;2011:2762–2767.

Schmid M, Pötter R, Brader P et al. Feasibility of transrectal ultrasonography for assessment of cervical cancer. *Strahlenther Onkol* 189;2013:123–128.

Tielens LKP, Damen RBCC, Lerou JGC, Scheffer GJ, Bruhn J. Ultrasound-guided needle handling using a guidance positioning system in a phantom. *Anaesthesia* 69;2014:24–31.

van Dyk S, Kondalsamy-Chennakesavan S, Schneider M, Bernshaw D, Narayan K. Comparison of measurements of the uterus and cervix obtained by magnetic resonance and transabdominal ultrasound imaging to identify the brachytherapy target in patients with cervix cancer. *Int J Radiation Oncol Biol Phys* 88;2014:860–865.

Whitaker M, Hruby G, Lovett A, Patanjali N. Prostate HDR brachytherapy catheter displacement between planning and treatment delivery. *Radiother Oncol* 101;2011:490–494.

III

Brachytherapy Suites

CHAPTER 19

Medical University of Vienna, Vienna, Austria

Christian Kirisits, Maximilian P. Schmid,
Nicole Nesvacil, and Richard Pötter

CONTENTS

19.1 STATUS IN THE 1980s AND 1990s (MÜNSTER, GERMANY, AND VIENNA, AUSTRIA)

Volumetric imaging, which was increasingly applied in clinical radiology, internal medicine, and gynecology in the 1970s, based on computed tomography (CT) of the brain and ultrasound (US) of the upper abdomen and uterus during pregnancy, led to the imaging revolution in radiotherapy, which started in the early 1980s. Axial CT scans were mainly used for external beam radiotherapy (EBRT) for increasing the number of body sites with two-dimensional (2D) treatment planning systems (TPSs) (based on "representative" image slices) and had to be linked to radiographic treatment planning, which was mainly fluoroscopic treatment simulation using the beam's eye view (BEV) with standard of individualized blocking of normal tissue. As information technology was limited, no direct link between volumetric and radiographic imaging and computerized dose planning was possible. Sporadic experience was collected with volumetric imaging (CT) for brachytherapy, for example, for gynecology and head and neck, and with transrectal ultrasound (TRUS) for prostate, which was challenging due to applicators with limited visibility and/or producing artifacts as they were constructed for radiographic imaging. Furthermore, computed 2D planning for brachytherapy was in its infancy.

With the coming up of magnetic resonance imaging (MRI) in the mid-1980s, with its superior soft tissue contrast and its capability of excellent imaging also in (para-) sagittal and (para-) coronal orientations, MRI was rapidly replacing CT in some body sites, such as the central nervous system (CNS) and pelvis (gynecology and prostate). This potential was used for the development of MRI simulation, linking fluoroscopic simulation with anterior–posterior and lateral radiographs, and sagittal and coronal imaging in order to select gross tumor volume (GTV) and clinical tumor volume (CTV) and adjacent organs at risk (OARs) in the BEV with image fusion through a camera fusion technology using a subtrascope. This planning technology could also be used for brachytherapy treatment planning, imaging the tumor with the applicator in place, for example, for uterine or vaginal cancer, for nasopharyngeal cancer, esophageal cancer, and pediatric malignancies with superimposition of isodoses from a 2D TPS in selected cases (Lukas et al. 1986; Nakano et al. 1987; Pötter et al. 1990; Kovacs et al. 1992). Major shortcomings were due to inappropriate hardware (e.g., metallic applicators) and due to limited planning technology (2D) with no direct possibilities of entering volumetric imaging.

In Vienna, with the opening of the New General Hospital (AKH) in 1993, there was more potential for linking volumetric imaging to brachytherapy applications. In 1996, a planning system became available that allowed for entering CT images, which was from then on used for image-based gynecologic applications for radical treatment and partly for other brachytherapy sites. The generation of dose volume histograms (DVHs) became possible based on both CTV and OAR contouring (Pötter 1997; Fellner et al. 2001).

In parallel, an open MRI was implemented in order to develop MRI-based treatment planning in radiotherapy in general (Pötter 1989) and for brachytherapy applications in particular (Pötter et al. 1990, 1991). However, up to the late 1990s, radiography-based dose planning was still the state of the art at the University Department of Radiotherapy, Vienna, for all tumor sites: breast, gynecology, prostate (US), head and neck, anus, esophagus, and bronchus. (Semi-) orthogonal radiographs were taken directly after applicator insertion. A conventional x-ray unit was available inside the operating theater and in front of the high-dose rate (HDR) treatment rooms. A target approach was followed as appropriate and combined with imaging methods, visualizing tumors and OARs in a nonsystematic approach. From 1998 onward, specific planning protocols were followed, developing a three-dimensional (3D) approach in gynecology using radiography, CT, and MRI; in prostate using US and MRI (CT); and in breast using radiography and CT/MRI on an occasional basis. For other sites such as head and neck, pediatric sarcoma, and esophagus, volumetric imaging was also used (Gerbaulet and Pötter 2002; Pötter 2002a,b).

For cervix cancer brachytherapy in the beginning, reference dose points according to ICRU 38 were defined and drawn on the two radiographs (ICRU 1985). The applicator and the dose points were reconstructed in three dimensions for a 3D radiography-based treatment plan with a digitalization device, which was, since 1996, complemented by referring applicator reference points to the planning CT with the applicator in place. While OAR doses were in the beginning estimated by calculating the dose at the ICRU reference points and then increasingly on brachytherapy treatment planning CT, the target dose was in the beginning prescribed based on Point A. Information about the target dimensions (width and thickness) was always available from the detailed clinical examination and to some degree later from volumetric imaging, which was first CT and then became MRI. Instead of normalizing/prescribing only to Point A, additional dose points were defined and the loading pattern and dwell times were adjusted accordingly. By that, the "prescription iso-dose" was moved up to 25 mm from the tandem in case of large tumors or to 15 mm in case of small tumors at the level of Point A instead of 20 mm with standard Point A loading. In addition to the clinical examination, the use of volumetric imaging became part of clinical routine. First CT and then MRI could show more reproducible and visible target struc-tures. The most optimal was imaging with the applicator in place as direct measurements of dimensions and distances related to the applicator became possible. A direct import of the volumetric images into the TPS was only possible from 1996 onward. The availability of CT- and MRI-compatible ring applicators became a breakthrough in the late 1990s.

Prostate HDR brachytherapy was performed with preplanning and TRUS and radiog-raphy-based needle implantation. With the use of volumetric image datasets (CT and later MRI), the location of the prostate could be defined on 3D radiography-based treatment plans. Individual optimization of dwell times could be performed taking into account dose volume constraints for the urethra and the rectum.

The highest patient numbers in brachytherapy in Vienna were due to breast cases. For the boost, in combination with EBRT, a straightforward single-fraction HDR technique is still standard. It is based on clinical information and mammography (Resch et al. 2002a). The target volume is implanted with fixed template geometry. For the more complex accel-erated partial breast irradiation (APBI) and locally recurrent breast cancer treatments with pulsed-dose rate (PDR) brachytherapy, individualized implants are used (Resch et al. 2002b, Strnad et al. 2016). The insertion is guided clinically with fixed templates and flex-ible catheters. The reconstruction of these catheters, radiopaque clips, and the fixation of the tubes at the skin was performed via semiorthogonal radiographs.

19.2 LEARNING PHASE WITH A DEDICATED MRI DEVICE

The decision for an MRI installed in a room within the brachytherapy ward had been taken by the head of the department based on his MRI experience and his vision for MRI-based treatment planning for radiotherapy in general and brachytherapy in particular. Instead of a C-arm x-ray unit dedicated to brachytherapy, the budget was reallocated in 1994 for a low-field open MRI located in the brachytherapy area, however, shared with the Department of Radiology. The commissioning of this device included detailed tests and fine-tuning of the MRI sequences with an MRI specialist physicist (Fransson et al. 2001). A new version of TPS allowed the export of a dose matrix based on the radiograph plan and imported into a separate module, which could for the first time visualize the isodoses directly on the monitor together with underlying MR images.

By that point, it was possible to go back to the radiography-based plan, perform changes in the loading pattern or dwell times, and repeat the export/import and visualization pro-cess. The dose grid was registered to the MRI via matching points related to the applicator but also catheters placed inside the patient (e.g., urinary bladder balloon, and rectal probe). This kind of registration had an intrinsic uncertainty of around half an MR slice thickness (which was 5 mm). Only one slice orientation was used; sagittal or coronal reconstructions were possible, but of limited quality due to the 5 mm slice thickness of the original axial slices. In addition, in the first versions, only strict axial slices could be imported. However, using this technique, a large cohort of patients was treated successfully (Pötter et al. 2007). In addition to the limitations using the whole potential of MRI with the commercially available brachytherapy TPSs, no reproducible dose reporting or prescription based on planning aims was performed.

Especially for gynecological brachytherapy but also for HDR prostate brachytherapy, which was then performed similar to gynecological with MRI-based planning, as well as PDR APBI with new implemented CT-based planning, the learning phase was driven to find appropriate dose and volume parameters.

19.3 ADVANCED PHASE OF MRI INTEGRATION, REPRODUCIBLE PLANNING AIMS, AND APPLICATOR OPTIMIZATION

Neither a target concept nor dose volume parameters were available in the early phase of MRI integration. Therefore, the development of the GEC-ESTRO recommendations during the early years of the 2000s was a major step forward (Haie-Meder et al. 2005; Kirisits et al. 2006). It marked the start of the "advanced phase" of MRI-based cervix cancer brachytherapy in Vienna (Kirisits et al. 2005; Pötter et al. 2011). The MRI was used routinely for all treatment plans, but also directly during the implantation in case of difficult topography. It is possible to shift the patients quickly in and out of the MR magnet. Dedicated MRI-compatible operation devices and anesthesia equipment in the MRI room allowed performing an interactive implantation process. Especially the use of free-hand-loaded needles could be guided by MRI.

In general, the use of needles became a logical consequence of MRI-based treatment planning. Of course, the need of increasing the treated volume in a conformal way was already evident from clinical examination for large tumors at the time of brachytherapy. But the visualization of the tumor and residual disease, the reproducible contouring of a CTV and OARs, and finally the dosimetric analysis with DVHs was big step forward but also showed the limitations of intracavitary techniques only. After the introduction of MRI into clinical routine in the advanced phase, the development became a main field of research in Vienna. Several applicators were individually designed for each patient using the workshop of the department (Kirisits et al. 2006). The main materials were different versions of thermoplastic. With a turning lathe, cylinders with individual shapes and diameters can be produced. Molding cutters allowed for the development of individual shapes, mainly used for perineal templates.

In addition to gynecologic applicators, individual applicator manufacturing was also done for other sites, especially for ano-rectal, head and neck, and pediatric cases. The main aim was to develop templates for applicator insertion, which was adjusted to the preimplant imaging or molds. For ano-rectal applications, the use of various balloons in combination with tubes provided suitable setups for endocavitary treatments as well.

The current imaging devices in the brachytherapy ward include an MRI (0.35 T Magnetom C!™, Siemens Healthcare, Erlangen, Germany), conventional x-ray unit (Optitop™, Siemens) for plain radiographs, a C-arm (ARCADIS Varic™, Siemens) in the operating theater, and US devices, including a B-K Medical System for clinical use und various additional systems for research and development. In the MR room, dedicated MR-compatible devices monitor the patient during full anesthesia. Treatment delivery is performed in four different rooms. Two of them contain a PDR microSelectron® device and two contain in total three HDR afterloaders: microSelectron® (Elekta Brachytherapy, initially Nucletron), Flexitron® (Elekta Brachytherapy, initially Isodose Control), and GammaMedPlus™ iX (Varian Medical Systems). In the treatment rooms, there is the possibility to perform *in vivo* dosimetry using the PTW VIVODOS system (PTB Freiburg). Outside of the brachytherapy ward, but in the vicinity of the department, there is access to CT, positron emission tomography (PET)/CT, and PET/MR systems.

The routine treatment workflow for cervical cancer patients includes usually two implantations. For each implantation, a full MRI dataset is acquired for treatment planning. One implant is used for two fractions. The patient is kept in the ward overnight and another MRI dataset is acquired for verification of the implant geometry and patient anatomy prior to the second treatment fraction. Immediate image inspection is done by a team of radiation oncologist and physicist. Based on a comparison of MR scans obtained for planning, the first brachytherapy fraction and images taken prior to the second fraction, dosimetric interfraction variations for this particular workflow have been analyzed (Lang et al. 2013). The mean dose differences between images from the two different time points were found to be of the order of a few percent of the fraction dose for target and OARs, but the reported standard deviations revealed that clinically significant changes of OAR doses can occur for individual patients. As a consequence, an additional tool was introduced in daily clinical routine, which allows better assessment of the dose distribution for the anatomy of the day, in cases where visual inspection of the MR images indicates substantial changes of organ shape or positions. This workflow is based on the Oncentra® GYN TPS (Elekta Brachytherapy, Veenendaal, The Netherlands) and requires a quick reconstruction of the intracavitary tandem/ring applicator by the radiation technologist (Boer et al. 2014). The planning images, with applicator, delineated structures, and dose distribution of the previous brachytherapy fraction, are automatically registered with the new MR image set of the day, based on the applicator as a reference. The dose distribution is inspected by the radiation oncologist, relative to the changed anatomy, and a decision for replanning or treatment with subsequent reevaluation of the DVHs can be taken.

19.4 FUTURE IN TECHNOLOGY

Ongoing and future developments will focus on improvements of online and in-room imaging devices in order to further enlarge the current spectrum of gynecologic brachytherapy as well as of brachytherapy in general. Online imaging can be used for the implantation procedure, treatment planning, and treatment verification. This will imply US (using transabdominal, transvaginal, and transrectal probes; see also Chapter 14), MRI, and new innovative imaging modalities (Schmid et al. 2016). Fast dynamic imaging can allow for real-time image guidance during applicator and/or needle insertion for individual adaptations of the implant within the operation theater, MRI room, or even directly at the HDR suite. The use of optical or electromagnetic tracking (Chapter 4) for applicator reconstruction is currently investigated for cervical cancer brachytherapy and can pave

(a)   (b)   (c)   (d)

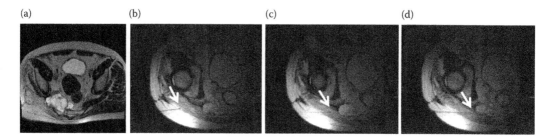

FIGURE 19.1   (a) T2-weighted MRI in a patient with a recurrent sarcoma at right pelvic wall. (b–d) Adapted fast sequences for interventional procedures using a multipurpose coil.

the way toward online treatment planning. This will reduce various current shortcomings of image-based brachytherapy such as uncertainties in treatment delivery and facilitate logistical challenges in case of poor implant geometry (of clinically guided insertions). For deeply located tumors in anatomically challenging regions, a cooperation with the Department of Interventional Radiology for MRI- and CT-guided approaches can be successfully initiated (Figure 19.1). This already enabled complex implants around or in close proximity to major nerves and vessels in the pelvis and head and neck region apart from standard brachytherapy indications. Based on these experiences, there is a vision to use the concept of an adaptive brachytherapy boost as developed for cervical cancer also for other indications. A close collaboration with the Department of Anaesthesiology is necessary to provide the basis for such extensive procedures. Therefore, all treatment rooms are also equipped with full monitoring possibilities. In selected cases, in particular the pediatric brachytherapy, the whole treatment (e.g., over a week in a PDR schedule) can be performed in general anesthesia with specific support by an intermediate care team.

Further, to improve patient transfer between different imaging and treatment facilities within the department, a custom-made, fully MR-compatible, combined operating table and trolley system has recently been introduced in the department. The trolley can be used directly next to the open 0.35 T scanner (Siemens Magnetom C!) in the MR suite. Using a rail system, the patient can be smoothly moved from the operating table onto the scanning table and back, and afterward be directly moved to the brachytherapy treatment room on the same trolley. The patient stays on the same table top during the whole procedure, which allows for a minimization of time-consuming patient lifting between different couches. Additional arm and leg support are optionally mounted on the trolley when needed. This flexible patient positioning and transfer system also allows for quick implant adaptations, such as needle repositioning, directly inside the MRI room. After verification MR scan, the patient can be immediately transferred to the HDR suite, which is located right next to the MRI room, for prompt irradiation.

REFERENCES

Boer, J., Koholka, K., Georg, P., Sturdza, A., Osztavics, A., Nesvacil, N., Kirisits, C., Pötter, R., Berger, D. 2014. PO-0975: Applicator-based image registration to support image-guided adaptive cervix brachytherapy in clinical routine. *Radiother Oncol.* 111(Suppl 1):136.

Gerbaulet, A., Pötter, R. 2002. Paediatric malignancies. In: Gerbaulet, A., Pötter, R., Mazeron, J. J., Meertens, H., and Van Limbergen, E., (eds), *GEC ESTRO Handbook of Brachytherapy*. ESTRO, Brussels, 611–632.

Fellner, C., Pötter, R., Knocke, T. H., Wambersie, A. 2001. Comparison of radiography- and computed tomography-based treatment planning in cervix cancer brachytherapy with specific attention to some quality assurance aspects. *Radiother Oncol.* 58:53–62.

Fransson, A., Andreo, P., Pötter, R. 2001. Aspects of MR image distortions in radiotherapy treatment planning. *Strahlenther Onkol.* 177(2):59–73.

Haie-Meder, C., Pötter, R., Van Limbergen, E., Briot, E., De Brabandere, M., Dimopoulos, J., Dumas, I. et al. 2005. Recommendations from Gynaecological (GYN) GEC ESTRO Working Group (I): Concepts and terms in 3D image-based 3D treatment planning in cervix cancer brachytherapy with emphasis on MRI assessment of GTV and CTV. *Radiother Oncol.* 74:235–245.

ICRU. 1985. International Commission on Radiation Units and Measurements. Dose and volume specification for reporting intracavitary therapy in gynaecology. *ICRU Report* 38. Oxford University Press, Oxford, United Kingdom.

Kirisits, C., Lang, S., Dimopoulos, J., Berger, D., Georg, D., Pötter, R. 2006. The Vienna applicator for combined intracavitary and interstitial brachytherapy of cervical cancer: Design, application, treatment planning, and dosimetric results. *Int J Radiat Oncol Biol Phys.* 65:624–630.

Kirisits, C., Pötter, R., Lang, S., Dimopoulos, J., Wachter-Gerstner, N., and Georg, D. 2005. Dose and volume parameters for MRI-based treatment planning in intracavitary brachytherapy for cervical cancer. *Int J Radiat Oncol Biol Phys.* 62:901–911.

Kovacs, G., Pötter, R., Prott, F. J., Lenzen, B., Knocke, T. H. 1992. The Münster experience with MRI assisted treatment planning used for HDR afterloading therapy of gynecological nasopharynx cancer. In: Breid (ed), *Advanced Tumor Therapy: Tumor Response Monitoring and Treatment Planning*, Springer, Berlin-Heidelberg, 662–665.

Lang, S., Nesvacil, N., Kirisits, C., Georg, P., Dimopoulos, J. C., Federico, M., Pötter, R. 2013. Uncertainty analysis for 3D image-based cervix cancer brachytherapy by repetitive MR imaging: Assessment of DVH-variations between two HDR fractions within one applicator insertion and their clinical relevance. *Radiother Oncol.* 107(1):26–31.

Lukas, P., Schröck, R., Rupp, N., Reiser, M., Allgayer, B., Feuerbach, S., Hofrichter, A., Heller, H. J. 1986. MR tomography of gynecologic diseases of the minor pelvis. *Rofo.* 144(2):159–65 [German].

Nakano, T. et al. 1987. Clinical evaluation of MRI applied to brachytherapy for cervical cancer. *Jap J Radiol.* 47(9):1181–1188.

Pötter, R. 1989. *Lokalisation mit Hilfe bildgebender Verfahren in der Strahlentherapie maligner Tumoren*. Habilitationsschrift, Universität Münster, Münster, Germany.

Pötter, R. 1997. Modern imaging methods used for treatment planning and quality assurance for combined irradiation of cervix cancer. In: Kovacs, G., (ed) *Integration of External Beam Therapy and Brachytherapy in the Treatment of Cervix Cancer: Clinical, Physical and Biological Aspects*. GEC ESTRO Workshop, May 1997, Course Book, 23–41.

Pötter, R. 2002a. Modern imaging. In: Gerbaulet, A., Pötter, R., Mazeron, J. J., Meertens, H., and Van Limbergen, E., (eds), *GEC ESTRO Handbook of Brachytherapy*. ESTRO, Brussels, 123–152.

Pötter, R., Dimopoulos, J., Georg, P., Lang, S., Waldhausl, C., Wachter-Gerstner, N., Weitmann, H. et al. 2007. Clinical impact of MRI assisted dose volume adaptation and dose escalation in brachytherapy of locally advanced cervix cancer. *Radiother Oncol.* 83:148–155.

Pötter, R., Georg, P., Dimopoulos, J. C., Grimm, M., Berger, D., Nesvacil, N., Georg, D. et al. 2011. Clinical outcome of protocol based image (MRI) guided adaptive brachytherapy combined with 3D conformal radiotherapy with or without chemotherapy in patients with locally advanced cervical cancer. *Radiother Oncol.* 100:116–123.

Pötter, R., Haie-Meder, C., Van Limbergen, E., Barillot, I., De Brabandere, M., Dimopoulos, J., Dumas, I. et al. 2006. Recommendations from Gynaecological (GYN) GEC ESTRO Working Group (II): Concepts and terms in 3D image-based treatment planning in cervix cancer brachytherapy: 3D dose volume parameters and aspects of 3D image based anatomy, radiation physics, radiobiology. *Radiother Oncol.* 78:67–77.

Pötter, R., Kovacs, G., Knocke, T. H., Heil, B., Haverkamp, U., Prott, F. J. 1990. MRI assisted localization of tumor and applicator for brachytherapy of cancer of nasopharynx, uterine cervix, and oesophagus. *GEC ESTRO Annual Meeting, Programme and Abstracts* May 21–23, Antwerp, Belgium, 29.

Pötter, R., Kovacs, G., Lenzen, B., Prott, F. J., Knocke, T. H., Haverkamp, U. 1991. Technique of MRI assisted brachytherapy treatment planning. *Activity J (Nucletron).* 5, 145–148.

Pötter, R., Van Limbergen, E. 2002b. Oesophageal cancer. In: Gerbaulet, A., Pötter, R., Mazeron, J. J., Meertens, H., and Van Limbergen E., (eds), *GEC ESTRO Handbook of Brachytherapy.* ESTRO, Brussels, 515–538.

Resch, A., Fellner, C., Mock, U., Handl-Zeller, L., Biber, E., Seitz, W., Pötter, R. 2002b. Locally recurrent breast cancer: Pulse dose rate brachytherapy for repeat irradiation following lumpectomy—A second chance to preserve the breast. *Radiology* 225(3):713–8.

Resch, A., Pötter, R., Van Limbergen, E., Biber, E., Klein, T., Fellner, C., Handl-Zeller, L. et al. 2002a. Long-term results (10 years) of intensive breast conserving therapy including a high-dose and large-volume interstitial brachytherapy boost (LDR/HDR) for T1/T2 breast cancer. *Radiother Oncol.* 63(1):47–58.

Schmid, M. P., Nesvacil, N., Pötter, R., Kronreif, G., Kirisits, C. 2016. Transrectal ultrasound for image-guided adaptive brachytherapy in cervix cancer—An alternative to MRI for target definition? *Radiother Oncol.* 120(3):467–472.

Strnad, V., Ott, O. J., Hildebrandt, G., Kauer-Dorner, D., Knauerhase, H., Major, T., Lyczek, J. et al. 2016. Groupe Européen de Curiethérapie of European Society for Radiotherapy and Oncology (GEC-ESTRO). 5-year results of accelerated partial breast irradiation using sole interstitial multicatheter brachytherapy versus whole-breast irradiation with boost after breast-conserving surgery for low-risk invasive and in-situ carcinoma of the female breast: A randomised, phase 3, non-inferiority trial. *Lancet.* 387(10015):229–38.

University Medical Center Utrecht, Utrecht, the Netherlands

Marinus A. Moerland, Astrid A. de Leeuw, Jochem R. van der Voort van Zyp, and Ina M. Jürgenliemk-Schulz

CONTENTS

20.1 INTRODUCTION

From its inception in the early 1970s, magnetic resonance imaging (MRI) has evolved into a technique with great potential in various subfields of medicine, including radiotherapy. This was acknowledged at the University Medical Center Utrecht (UMC Utrecht). Investigations concerning the use of different MRI techniques for radiotherapy and hyperthermia were started (Bakker and Vriend 1983; Sijens et al. 1988; Moerland et al. 1994). Nowadays, MRI is widely applied in planning and evaluation of external beam radiotherapy and brachytherapy (Moerland et al. 1997; Jürgenliemk-Schulz et al. 2009; Pötter 2009; Lagendijk et al. 2014). The transition from "finger and eye guided" brachytherapy to three-dimensional (3D) image-guided brachytherapy has improved results and the application of MRI-guided brachytherapy may further improve clinical outcome including morbidity (Pötter 2009; Hinnen et al. 2010; Nomden et al. 2013). This chapter describes the development toward MRI-guided brachytherapy at the UMC Utrecht such as the MRI-brachytherapy suite, the development of an MRI-compatible afterloader, MRI visualization of the ^{192}Ir source, the development of a robotic prostate implant device, MRI-based gynecology, prostate and nasopharynx applications, and safety issues.

20.2 MRI-BRACHYTHERAPY SUITE

The basic layout of the MRI-brachytherapy suite at the UMC Utrecht is depicted in Figure 20.1a. It is designed as a treatment room with a maze and 50 cm of concrete for shielding. The neighboring walls of the linear accelerator rooms have 150 cm concrete walls. The entrance wall and doors contain 2 mm Pb for shielding of scattered radiation. Another main part of the treatment room is the Faraday cage for shielding of radio frequency (RF) noise. The dimensions of the room are 7.5×6.7 m^2 (including technical/storage room). The MRI-brachytherapy suite is equipped with a 1.5 T wide bore Ingenia MRI scanner (Philips, The Netherlands), a microSelectron® high-dose rate (HDR) afterloader (Elekta Brachytherapy, Veenendaal, The Netherlands), and an MRI-compatible Aestiva anesthesia system (GE Healthcare, USA). Furthermore, Figure 20.1a shows a preparation room, a control room with operator consoles for the MRI scanner and the afterloader, and a technical room with MRI scanner cabinets for the gradient amplifier, spectrometer, and so on, and holes in the concrete wall for afterloader and scanner cables.

The afterloader is not MRI compatible. Firstly, some parts of the afterloader are ferromagnetic and it is attracted to the 1.5 T MRI scanner if it comes too close. During treatment inside the suite, the afterloader is secured to the wall so that it remains at a safe distance from the MRI scanner. Second, the afterloader is not RF shielded and therefore would contribute noise to MR images if imaging is performed with the afterloader inside the suite. For this reason, MRI scanning and irradiation are performed in a sequential manner. The procedure starts with MR imaging, reconstruction of the implant, and dose planning. Next, an MRI for position verification is performed and the patient is shifted out of the scanner. Finally, the transfer tubes from the afterloader are connected to the needles or catheters and the dose is delivered (Figure 20.1b, patient in irradiation position). An MRI-compatible afterloader is being developed in cooperation with Elekta Brachytherapy, The Netherlands. The advantage would be that the patient can be irradiated in imaging position, which potentially reduces catheter or needle displacements. With MRI source visualization, actual dwell positions can be verified and dose delivery adjusted if necessary due to changes in anatomy. With special MR acquisition and reconstruction techniques, it is possible to localize the HDR ^{192}Ir source with millimeter accuracy (De Leeuw et al. 2013). Clinical introduction awaits the availability of an MRI-compatible afterloader.

20.3 GYNECOLOGY

For the treatment of patients with advanced cervix cancer, the workflow in the combined MRI-brachytherapy suite is now fully adapted to exploit MR imaging immediately before dose delivery. Our standard treatment consists of external beam radiotherapy combined with two brachytherapy applications, with two HDR fractions each. MRI investigation is performed prior to the first application to evaluate tumor regression and in order to do a qualitative exploration/estimation for the need for interstitial needles and their eventual

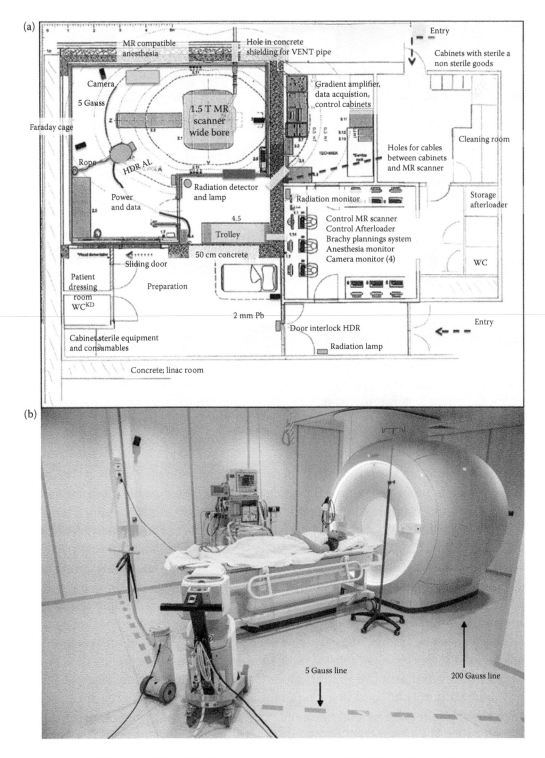

FIGURE 20.1   (a) Layout of the MRI-brachytherapy suite at the UMC Utrecht and (b) patient on the MR trolley in irradiation position.

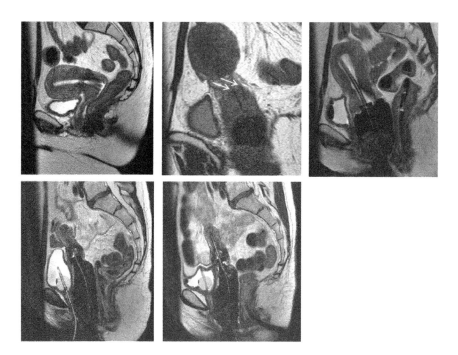

FIGURE 20.2 (Top row) Sagittal T2-weighted TSE images of a cervix cancer patient: image immediately before the application, image with interstitial needles (arrows) during application, and image with applicator (MRplan), respectively. (Bottom row) Example of another patient: sagittal MRplan and MRprefraction. In this situation, different bladder fillings lead to plan adjustment before dose delivery.

position in addition to the intracavitary parts of the applicator (Figure 20.2). Thereafter, the application is performed in the MRI-brachytherapy suite on the MR trolley with additional leg supports to ensure adequate positioning for the application procedure. If needles are placed, a short MRI sequence can be scanned to evaluate the positioning and depth of the needles (Figure 20.2). The position of the needles may be adjusted accordingly, followed by a full scan (with T2-weighted turbo spin echo (TSE) transversal, coronal, and sagittal sequences) taken for treatment planning (MRplan) (Figure 20.2).

After reconstruction of the applicator, contouring, plan optimization and approval, the patient is again placed on the MR trolley. A sagittal MRI scan is taken for visual inspection. Adaptations for bladder and rectum may be performed in case of unfavorable organ changes, for instance trying to deflate gas, or adjusting bladder filling (Figure 20.2, bottom row). The transversal scan (MRprefraction) is then registered to the transversal MRplan, using mutual information on a box around the applicator. In this way, the anatomy relative to the applicator can be evaluated and compared with the contours used for treatment planning. If visual inspection reveals that the positions of the needles and organs at risk (OARs) are acceptable compared with the treatment plan, then the dose can be delivered. However, if OARs have changed positions relative to the planned positions, the contours can be adapted and a new dose volume histogram (DVH) analysis is performed. Either the dose is then delivered, but the new estimated dose is taken into account for the constraints for the second application, or the treatment plan can be adapted in case of major changes. The procedure is repeated for the second fraction which is usually delivered the next morning.

For the first 15 patients treated in the MRI-brachytherapy suite, the procedure including as well post treatment scan has been described by Nomden et al. (2014). In clinical practice during the last 2 years (34 patients), the OARs situation was adapted 14 times and delivered dose was recalculated 17 times, and plan adaptation was considered necessary for the fourth fraction in one patient.

20.4 PROSTATE

The potential of focal MRI-guided HDR brachytherapy in patients with localized prostate cancer is investigated in an Institutional Review Board (IRB) approved feasibility study. Patients receive spinal anesthesia and are placed in a supine position with legs spread on the MR trolley with leg support. Since manual MRI-guided insertion of needles is not feasible due to limited access to the patient in the MRI scanner bore, ultrasound (US)-guided insertion and use of MR information is applied. After fusion of US images with diagnostic multi-parametric MRI, the catheters are inserted in the preparation room. Hereafter, an intra-operative MRI is performed and fused with the diagnostic MRI for delineation of the regions of interest. Delineation adjustments are made for intra-operative prostate deformation and movement of the OARs. Following this procedure, catheter reconstruction and dose planning are performed on the intra-operative MRI. Once there is an optimal dose plan, a single fraction of 19 Gy is administered to the clinical target volume (CTV) (Figure 20.3).

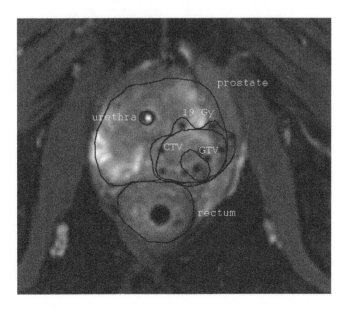

FIGURE 20.3   Transverse MR image of the prostate depicting prostate, GTV, clinical target volume (CTV), rectum, urethra catheter, and the signal voids of the inserted catheters with projection of the 19 Gy isodose line. The image is acquired with a 3D balanced turbo gradient echo sequence with spectral attenuated inversion recovery (SPAIR).

Catheter reconstruction and position verification just before dose delivery are performed on transversal T2-weighted TSE images and transversal and sagittal 3D turbo gradient echo images with 1 mm isotropic resolution. A robotic device is under development for MRI-guided needle insertion and dose delivery (Van den Bosch et al. 2010).

20.5 NASOPHARYNX

MRI guidance for brachytherapy has potential for different body sites. We explored MRI for delivery of a boost dose using the Rotterdam nasopharynx applicator (Levendag et al. 2013). After insertion of the applicator, computed tomography (CT) and MRI are performed. CT is still used as the gold standard for applicator reconstruction and MRI is used for reconstruction and delineation of the boost volume and OARs. For applicator reconstruction on CT, the two inner tubes are filled with a thin metal wire (Figure 20.4a), for reconstruction on MRI, the two inner tubes are filled with a saline solution (Figure 20.4b).

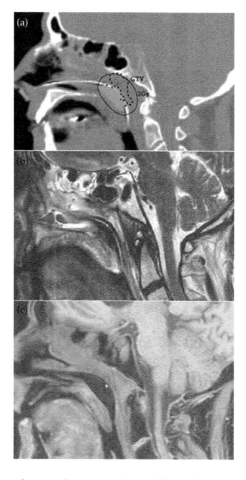

FIGURE 20.4 (a) CT image of a nasopharynx patient with applicator *in situ*, inner tubes filled with metal wires, pretreatment GTV, and 3 Gy isodose line. (b) Sagittal T2-weighted TSE image with inner tubes filled with saline solution and (c) fast gradient echo with Dixon fat suppression image for position verification.

Before irradiation, the inner tubes are replaced by new tubes in order not to pollute the afterloader. The brachytherapy boost protocol consists of six fractions of 3 Gy. The pre-treatment gross tumor volume (GTV) is considered to be the boost volume (Figure 20.4a). For all fractions, MRI is used to verify the correct position of the applicator (Figure 20.4c), which in practice proves to be time efficient and more accurate than the conventional method using C-arm fluoroscopy.

20.6 ROBOTIC IMPLANT DEVICE

As mentioned above, access to the patient for MRI-guided prostate interventions is restricted due to the 70 cm diameter bore of the MRI scanner. To overcome this limitation, an MRI-compatible robotic device has been developed. It is built of polymers and non-ferromagnetic materials, such as copper, titanium, and aluminum (Figure 20.5). Piezoelectric motors position the needle entry point and the tapping part is driven by pneumatics. The device is fixed to a wooden base plate that can slide over the MR table top and is placed between the legs of the patient. The robot taps the needle stepwise from a single entry (rotation) point just underneath the perineal skin into the prostate. By choosing different angles of needle insertion, the whole prostate can be reached. Stepwise tapping reduces tissue deformation (Lagerburg et al. 2006). In a first clinical test, the device has been used for insertion of fiducial gold markers in the prostate of patients for external beam radiotherapy (Van den Bosch et al. 2010). After redesigning the robot, needle insertion for prostate HDR brachytherapy will be tested in a clinical study, which is now awaiting approval by the IRB. For robotic developments, see also Chapter 12.

20.7 MRI VISUALIZATION OF THE ^{192}Ir SOURCE

The ultimate goal of our research is MRI-guided needle insertion and dose delivery, for which an MRI-compatible afterloader and special MR imaging and reconstruction

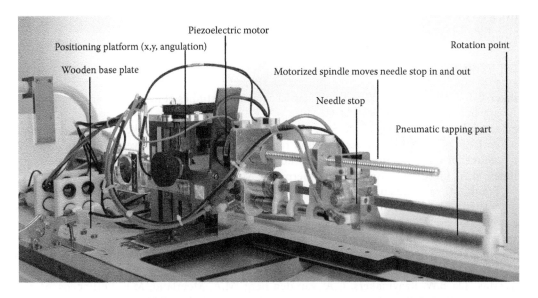

FIGURE 20.5   Robotic implant device for HDR prostate brachytherapy with main parts depicted.

techniques are being developed. De Leeuw et al. (2013) developed an MR imaging method that depicts an ^{192}Ir source on MR images. In this so-called center-out RAdial Sampling with Off-resonance Reconstruction (co-RASOR) technique, frequency offsets are applied during signal reconstruction in a center-out radial MRI acquisition. With optimal frequency offset tuning, a high positive contrast is generated at the location of the magnetic field disturber, that is, the ^{192}Ir source. It was shown that with two orthogonal two-dimensional (2D) co-RASOR images, the ^{192}Ir source can be located with millimeter accuracy and with a frame rate of approximately 0.25 Hz. Current research focuses on faster reconstruction and acquisition methods to increase the temporal resolution.

20.8 SAFETY ISSUES

An important safety measure is the screening of patients regarding contraindications for MRI, which is not different from the screening protocols as applied in radiology departments. The special issue for a MRI-brachytherapy suite is the application of interventional instruments like needles, scissors, clamps, and obturators, and also the use of an US machine, for example, to check the insertion of a uterine tube. In the past years, the small instrumentations were replaced by titanium or plastic instruments which are safe to use in an MR environment. Other devices like the US machine, the HDR afterloader, and emergency container, which are not MR safe, are used in the MRI-brachytherapy suite under strict regulations. Our experience is that these devices can be safely used if they are placed behind the 200 Gauss line (marked on the floor with a dashed line, see Figure 20.1b) and securely attached with a rope to the wall to prohibit the device being placed too close to the scanner by accident. Generally, the safe area is behind the 5 Gauss line (marked on the floor with a dashed line, see Figure 20.1b), where we also treat patients with contraindications for MRI (e.g., patients with a pacemaker or implantable cardioverter defibrillator). Furthermore, all personnel that works at the MRI-brachytherapy suite undergoes MRI safety training in combination with an HDR emergency procedure, which is repeated annually and is required to gain entry to the suite.

REFERENCES

Bakker, C.J., J. Vriend. 1983. Proton spin–lattice relaxation studies of tissue response to radiotherapy in mice. *Phys Med Biol.* 28:331–40.

De Leeuw, H., M.A. Moerland, M. van Vulpen, P.R. Seevinck, C.J. Bakker. 2013. A dual-plane co-RASOR technique for accurate and rapid tracking and position verification of an Ir-192 source for single fraction HDR brachytherapy. *Phys Med Biol.* 58:7829–39.

Hinnen, K.A., J.J. Battermann, J.G. van Roermund, M.A. Moerland, I.M. Jürgenliemk-Schulz, S.J. Frank, M. van Vulpen. 2010. Long-term biochemical and survival outcome of 921 patients treated with I-125 permanent prostate brachytherapy. *Int J Radiat Oncol Biol Phys.* 76:1433–8.

Jürgenliemk-Schulz, I.M., R.J. Tersteeg, J.M. Roesink, S. Bijmolt, C.N. Nomden, M.A. Moerland, A.A. de Leeuw. 2009. MRI-guided treatment-planning optimisation in intracavitary or combined intracavitary/interstitial PDR brachytherapy using tandem ovoid applicators in locally advanced cervical cancer. *Radiother Oncol.* 93:322–30.

Lagendijk, J.J., B.W. Raaymakers, C.A. van den Berg, M.A. Moerland, M.E. Philippens, M. van Vulpen. 2014. MR guidance in radiotherapy. *Phys Med Biol.* 59:R349–69.

Lagerburg, V., M.A. Moerland, M. van Vulpen, J.J. Lagendijk. 2006. A new robotic needle insertion method to minimise attendant prostate motion. *Radiother Oncol.* 80:73–77.

Levendag, P.C., F. Keskin-Cambay, C. de Pan, M. Idzes, M.A. Wildeman, I. Noever, I.K. Kolkman-Deurloo, A. Al-Mamgani, M. El-Gantiry, E. Rosenblatt, D.N. Teguh. 2013. Local control in advanced cancer of the nasopharynx: Is a boost dose by endocavitary brachytherapy of prognostic significance? *Brachytherapy* 12:84–89.

Moerland, M.A., A.C. van den Bergh, R. Bhagwandien, W.M. Janssen, C.J. Bakker, J.J. Lagendijk, J.J. Battermann. 1994. The influence of respiration induced motion of the kidneys on the accuracy of radiotherapy treatment planning, a magnetic resonance imaging study. *Radiother Oncol.* 30:150–4.

Moerland, M.A., H.K. Wijrdeman, R. Beersma, C.J. Bakker, J.J. Battermann. 1997. Evaluation of permanent I-125 prostate implants using radiography and magnetic resonance imaging. *Int J Radiat Oncol Biol Phys.* 37:927–33.

Nomden, C.N., A.A. de Leeuw, J.M. Roesink, R.J. Tersteeg, M.A. Moerland, P.O. Witteveen, H.W. Schreuder, E.B. van Dorst, I.M. Jürgenliemk-Schulz. 2013. Clinical outcome and dosimetric parameters of chemo-radiation including MRI guided adaptive brachytherapy with tandem-ovoid applicators for cervical cancer patients: A single institution experience. *Radiother Oncol.* 107:69–74.

Nomden, C.N., A.A. de Leeuw, J.M. Roesink, R.J. Tersteeg, H. Westerveld, I.M. Jürgenliemk-Schulz. 2014. Intra-fraction uncertainties of MRI guided brachytherapy in patients with cervical cancer. *Radiother Oncol.* 112:217–20.

Pötter, R. 2009. Image-guided brachytherapy sets benchmarks in advanced radiotherapy. *Radiother Oncol.* 91:141–6.

Sijens, P.E., H.K. Wijrdeman, M.A. Moerland, C.J. Bakker, J.W. Vermeulen, P.R. Luyten. 1988. Human breast cancer in vivo: H-1 and P-31 MR spectroscopy at 1.5 T. *Radiology.* 169:615–20.

Van den Bosch, M.R., M.R. Moman, M. van Vulpen, J.J. Battermann, E. Duiveman, L.J. van Schelven, H. de Leeuw, J.J. Lagendijk, M.A. Moerland. 2010. MRI-guided robotic system for transperineal prostate interventions: Proof of principle. *Phys Med Biol.* 55:N133–40.

CHAPTER 21

University Hospital Erlangen, Erlangen, Germany

Vratislav Strnad and Michael Lotter

CONTENTS

21.1 HISTORICAL INTRODUCTION

Radiation therapy in Erlangen has a long history that dates back to the beginning of the twentieth century. It was already during World War I and in the following decade, as part of the Department of Gynecology, that Erlangen's radiation therapy developed into an internationally renowned center. This story is associated with the names of the directors Ludwig Seitz (1872–1961) and Hermann Wintz (1887–1945). One of the most famous discoveries at this time was a special form of radiation treatment (the so-called "Röntgen-Wertheim-Fernfeldmethode") in which by analogy to the radical surgery of cervical carcinoma, developed by the gynecologist Ernst Wertheim, treatment was directed not only against the tumor itself but also against adjacent tissues at-risk that were included within the safety margin. Things then lay dormant for around 20 years following World War II until finally a period of renaissance for the development of radiation therapy was initiated by Rolf Sauer, who was director of the Department of Radiation Oncology from 1977 to 2008 and who inaugurated a new modern age that gave rise to the further rapid development of radiation therapy. He was the originator of the so-called "Erlangen School"—a novel way of thinking that implements radiation therapy within the concept of "radiation oncology." Within this new concept, all methods of radiation oncology—external beam radiation therapy, brachytherapy, hyperthermia, simultaneous chemotherapy, radiobiology, and onco-psychology—were to be used in a synergistic way aiming at restoring the patient's health. At this time, brachytherapy was already an integral part of the Department of Radiation Therapy: high-dose rate (HDR) and low-dose rate (LDR), followed by pulsed-dose rate (PDR) treatment in the late 1990s, were performed in the old "Radium Department" in the rooms located in the basement of the Women's Hospital in Erlangen. Rolf Sauer's successor Rainer Fietkau (2008) continued the established way of thinking and even amplified some aspects—particularly simultaneous chemotherapy, brachytherapy, and hyperthermia. His appointment as director of the Department of Radiation Oncology at Erlangen was tied to the construction of a self-contained unit dedicated solely to brachytherapy—the "Division of Interventional Radiation Therapy" which is the major provider of brachytherapy treatments in Germany.

21.2 STRUCTURAL CONCEPT OF THE BRACHYTHERAPY SUITE AS A SELF-CONTAINED BUILDING

The self-contained unit is dedicated solely to the use of brachytherapy. The construction encompasses two floors with an area of 682 m² (Figures 21.1 and 21.2). While the second floor hosts office rooms for the physicians and the physicist as well as a conference room and changing rooms, the first floor offers the complete infrastructure for all indications of brachytherapy: one surgical theater (Figure 21.3) with corresponding environment (one room for induction of general anesthesia, one room for recovery from general anesthesia, two storage rooms for devices and for instruments) that allows most surgical procedures to be performed on site, one intervention room (HDR-suite), four patient rooms (PDR-suites, Figure 21.4) with full equipment for intensive medical care and central monitoring, nurse point, pharmacy room, storage room for adjuvant devices and additional materials,

FIGURE 21.1 Edge-on view of the brachytherapy unit.

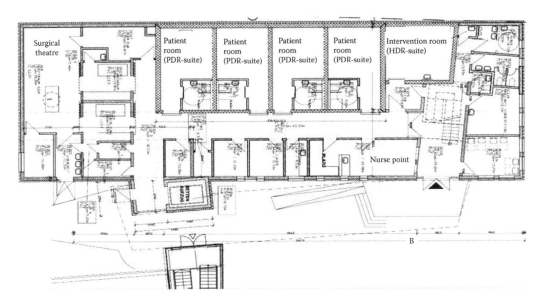

FIGURE 21.2   Layout of first floor.

disinfection and preparation room, patients' kitchen, physician's room, and waiting room (Figure 21.3).

This infrastructural arrangement together with a specialized team of eight nurses, one radiotherapy assistant, two physicists, and 2.5 physicians in service 24 hours per day, makes it possible to deliver any kind of brachytherapy treatment and ensures a maximum of care during the entire brachytherapy procedure.

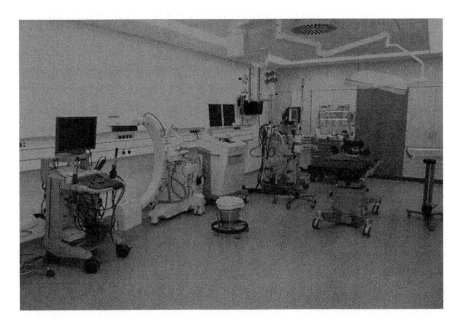

FIGURE 21.3   Surgical theater with 3D ultrasound (US) device (B&K) including a "HistoScanning™ tool" and with x-ray 3D C-arm (Ziehm).

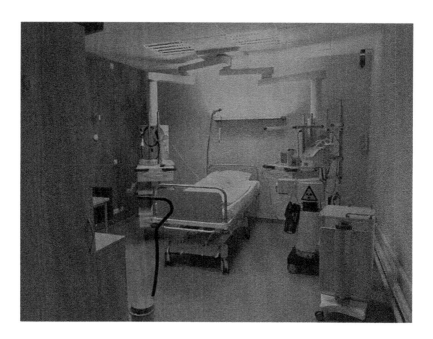

FIGURE 21.4   Patient room as intensive care unit.

21.3 BRACHYTHERAPY EQUIPMENT AND REVIEW OF INDICATIONS

The Division of Interventional Radiation Therapy at Erlangen uses four afterloading devices (one micro-Selectron® HDR- and three micro-Selectron® PDR (Elekta, Stockholm)) in clinical routine. A newly developed Flexitron® (Elekta, Stockholm) is available for experimental use. This broad spectrum of afterloading devices makes it possible to offer to each individual patient the most convenient schedule of brachytherapy treatment depending on his or her illness, co-morbidities, and personality. With regard to the applicators and instruments for brachytherapy procedures, the Erlangen unit offers the whole spectrum of surgical instruments and applicators (needles and catheters, templates, auxiliary instruments to be able to treat each of the anatomical tumor sites, both with temporary as well as with permanent brachytherapy). All applicators are manufactured by Elekta while accessory equipment from other companies like Best Medical, Varian, or Bebig is also used to be able to choose the most appropriate instrument and applicator for each individual patient.

We treat around 400 patients per year, the vast majority (>90%) of them with interstitial brachytherapy. The main indication for brachytherapy in our division is breast cancer both as sole partial breast treatment (accelerated partial breast irradiation [APBI]) and as a boost after whole breast irradiation (~40% of brachytherapy treatments). The second and third most common indications are low-, intermediate-, and high-risk prostate cancer (~20%) and head and neck (H&N) cancer (~17%). For both indications, we use both HDR- and PDR-brachytherapy. For H&N cancer, HDR-brachytherapy is still rarely used—and restricted to palliative indications. For patients with prostate cancer, in case of a salvage-brachytherapy (re-irradiation), we prefer PDR-brachytherapy. In addition, for low-risk prostate cancer patients, we also offer LDR-brachytherapy using I-125 sources. Patients with recurrent tumors (H&N cancer) are treated with interstitial PDR-brachytherapy with simultaneous chemotherapy (in case of H&N recurrences) and interstitial hyperthermia. The fourth most common indication (~15%) is represented by gynecological tumors. Here typically a combination of intracavitary and interstitial brachytherapy is performed. Typical indications are also unusual cancers as penile carcinoma, anal carcinoma, vulvar carcinoma, and other similar.

All steps of the brachytherapy procedure—(pre-planning, insertion of applicators, post-planning and quality assurance [QA])—are image guided. We are convinced that "image-guided brachytherapy"—in the true sense of the word is a "conditio sine qua non" without exception. For adequate and appropriate imaging—pre-interventional, online (intraoperative), post-interventional—a broad spectrum of imaging equipment is available as described below. Typically for brachytherapy of the female breast, having the knowledge of all "pre-brachytherapy information" like the primary surgical procedure (type of surgery, number and location of surgical clips, and position of the skin scar), all details of the pathology report including size of resection margins in six directions, and preoperative imaging (mammography and/or magnetic resonance imaging (MRI) and/or ultrasound), we use an x-ray C-arm (with a cone-beam computer tomography [CT] tool) for visualization surgical clips and needles during the insertion of catheters. If necessary and helpful, US imaging is also applied. Of course the detailed target definition and target delineation for both APBI, and boost irradiation is CT-based according to GEC-ESTRO recommendations. For prostate brachytherapy, independently whether it is temporary or permanent brachytherapy, we perform solely US-based brachytherapy in all steps—pre-interventional, online (intraoperative), post-interventional, whereby the online verification of needle localization together with the online dose calculation is the typical arrangement. In brachytherapy for H&N cancer only pre- and post-interventional imaging is used, we perform the insertion of needles and catheters solely based on palpation and inspection, supported with an endoscopy instrument if necessary. For brachytherapy of cervical carcinoma, we use pre- and post-interventional MRI according to GEC-ESTRO recommendations for target definition and delineation. For reconstruction of applicator and needles, we prefer to use CT-data and at the end to fuse CT- and MR-data in an appropriate way. For other inoperable gynecologic tumors, we prefer using only CT-based target definition. The online (intraoperative) imaging for verification and guidance (if necessary) of needle, catheter, and applicator insertion in all gynecologic tumors is strictly US based. In the postoperative situation of endometrium cancer, we perform the entire planning only with US based—individual estimation and the thickness of vaginal mucosa and distances of organs at risk (OARs) such as the rectum and bladder. Of course, as well as for brachytherapy of unusual cancers, the imaging remains an integral part of all brachytherapy procedures.

21.4 IMAGING EQUIPMENT

Both inside the surgical theater (Figure 21.4) as well as inside the intervention room ("small surgical theater"), the department has a vast range of imaging equipment available as follows: two three-dimensional (3D) ultrasound devices (B&K), including a "HistoScanning™

tool" and full DICOM-based connectivity to all planning systems, and x-ray C-arm with a cone-beam CT tool (Ziehm). Both are used for online planning or online verification of brachytherapy procedures. Moreover, in the immediate vicinity of the "brachytherapy building" within a 15 m radius approximately (across the corridor or alternatively accessible via lift in the basement of the building), we also have access to a CT as well as an MRI scanner—which are for the exclusive use of the Department of Radiation Oncology and are available also for brachytherapy treatment planning.

21.5 EQUIPMENT FOR TREATMENT PLANNING, DOSIMETRY, AND QA

21.5.1 Treatment Planning Systems

Physical treatment planning for nearly all brachytherapy techniques—for example, intracavitary treatment of the cervix, intraluminal treatment of esophagus, interstitial treatment of the female breast, tumors of the head and neck or prostate, as well as superficial lesions is performed with Oncentra® Brachy V.4.3 (Elekta—Nucletron). Moreover, a dedicated system—Oncentra® Prostate V.4.2.2.1 (Elekta—Nucletron) is used for ultrasound-based online planning of interstitial brachytherapy with I-125 Seeds for patients with low-risk carcinoma of the prostate. In case of treatment planning where structures like clinical target volume (CTV) and OARs need to be identified and defined on an MRI data set whereas catheter reconstruction is done on a CT image, image fusion and registration as well as outlining of the contours is performed with the software Velocity Al V.2.7.0 (Velocity medical solutions).

21.5.2 Dosimetric Equipment

To measure the source strength of the Ir-192 sources for the PDR and HDR afterloading devices, a well-type chamber in combination with an electrometer Unidos (PTW Freiburg) is used. In case of I-125 seeds, the source strength of one seed out of the whole batch is determined with the tool Sourcecheck also together with the electrometer Unidos (PTW Freiburg). With a MOSFET detector system (Best Medical), especially in an experimental setup, comparing results from the treatment planning systems (TPS) with online measurements, for example, during interstitial treatment of the female breast, *in vivo* dosimetry can be performed.

For measurements relating to radiation protection, the department is equipped with H*-10 calibrated measuring tools like Szintomat® 6150AD (Automess GmbH Ladenburg) or RadEye™ PRD-ER (Thermo Fisher Scientific Messtechnik GmbH).

21.5.3 QA Equipment (Basic, Research, and Emergency Training)

Checking the source position via autoradiographs is done with a homemade phantom as well as with QA tools like the "check ruler" from the developer (Nucletron that is an Elekta Company) where AgBr or self-developing Gafchromic films can be inserted.

As a research project in collaboration with Elekta-Nucletron, an electromagnetic tracking system (NDI System Aurora) is used for catheter tracking. Comparing the catheter coordinates received from the electromagnetic tracking system with those from the TPS, potentially interfractional catheter movement or shift during fractionated HDR irradiation of breast cancer patients could be detected.

For emergency training (retrieving the source in case of problems), a PDR classic afterloader that uses only a dummy source (no Ir-192 source) is available. In this way, an array of different problems involving applicators and source wire can be simulated. Safe from the hazard of ionizing radiation, the staff can easily obtain experience in drawing back the (dummy) source with the hand crank and get familiar with the workflow in case of an emergency situation.

CHAPTER 22

Charles LeMoyne Hospital, Montreal, Canada

Marjory Jolicoeur, Talar Derashodian, Marie Lynn Racine, Georges Wakil, Thu Van Nguyen, and Maryse Mondat

CONTENTS

22.1 INTRODUCTION

Brachytherapy was always considered more precise and more conformal than external beam radiotherapy (EBRT). The conformity was assured by careful placement of radioactive source(s) within the tumor, making it highly operator-dependent. The use of three-dimensional (3D) imagers such as computed tomography (CT) and magnetic resonance imaging (MRI) in brachytherapy has led to the image-guided brachytherapy (IGBT) concept, a more precise and less operator-dependent form of the technique. The first reports of the use of MRI in brachytherapy were in 1988 by Wijrdeman and Bakker (1988) and, in 1992, by Schoeppel et al. (1992). One of the difficulties for IGBT is access to a 3D imager. Usually, following implantation of the radioactive source(s), patients have to be moved to a CT scanner located in an EBRT area. For MRI-guided IGBT it is even more difficult, as MRIs are usually at some remove from the brachytherapy area.

The goal behind Charles LeMoyne's advanced multimodality integrated brachytherapy suite is to have all of the imaging modalities in close proximity to achieve accurate positioning of the implant and accurate definition of the target, while minimizing the potential for displacement of the implants within the patient.

22.2 REQUIREMENTS FOR CHARLES LEMOYNE HOSPITAL'S BRACHYTHERAPY SUITE

While designing our brachytherapy program, we had three priorities: first, unlimited access to 3D imaging relevant to all forms of brachytherapy; second, minimal moving of the implanted patient; and, third, a workflow permitting rapid turnover of the heavy load of patients expected at our institution. With all that in mind, we decided that the best design approach would be to provide a setting in which patient preparation, applicator placement, imaging with the implant in place, treatment planning, and treatment delivery could all be performed in one area.

22.2.1 Equipment

In brachytherapy, different 3D imagers can be used for different tumor sites and for different steps within the workflow preplanning, implantation, contouring, reconstruction, and dosimetry. By 2005, the American Image-Guided Brachytherapy Working Group and the Groupe Européen de Curiethérapie—European SocieTy for Radiotherapy and Oncology (GEC-ESTRO) had proposed that MRI should be used for imaging in cervical cancer brachytherapy (Haie-Meder et al., 2005; Pötter et al., 2006). Given the growing literature of reports regarding multiple tumor sites (Jolicoeur et al., 2011; Register et al., 2013), we decided to acquire, along with a CT scanner and an ultrasound system, an MRI scanner for our new suite. We chose a closed magnet scanner because its higher magnetic field offers better image quality. This equipment was intended to guide, verify, image, and aid in creation of a treatment plan. For that imaging armada to be of efficient use in IGBT, we also required applicators compatible with CT and MRI scanners.

22.2.2 Safety Features

22.2.2.1 For Patients

From the patients' perspective, safety features of our program include: limiting the risk of infection, ensuring an adequate and rapid response in case of emergencies, and delivering the best treatment achievable with less pain and greater comfort. Therefore, we conceived the brachytherapy suite environment to be completely sterile, respecting the strictest and most modern safety norms for an operating room. Patients would remain in the suite under the care of a radiation oncologist, an anesthesiologist, a respiratory therapist, a radiation therapist (RTT), and nurses.

In addition, all staff present in the suite, including the orderlies and physicists would be highly trained to respond to emergencies, either medical or radiation related, at their respective levels. Careful care was given to the choice of monitoring equipment to make sure all items were compatible with all our imaging modalities, including the MRI scanner. The monitoring equipment and a live video feed of the patient had to be accessible throughout the suite.

22.2.2.2 For Staff

Safety implies minimizing injuries and adequate response to emergencies. Therefore, we ensured the maneuverability and interchangeability of stretchers and operating tables. All three operating rooms were also adequately shielded for radiation. A high-dose rate (HDR) afterloader was chosen to reduce the risk of radiation exposure. Adequate training on the safe use of each modality was to be guaranteed, along with regular refresher courses.

22.2.3 Description of Charles LeMoyne Hospital's Brachytherapy Suite

The IGBT suite (Figure 22.1), consisting of one confined sterile environment, includes:

- One preparation and recovery room with X-ray capacity (mobile X-ray), emergency cart, pharmacy, access to electronic files, and personalized monitoring.

- One shielded procedure room equipped with a dedicated large-bore CT on-rail, a Siemens SOMATOM® Sensation with a custom-made Trumpf operating room table.

- One shielded procedure room with a Faraday cage. The room can be split in two by double Faraday cage sliding doors. The equipment is a large-bore 1.5T MRI scanner, a Siemens MAGNETON® Espree.

- One shielded room reserved to accommodate multi-fractionated courses of treatment such as breast and H&N, and for implant removal.

- The last imaging device consists of a mobile BK medical ultrasound used for implant guidance in all cases.

- Other equipment used for implant localization includes several scopes from Storz, all compatible with our audiovisual network.

- One HDR mobile afterloader (Flexitron® V3.2, Elekta Brachytherapy), that can treat in all three rooms.

- The electronic files and images are readily available on monitors in all areas.

- Planning systems Oncentra and Pinnacle are readily accessible in all areas. Pinnacle is used for contouring.

- A network of monitoring and video feeds for all patients under or recovering from anesthesia, as well as patients in any treatment room (Figure 22.1).

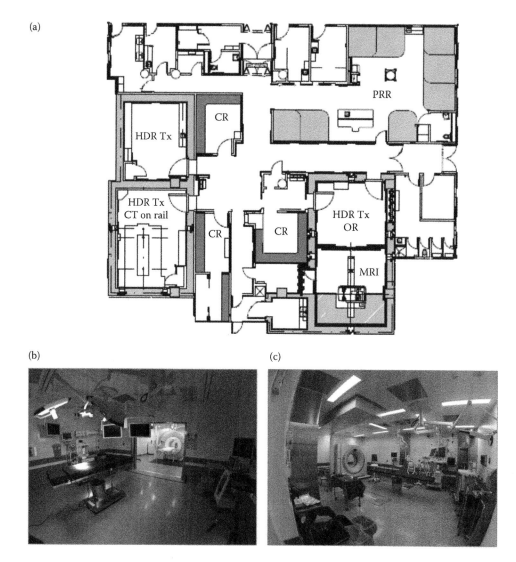

FIGURE 22.1   Images of the suite: (a) floor plan, panoramic view of (b) MRI intervention room and (c) CT intervention room.

22.3 MRI IN BRACHYTHERAPY: IMPLEMENTATION CHALLENGES

Once we decided to have MRI in our brachytherapy suite, we had no model to follow. In order to meet all challenges that came along with our decision, we needed the involvement and close collaboration of both radiation oncologists and physicists in the design process for the facility.

22.3.1 Construction

All major equipment including the afterloader was purchased in advance (in 2008), ensuring the most accurate structural plan, before starting the complex construction phase. The selection of the MRI scanner was a challenge: it had to be compatible with the air exchange and the negative pressure required in a sterile environment, without creating any additional pressure on the mechanical room. The next major impediment impediment was the question of how to design a shielded procedure room, including a Faraday cage, in a sterile environment in which anesthetized patients were to be kept under monitoring.

22.3.2 Compatible Materials and Applicators

The challenge was to find MRI-compatible equipment. We acquired an MRI-compatible anesthesia cart, stretchers, and operating room materials, as well as wireless monitoring equipment that would permit a strict monitoring of patients through shielding and a Faraday cage. MRI-compatible applicators for brachytherapy are available for gynecological cancers since the acceptance of the practice (Haie-Meder et al., 2005; Pötter et al., 2006; Viswanathan et al., 2012). For other sites, we had to be more creative and the search for suitable applicators remains an ongoing task.

22.3.3 Functionality

One important consideration was the fact that our afterloader was not MRI-compatible. Therefore, in order to deliver treatment in the MRI room, we had to conceal the magnet behind special doors, which not only splits the room but also serve as a noise barrier. Furthermore, operating MRI scanners and interpreting the images requires either additional personnel or expertise. The choice was made to provide training to RTTs and medical physicists in MRI technology. This turned out to be a sensible choice because we now have highly skilled RTTs. Nevertheless, adoption of MRI-guided IGBT may place additional pressure on the number of patients treated in the brachytherapy suite because of the additional time it takes to acquire images compared to conventional IGBT. Since the intent was to have a pace of four to five patients per day at maturity, we had to be creative in mapping the process, optimizing each imaging device at each step for maximum patient throughput (Figure 22.3).

22.3.4 Finding the Right MRI Sequences for Each Situation

Because electron density does not play a significant role in brachytherapy dose calculations at the ^{192}Ir energy level, MRI-based dosimetry per se is possible. At this point, the challenge is image quality and applicator reconstruction. Identification of applicators on MRI images is not as simple as on CT images: cardiac, bowel, and respiratory motions, and the presence of applicators, may cause distortion in the magnetic field and alter image quality. We acquire T2-weighted sequences for contouring, whereas T1-weighted sequences are most helpful with applicator reconstruction. For gynecological cases, applicators can be reconstructed with the Oncentra library. Work is ongoing on the accurate reconstruction of other implant devices. We are currently working on alternate sequences that could help with both contouring and reconstruction (Figure 22.2).

22.4 3D IGBT WORKFLOW

3D imaging is used at every step of our brachytherapy planning. First, at preplanning, we use MRI for the localization and definition of the target volume. In the interest of keeping

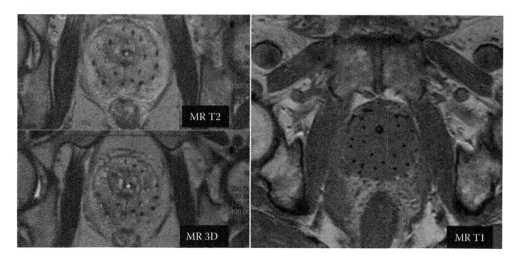

FIGURE 22.2   MRI sequences used in prostate brachytherapy: MR T2 are T2-weighed images used for the purpose of contouring, MR T1 are used for reconstruction but the catheter tips are sometimes difficult to see. MR 3D could help with both reconstruction and contouring.

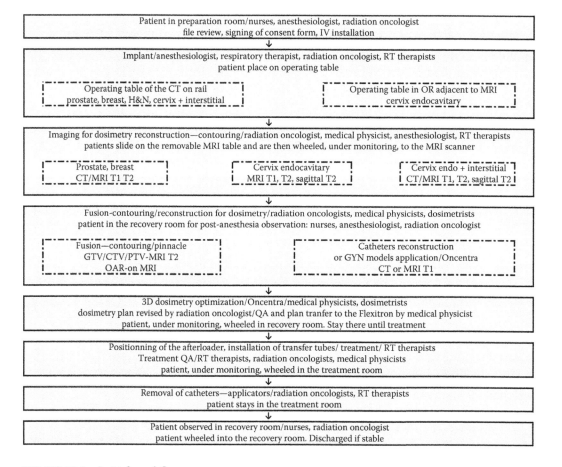

FIGURE 22.3   Initial workflow.

the procedure time as short as possible, all implants are guided by ultrasound. Subsequently, the patients undergo CT or MR imaging, depending on the treatment site. Since patients are scanned daily, during or after the implant, the overall increase in the procedure time is substantial. Another area of concern for interstitial implants is the uncertainty associated with needle-tip localization on MRIs. As a result, our initial workflow relied on CT/MRI co-registration for most sites (Figure 22.3).

22.5 EXAMPLES OF STANDARD PROCEDURES FOR CLINICAL IMAGE-GUIDED ADAPTIVE HIGH-DOSE BRACHYTHERAPY

22.5.1 Image-Guided Interstitial Breast Brachytherapy

Our indications include breast re-irradiation, tumor cavity boost in cases with positive margins, and accelerated partial breast irradiation (APBI). We integrated 3D imaging techniques throughout the steps. In the 4–6 weeks following surgery, preplanning the CT scan and MRI in the supine position are done to localize the surgical bed. Details of the procedure have been described elsewhere (Jolicoeur et al., 2011). The brachytherapy implantation is performed under local anesthesia and conscious sedation. The implant is ultrasound guided using a surface probe. Once the catheters are in place, a CT and an MRI scan are acquired. Images are then co-registered and contouring is done on the MRIs. The CT scan is used for catheter reconstruction. After dosimetric planning, patients receive several fractions. The implant is removed after the last treatment.

22.5.2 Image-Guided Interstitial Prostate Brachytherapy

HDR brachytherapy is one of the safest techniques used for prostate dose escalation (Martinez et al., 2002). MRI gives better visibility of the prostate gland compared to CT and ultrasound (McLaughlin et al., 2005). We use HDR prostate brachytherapy as a boost to EBRT, for recurrences of previous irradiation and for monotherapy. All patients have a preplanning multi-parametric MRI to evaluate the size of the prostate, the presence of a nodule, or capsular involvement. After regional anesthesia, the transperineal implant is placed in the classical manner under transrectal ultrasound (TRUS) guidance. The catheters are secured using an MRI-comaptible solidifying material, which is then sutured to the perineum. CT and MRI images are acquired. The images are then co-registered and contours are drawn on the MRI whereas catheter reconstruction is done on CT images.

22.5.3 Image-Guided Cervical Cancer Brachytherapy

Performed one day before the procedure, the preplanning MRI helps to adequately assess the extent of the tumor and select the optimal type of brachytherapy technique (e.g., intra-cavitary versus interstitial). The patient is prepared under general or regional anesthesia. Then the implant is put in place under transabdominal ultrasound guidance. We often use a combination of intracavitary and interstitial brachytherapy. A post implant MRI is performed with legs down. Contouring is done on the T2 images, while reconstruction is done

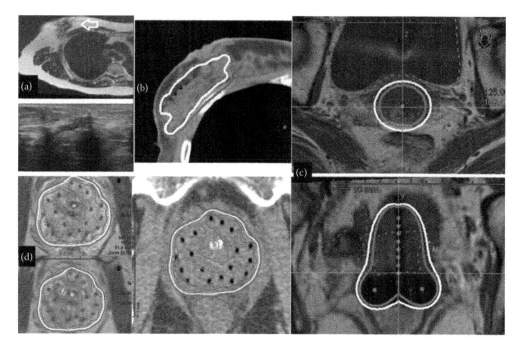

FIGURE 22.4 Brachytherapy images for different sites: (a) breast implant image: preplanning MRI and ultrasound for implant guidance, (b) breast implant dosimetry, (c) cervical intracavitary dosimetry, and (d) prostate dosimetry.

on the T1 images using the Oncentra applicator library. Dose calculations are performed and overlaid on the MRI images (Figure 22.4).

22.6 CONCLUSION

3D imaging is the cornerstone of modern IGBT. We postulate that multimodality imaging devices, when used in brachytherapy, will level out the high dependence of brachytherapy on specific physician skills. Furthermore, MRI-guided IGBT improves outcomes for cervical cancers, and it has great potential for the treatment of prostate and breast cancers. Altogether, these three count for more than a third of brachytherapy practice. Therefore, an MRI scanner should be part of the 3D imaging capacity of a brachytherapy suite built in the twenty-first century. We acknowledge the fact that implementing MRI-guided IGBT is time-consuming and requires a considerable amount of planning. Our comprehensive brachytherapy suite was designed to accommodate a heavy workload. Although we started at a slow pace, ramping up has not been an issue, and we can now perform three to four implants per day with their associated imaging, planning, and treatment. We treated 650 patients between January 2013 and October 2015.

In parallel, we are working on developing MRI-based brachytherapy for diseases other than cervical cancer. In the near future, we will be able to perform reconstruction and dose planning on MRI images for other tumor sites.

REFERENCES

Haie-Meder C., Potter R., Van Limbergen E. et al. Recommendations from Gynaecological (GYN) GEC-ESTRO Working Group *. *Radiother Oncol.* 2005;74:235–245

Jolicoeur M., Racine M.L., Trop I. et al. Localization of the surgical bed using supine magnetic resonance and computed tomography scan fusion for planification of breast interstitial brachytherapy. *Radiother Oncol.* 2011;100(3):480–484.

Martinez A.A., Gustafson G., Gonzalez J. et al. Dose escalation using conformal high-dose-rate brachytherapy improves outcome in unfavorable prostate cancer. *Int. J. Radiat. Oncol. Biol. Phys.* 2002;53(2):316–327.

McLaughlin P.W., Troyer S., Berri S. et al. Functional anatomy of the prostate: Implications for treatment planning. *Int. J. Radiat. Oncol. Biol. Phys.* 2005;63(2):479–491.

Pötter R., Haie-Meder C., van Limbergen E., Barillot I. et al. Recommendations from Gynaecological (GYN) GEC-ESTRO Working Group * (II). *Radiother Oncol.* 2006;78(1):67–77.

Register S.P., Kudchadker R.J., Levy L.B. et al. An MRI-based dose-response analysis of urinary sphincter dose and urinary morbidity after brachytherapy for prostate cancer in a phase II prospective trial. *Brachytherapy.* 2013;12(3):210–216.

Schoeppel S.L., Ellis J.H., LaVigne M.L. et al. Magnetic resonance imaging during intracavitary gynaecologic brachytherapy. *Int. J. Radiat. Oncol. Biol. Phys.* 1992;23:169–74.

Viswanathan A.N., Thomadsen B. et al. American Brachytherapy Society Cervical Cancer. *Brachytherapy.* 2012;11:33–46.

Wijrdeman H.K., Bakker C.J.G. Multiple slice MR imaging as an aid in radiotherapy of carcinoma of the cervix uteri: A case report. *Strahlenther Onkol.* 1988;164:44–47.

Sunnybrook Health Sciences Centre, Toronto, Canada

Amir Owrangi, Geordi Pang, and Ananth Ravi

CONTENTS

23.1 INTRODUCTION

The Odette Cancer Centre (OCC), of the Sunnybrook Health Sciences Centre, is home to one of the largest mono-institutional brachytherapy programs in North America. The volume of patients treated at OCC in 2015 was 300 high-dose rate (HDR) prostate cancer brachytherapy patients, 75 low-dose rate (LDR) prostate cancer patients, 35 cervical cancer patients, 100 endometrial cancer patients (35 of the gynecological patients were treated with interstitial brachytherapy), 5 penile cancer patients, and 1 anal canal cancer patient. The American Brachytherapy Society (ABS) has identified the OCC program as a center of excellence for brachytherapy training; as such, the institution plays host to several ABS-sponsored fellows annually. The highlights of the OCC brachytherapy program are the incorporation of transrectal ultrasound (TRUS), based intraoperative treatment planning for prostate HDR brachytherapy. With this initiative, the throughput of the program increased 10-fold, and is now in-line with ABS recommendations of performing real-time treatment planning for prostate HDR (Nag et al., 2001). Another area of major development within the program is in gynecological brachytherapy. Up until 2013, the gynecological brachytherapy program at the OCC was limited to two-dimensional (2D) brachytherapy with restricted access to magnetic resonance imaging (MRI), as was the case with the majority of centers in the Province of Ontario, Canada. As such, gynecological community of practice within Ontario advocated for improved access to MRI and wrote a recommendation document that was endorsed by the provincial funding body (Ravi et al., 2014). Owing to the large patient volumes at OCC, the institution was selected as one of the sites to receive a dedicated MRI system for brachytherapy image guidance. In addition, the institution decided to capitalize on this opportunity and through the generous support of donors, it was able to supplement the contributions by the province to create a unique MR suite that would combine an interventional MR suite, a brachytherapy treatment bunker, and an MR simulator all in one space. This novel suite will enable the brachytherapy program to investigate four fascinating new areas: (1) efficient MR-guided treatment planning—by leveraging an MR-compatible anesthetic delivery system, patients can be imaged while they are anesthetized, thereby, dramatically reducing the time between implantation and treatment plan creation; (2) MR-guided real-time applicator guidance—by using an MR-compatible heads-up display, it is hoped that applicator placement and needle implantation can be performed with greater accuracy avoiding critical organs while being optimally placed within the target; (3) delivery monitoring—with a shielded bunker, it may be possible in the future with advances in afterloader technology to be able to image the patient in the MR bore as they are being treated with the HDR source; and (4) tumor response—with the magnet adjacent to the existing brachytherapy suite and it being situated in a radiation-shielded bunker itself, it is possible to perform functional assessment of the tumor burden immediately after (e.g., within an hour) and during treatment delivery to assess efficacy/functional changes.

23.1.1 Gynecological Brachytherapy at OCC

Currently, at the OCC, limited disease cervix cancer is treated with 5 weekly HDR fractions of 550 cGy. Alternately, if patients have extensive disease, they undergo interstitial brachytherapy. The choice of technique and applicator for all gynecological brachytherapy is dependent on geometry and extent of the disease. The OCC workflow will utilize the dedicated MR suite to provide unprecedented guidance and dosimetric accuracy while also limiting the impact on patient throughput. This will be accomplished by implanting the patients in the MR suite, while they are under general anesthetics. Once the images have been acquired, treatment planning can immediately commence while the patient recovers from anesthetics; thereby, improving the efficiency of the workflow by at least 2 hours. This will facilitate MR-guided brachytherapy for each fraction of brachytherapy. This workflow will also enable MR-guided applicator placement, which in turn has the desired consequence of facilitating planning and achieving optimal dosimetry. With the patient in the lithotomic position on an MR-compatible Zephyr patient transfer system

(Diacor, USA), in conjunction with in-house fabricated MR-compatible stirrups, the applicator is to be inserted just outside the MR bore. Immediately after applicator insertion, the placement can be evaluated using MRI by lowering the patient's legs and sliding the patient from the end of the couch until they are in an ideal imaging position.

23.1.2 Prostate Brachytherapy at OCC

Prostate cancer patients have the option to be treated with LDR or HDR brachytherapy. Currently, LDR at OCC is practiced using a TRUS-based preplanning approach, and postimplant dosimetry is conducted using computed tomography (CT) imaging. With access to a dedicated brachytherapy MR suite, postimplant dosimetry will be transitioned to MR-based evaluation (D'Amico et al., 2000; Van Gellekom et al., 2004; Bomers et al., 2012). HDR brachytherapy at OCC is performed using intraoperative TRUS-based treatment planning. The use of MR in the HDR brachytherapy workflow is primarily for the identification of the dominant intraprostatic lesion (Chung 2000–2015a,b; Morton 2000–2015; Niranjan et al., 2013). With the novel suite at OCC, patients treated with prostate HDR can be easily transferred (using the Zephyr patient transfer system) from the standard operating room where TRUS-based catheter insertion will be completed, through a shielded pass-through door into the MR suite with only one transfer to a movable MR couch. Areas of further investigation will focus on determining the value of MR-based treatment planning for prostate brachytherapy in both the HDR and LDR settings. Additional areas of exploration will focus on monitoring treatment response on timescales that have been difficult to achieve previously, within an hour, to just minutes after treatment.

23.2   LAYOUT AND DESIGN

In Canada, the use of nuclear energy is governed by the Federal laws, that is, the Nuclear Safety and Control (NSC) Act and its regulations (http://laws-lois.justice.gc.ca). Based on the regulations, an HDR unit is a Class II prescribed equipment and the facility housing the unit is a Class II Nuclear Facility. Licenses are required to build and operate a Class II Nuclear Facility.

23.2.1   General Layout

A general layout of an MR-HDR room is shown in Figure 23.1 (i.e., the MR-brachytherapy suite, Zone 4). Compared to a conventional HDR room, special considerations are required here due to the addition of the MRI machine inside a HDR room. These include: (1) the MR-HDR room should have a sufficient size to accommodate an MRI machine, that is, a minimum of ~9 m × 5.5 m is typically required. (2) The weight of the MRI machine should be considered as well, especially when there is an occupancy directly below the MR-HDR room. The structure of the building should be assessed by a structural engineer to ensure the structure can support the extra weight of the MR machine and the materials used for shielding. The MRI system that is going to be installed in OCC MR-brachytherapy suite is a Philips Ingenia® 1.5 T MR-RT system with 70 cm bore size. (3) The room should be shielded for both radiation (from the HDR source) and magnetic field (from the MRI machine). The spatial order of the shielding is also important. Practically, it makes more sense to place

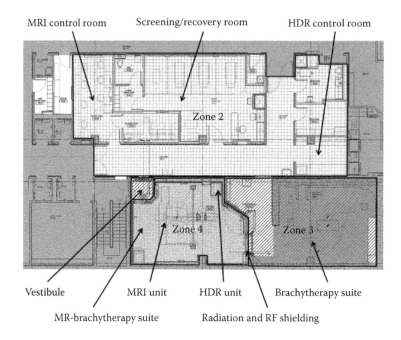

FIGURE 23.1 Overall facility layout showing the MR-brachytherapy suit and the ACR zone definitions used in this facility. Zone 1 (public) is not shown, but it is understood to be the outside space surrounding the current diagram. Zone 2, which is for patient check in, change room, waiting area, lockers, screening, and recovery has two access point from Zone 1 and two access points to Zone 3, which is the small vestibule in the top left corner of the MR suit and the opening between the brachytherapy suit and MRI suit as well as the whole brachytherapy suit. Zone 4 is the whole MRI suite.

the radiofrequency (RF) shielding (a Faraday cage) inside the radiation shielding since the materials used for radiation shielding would have to be MRI-compatible if placed inside an RF cage. (4) Owing to the presence of an MRI machine, there are a lot more cables and pipes that need to penetrate the radiation shielding from outside to inside the MR-HDR room, which requires special attention. (5) Both radiation safety and MR safety systems need to be in place.

23.2.2 Radiation Shielding

Radiation shielding design for the MR-HDR room is not much different from that for a conventional HDR room except that there are many more cables and pipes in the present case. Table 23.1 shows the effective dose limits in Canada. The ALARA design limits (which

TABLE 23.1 Effective Dose Limits in Canada

Personnel Classification	Legal Limits (mSv)	ALARA Design Limits (mSv)
Nuclear energy worker	50 for 1 year 100 over 5 years	1 per year
Public	1	0.05 per year

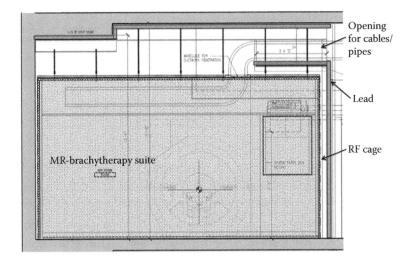

FIGURE 23.2 Schematic diagram of a cross-section showing the shielding design surrounding a large opening in the wall to let cables, pipes, and ducts enter an MR- HDR room.

are much lower than the legal limits) should be used for radiation shielding design. This is based on the ALARA principle (exposures and likelihood of exposure should be kept as low as reasonably achievable, economic and social factors being taken into account), one of the most important principles in radiation protection (ICRP, 1990).

Owing to the presence of the MR machine, there are many more cables that need to penetrate the shielded walls than a conventional HDR room. In addition, these cables should be accessible for maintenance and repair. Thus, large openings in shielded walls are needed, which can be a challenge for shielding design. Figure 23.2 shows a feasible solution. The wall at the control console is cut open at the top, close to the ceiling with the height of the opening of ~1 m. A second layer of lead is placed under the ceiling shielding to block the gamma rays that would pass through the opening without being attenuated.

23.2.3 Radiation Safety Systems and Security Issues

Based on the Class II Nuclear Facilities and Prescribed Equipment Regulations, there are a number of radiation safety systems that shall be installed in an MR-HDR room to prevent accidental exposures. These include door/entrance interlocks, warning lights, independent radiation monitoring system, emergency off buttons, viewing system, warning signs, and tools and equipment for source emergencies.

In addition, there are new source security regulations in Canada (CNSC's Security of Nuclear Substances: Sealed Sources). Different levels of security measures are required for different source categories. A single-source HDR unit (with ~10 Ci or less of ^{192}Ir) belongs to category 3 (a dual-source HDR would be of category 2). All security measures required by the regulations shall be in place. In particular, there shall be a minimum of two different physical barriers, to prevent unauthorized access to the HDR unit in storage, and the trustworthiness and reliability of all persons who require access to the HDR unit shall be verified.

23.3 MRI SYSTEM SAFETY

The safety considerations for separate MRI suites and brachytherapy suites are very well understood and developed. However, there is not much prior knowledge and documentation regarding the integration of the two suites. Proper implementation of the MRI safety zones and enforcing strict personnel access after going through proper safety training are vital for safe operation of MR-brachytherapy suites.

23.3.1 MR Zone Definitions

To be consistent with the American College of Radiology (ACR) recommendations, standard MRI zone definitions were applied (Kanal et al., 2002, 2013). The brachytherapy facility described here is part of the larger program for MRI guidance in radiation therapy within the OCC. A schematic of the MR-brachytherapy suite within the brachytherapy facility is shown in Figure 23.1. Zone 1 (not shown in Figure 23.1) is any area outside the brachytherapy facility that is open to general public access. Zone 2 is the area that is the connection between Zone 1 and the MR-brachytherapy suite and needs to be restricted with card key access. This includes the main brachytherapy hallway, patient recovery area, nursing station, patients change rooms/washrooms, and MRI/brachytherapy control areas. All persons, including patients, visitors, attending physicians and others, must be accompanied by safety-trained personnel to enter this area. The main purpose of this zone is to restrict further public access to the MR-brachytherapy suite until they are completely prescreened and escorted by the MRI staff. Final preparations and instructions are given to scanning participants prior to entry into Zone 3. This area is the last barrier to prevent any potential high magnetic field-related incidents; final checks in this zone ensure that any ferromagnetic materials do not accidentally enter the MR-brachytherapy suite. This zone includes a small vestibule that is shown in the top left corner of the MR-brachytherapy suite in Figure 23.1 and a larger vestibule that is the connection between MR-brachytherapy suite and the conventional brachytherapy operating room. As per ACR recommendations, all of the entrances into Zone 4 have ferromagnetic detection systems that alarm when metallic materials cross the barrier. However, it should be reiterated that their use is in no way meant to replace a thorough screening program, but rather act as a supplement (Kanal et al., 2002, 2013). Zone 4 is the MRI scanning room where the 5-Gauss line should be clearly delineated on the floor. The 5-Gauss line demarcates the regions where MRI-safe equipment can be used and beyond which MRI-conditional equipment can be placed. In our case, nonferrous IV poles, wheelchairs, and medical supplies that have MRI-safe or MRI-conditional sign can enter Zone 4.

23.3.2 Personnel Authorizations

According to the ACR recommendations, all individuals working in Zones 3 and 4 should go through MR safety courses at least once a year and their successful completion of the safety courses needs to be documented. Upon successful completion of these courses, they will be referred to as "MR personnel." Based on the ACR recommendations, there are two levels of MR personnel. Level 1 is referred to those who finished minimal safety education to ensure their own safety as they work in Zones 3 and 4. Level 2 personnel are those who have been more extensively trained and educated in the broader aspects of MR safety issues, including issues relating to thermal loading or burns, or direct neuromuscular excitation from MR gradients. All those that have not successfully complied with this MR safety defined by the MR safety director of that facility within the previous 12 months shall be designated as "non-MR personnel."

23.4 SPECIALIZED AUXILIARY EQUIPMENT

In order to enable image-guided implantation and MR treatment monitoring for brachytherapy, there are a number of specialized systems and equipment that are required.

23.4.1 Anesthesia Cart

It is the intention to perform applicator insertions in the suite, as such an MR-compatible solution to provide anesthesia is required so that real-time image guidance can be achieved. This was addressed by accounting for the necessary gas supplies into the MR-brachytherapy suite, and also through the selection of an MR-compatible anesthesia cart: the Aestiva/5 MRI (GE Healthcare, Madison, Wisconsin). The Aestiva system is safe for use in magnetic environments up to 3 T. On a pragmatic level, the Aestiva system is compatible with the fleet of GE anesthetic carts used on the Sunnybrook campus, facilitating its adoption by the respiratory technologists and anesthesiologists.

23.4.2 MR-Safe Monitor

Real-time guidance of applicator placement will not be possible without the visual feedback of the applicator's position relative to the patient's anatomy. While the patient is in the MR bore, images can be acquired in real time; however, typically, they are reconstructed and displayed on viewing stations outside the MR-brachytherapy suite, making it impossible to use this information to guide the implant efficiently. To overcome this, OCC will use an MR-compatible heads-up display SensVue (Invivo, Florida), a 32″ high-resolution display safe in magnetic environments up to 3 T. It was initially intended for fMRI studies, but has been repurposed to provide imaging feedback during implantation.

23.4.3 HDR Afterloader

Currently, there are no HDR afterloaders that are MR-compatible. However, if the Flexitron® afterloader is placed outside the 5-Gauss line and is physically tethered to the ground, there is limited risk of using the HDR unit in the MR suite. OCC has proactively designed the new MR suite to accommodate the afterloader in two locations with physical anchor points. Additionally, all the radiation safety precautions outlined earlier will enable safe brachytherapy treatments in the novel suite.

REFERENCES

Bomers JG, Sedelaar JP, Barentsz JO et al. MRI-guided interventions for the treatment of prostate cancer. *Am J Roentgenol.* 2012;199(4):714–720.

Chung H. Focal salvage HDR brachytherapy for the treatment of prostate cancer. In: *ClinicalTrials. gov (Internet)*. Bethesda, MD: National Library of Medicine (US). 2000–2015a. Available from: NCT01583920.

Chung H. Pilot Study of Whole Gland Salvage HDR Prostate Brachytherapy for Locally Recurrent Prostate Cancer. In: *ClinicalTrials.gov (Internet)*. Bethesda, MD: National Library of Medicine (US). 2000–2015b. Available from: NCT02560181.

D'Amico A, Cormack R, Kumar S, Tempany CM. Real-time magnetic resonance imaging-guided brachytherapy in the treatment of selected patients with clinically localized prostate cancer. *J Endourol.* 2000;14:367–370.

ICRP. *International Commission on Radiological Protection.* 1990 Recommendations of the International Commission on Radiological Protection, ICRP Publication 60, Annals of the ICRP 21 (1–3), Elsevier Science, New York, 1991.

Kanal E, Borgstede JP, Barkovich AJ et al. 2002. American College of Radiology white paper on MR safety. *Am J Roentgenol.* 178:1335–1347.

Kanal E, Barkovich AJ, Bell C et al. ACR guidance document on MR safe practices: 2013. *JMRI* 2013;37:501–530.

Morton G. A phase II trial of high dose-rate brachytherapy as monotherapy in low and intermediate risk prostate cancer. In: *ClinicalTrials.gov (Internet)*. Bethesda, MD: National Library of Medicine (US). 2000–2015. Available from: NCT01890096.

Nag S, Ciezki JP, Cormack R et al. Intraoperative planning and evaluation of permanent prostate brachytherapy: Report of the American Brachytherapy Society. *Int J Radiat Oncol Biol Phys.* 2001;51(5):1422–1430.

Niranjan V, Loblaw A, Ravi A. Real-time integration of MR imaging for prostate high-dose-rate brachytherapy using 3D slicer. *Brachytherapy.* 2013;12:S34.

Ravi A, Fyles A, Surry K et al. Cancer Care Ontario; Imaging Strategies for Definitive Intracavitary Brachytherapy of Cervical Cancer. 2014; https://www.cancercare.on.ca/common/pages/UserFile.aspx?file-Id=309545.

Van Gellekom MP, Moerland MA, Battermann JJ, Lagendijk JJ. MRI-guided prostate brachytherapy with single needle method: A planning study. *Radiother Oncol.* 2004;71:327–332.

CHAPTER 24

Princess Margaret Cancer Centre, Toronto, Canada

Marco Carlone, Teo Stanescu, Tony Tadic, Kitty Chan, Colleen Dickie, Stephen Breen, Hamideh Alasti, Kathy Han, Michael Milosevic, Cynthia Ménard, Alexandra Rink, Anna Simeonov, and David Jaffray

CONTENTS

24.1 CLINICAL OBJECTIVES OF THE SUITE

Three-dimensional treatment planning is the modern standard for brachytherapy as computed tomography (CT) imaging has widely replaced projection imaging in brachytherapy practice. This has led to volume imaging-based planning as a standard for gynecologic, prostate, and other types of brachytherapy plans. Despite the clear improvements in planning techniques and protocols, CT-based imaging for brachytherapy still lack the necessary soft-tissue contrast required to visualize target areas appropriately. This is especially true for gynecological cancers (Haie-Meder et al., 2005; Pötter et al., 2006; Hellebust et al., 2010; Dimopoulos et al., 2012). For prostate brachytherapy, where focal disease is increasingly targeted, CT imaging is also unable to sufficiently discriminate subglandular structures.

The facility described here was designed to advance brachytherapy into the era of magnetic resonance (MR)-guided adaptive brachytherapy. The suite integrates a radiation-shielded brachytherapy delivery room with fully functional diagnostic quality magnetic resonance imaging (MRI) as well as C-arm and ultrasound imaging. It was conceived to allow rapid assessment of applicator placement with exquisite MRI so that instant feedback can be utilized to readjust both applicator and needle placement. Further, this suite also allows treatment delivery within the same space as applicator insertion so as to minimize patient transfers and movement between applicator insertion and imaging. Finally, the facility is meant to also allow treatment delivery monitoring and response, which we hope will lead to further innovations in brachytherapy practice.

24.2 LAYOUT AND DESIGN

The brachytherapy facility described here is a component of a larger MRI guidance facility built within the Princess Margaret Cancer Centre. A schematic of the entire MRI guidance facility is shown in Figure 24.1. The facility is made up of three collinear rooms linked by ceiling-mounted rails, which permits a 1.5 T IMRIS MRI to be moved to any of the three adjacent rooms. The middle room is the MRI simulation room, which is adjacent to a linac bunker and a brachytherapy treatment suite. These last two rooms have radiation shielding such that normal patient treatments can be performed within them. This chapter will describe the configuration and design of the brachytherapy room when it is configured to interface with the MR system.

When the MR-brachytherapy doors (Figures 24.1 and 24.2), which provide both radiation and radiofrequency (RF) shielding, are closed, the adjacent MR room can be used as a standalone MR simulation suite. The radiation-shielding doors are unique in that a pair of sliding doors was required so that there was sufficient space for the doors to open into wall pockets between the brachytherapy and MR simulation room. Because of this nonstandard configuration, S-shaped door ends provided by Nelco (Boston, Massachusetts, USA) were used to minimize leakage radiation when the doors were fully closed.

The brachytherapy room features sufficient RF shielding so that the space can also be used for MRI when the MR-brachytherapy doors are opened to allow the MR travel into the brachytherapy space. In this configuration, MRI can be completed without moving the patient outside of the treatment suite; however, brachytherapy delivery is not possible since the MR-brachytherapy doors between the brachytherapy suite and the MR suite are open. After the brachytherapy applicator has been positioned as desired (which can be verified by MRI), the MR scanner is moved back into the middle (MR) simulation room and the radiation-shielded doors are then closed, thus completing the radiation shield and enabling brachytherapy treatment delivery.

One of the design challenges was to accommodate non-MR-compatible equipment in the brachytherapy treatment area. The remote afterloader and ancillary imaging equipment are not available in MR-safe configurations, and so the facility design had to accommodate this. A further complication of this requirement was the need to have this equipment in an operating state while the MR scanner may be employed in the brachytherapy room. For instance, this is a requirement of radiotherapy quality assurance (QA) programs that any equipment that goes through morning QA procedures be left in an operating state until used clinically. The solution was to build an equipment closet in the brachytherapy treatment room behind RF-shielded doors. When not needed, the remote afterloader, C-arm, or

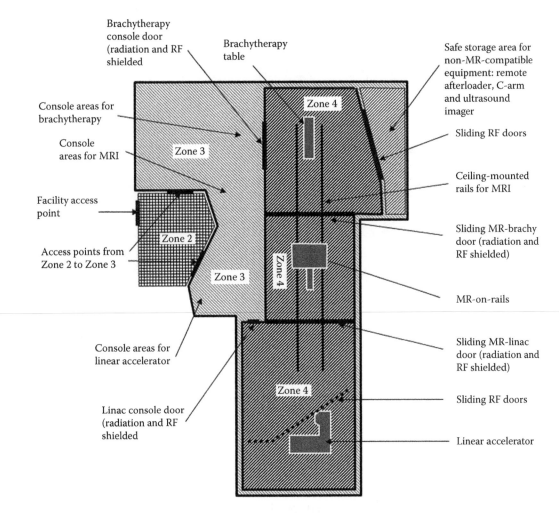

Brachytherapy console door (radiation and RF shielded

Brachytherapy table

Safe storage area for non-MR-compatible equipment: remote afterloader, C-arm and ultrasound imager

Console areas for brachytherapy

Zone 4

Console areas for MRI

Zone 3

Sliding RF doors

Facility access point

Zone 2

Ceiling-mounted rails for MRI

Access points from Zone 2 to Zone 3

Zone 3

Zone 4

Sliding MR-brachy door (radiation and RF shielded)

MR-on-rails

Console areas for linear accelerator

Zone 4

Sliding MR-linac door (radiation and RF shielded)

Sliding RF doors

Linac console door (radiation and RF shielded

Linear accelerator

FIGURE 24.1 Overall facility layout showing the ACR zone definitions used in this facility. Zone 1 (public) is not shown but it is understood to be the outside space surrounding the current diagram. Zone 2, which is for patient check in, change room, waiting area, lockers, and screening, has a single access point from Zone 1 and two access points to Zone 3, which is where the brachytherapy, MRI, and linac consoles are located. Zone 4 follows a nonstandard definition in that all three suites, the brachytherapy, MRI, and linac suites, are always Zone 4, regardless of the location of the MRI and the open/close state of the MR-brachytherapy and MR-linac doors.

ultrasound unit is stored in a designated location in this closet. With the RF doors closed, the equipment is both RF quiet to the MR scanner and far enough away so as to not pose any safety or interference problems to the MR scanner or imaging.

24.3 SYSTEM SAFETY

Patient and staff safety was the principal concern of the design of the brachytherapy suite described here. Safety considerations for radiation therapy and MRI are both well developed, with well-known best practices. When integrating these safety systems, it became apparent that there was very little overlap in the safety principles and regulatory requirements of radiotherapy

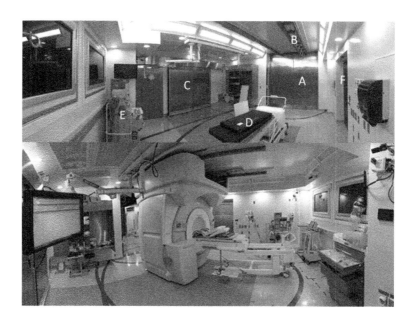

FIGURE 24.2   Top: A wide-angle view of the brachytherapy suite without the MRI. Bottom: View of the brachytherapy facility with the MRI moved into the brachytherapy suite. The labeling of the room for both images is as follows: (A) the MR-brachytherapy sliding doors, separating the MRI simulation and brachytherapy suites, (B) the ceiling rails for MRI support and movement, (C) the doors to the equipment cabinet RF shielded, (D) the procedure table, (E) anesthesia equipment, and (F) the door to the console area.

and MRI, which complicated the design. Further, many brachytherapy procedures require equipment and accessories that are not available with MR-safe devices. The solutions that were developed were to incorporate a systems approach to safety involving high-level principles for both brachytherapy and MRI. This involved applying conventional MR safety zone definitions and implementing strict control personnel access. Each of these is addressed below.

24.3.1 MR Zones Definitions

Standard MR zone definitions (Kanal et al., 2002, 2013) were employed, other than for Zone 4. In this case, we instead chose the conservative approach where all three rooms where the MRI is able to be located, regardless of the actual location of the MRI, are considered Zone 4. In other words, the brachytherapy treatment room is classified as Zone 4 even when the MRI is in the MR simulation room, and the doors between the two rooms are closed. The reason was to ensure that any MR-unsafe equipment or devices would be controlled when used in the brachytherapy suite, and that the suite would be returned to an MR-safe state when the brachytherapy procedure or activity (e.g., QA, servicing) is completed, as would be the case if the MRI was physically located in the room.

24.3.2 Personnel Authorizations

The 2013 American College of Radiology (ACR) Guidance Document on MR Safe Practices (Kanal et al., 2002) was used to define different levels of MR safety trained staff (Level 1 and Level 2) as well as non-MR personnel, and supervision privileges. Since the magnet can be placed in three different rooms, it was decided to use different terms to describe room states. "High Field State" refers to when the MR scanner is present in the brachytherapy or linac room. "Non-high-field state" refers to the brachytherapy room/linac room without the presence of the MR scanner. Along with differentiating the imaging states of the rooms, personnel authorizations differed according to the level of MR safety training as defined by the ACR guidance document. Level 1 MR personnel are those who have accomplished minimal safety educational requirements to ensure their own safety as they work within Zones 3 and 4. Level 2 MR personnel are those with extensive training and education in the broader aspects of MR safety issues, including issues relating to thermal loading or burns, or direct neuromuscular excitation from MR gradients. Various institutional authorizations were granted according to staff involvement in daily QA activities, some of which include accessing the magnet room to perform certain tests. The staff involved in the magnet room QA activities was designated "authorized users," whereas staff only required to work within the facility but not enter the magnet room for QA procedures were given the distinction "authorized access." All staff expected to work within the facility required level 1 safety training before they were granted authorized access to the facility. Authorized users required additional QA-related and safety training. The "MRgRT (MR-guided radiotherapy) duty officer" role was designed to oversee all operations involving the MR scanner and MR safety in the MRgRT facility for the day. The MRgRT duty officer must be an MR-certified staff (Level 2 MR safety training). A "Radiation Medicine Program MR Safety Officer" has the responsibility to ensure MR safety for all Radiation Medicine Program MR-related activities.

24.4 AUXILIARY EQUIPMENT

A C-arm with flat-panel detector is necessary for non-MR-guided brachytherapy procedures such as high-dose rate (HDR) brachytherapy to the lung and esophagus. When the magnet is in the MR suite with MR-brachytherapy doors closed, images from an image intensifier acquired close to the fringe field line (5G) yielded significant distortion. Flat-panel-based C-arms do not exhibit any distortion at the 5G line and so were employed in this facility. An ultrasound machine was also included in the suite for standard transabdominal and transrectal imaging for gynecologic and prostate brachytherapy procedures. These are stored in the RF storage area. Other MR-safe auxiliary equipment was also incorporated in the suite to support interventional MR-guided brachytherapy procedures such as HDR prostate brachytherapy. This included an MR-safe physiological monitor, MR-safe anesthesia machine, and MR-safe camera system for patient monitoring during anesthesia; MR-safe display monitor for intraoperative image guidance; and MR-safe brachytherapy applicators, MR-safe procedural table, and MR conditional operation room (OR) instruments. Additional MR coils such as the endorectal coil were purchased to enhance image quality.

A novel patient transfer system allows for seamless/effortless patient transfer between the surgical operating bed, MRI simulation couch, CT simulation couch, and brachytherapy treatment couch using an air-based hovercraft system (Zephyr System, Diacor, USA).

This system is compatible with the brachytherapy immobilization and treatment applicators, as well as the external beam immobilization devices. The patient is positioned and immobilized on a patient transfer sled once, which then hovers between different treatment beds for various aspects of the treatment process. For brachytherapy-related treatment, this system eliminates patient transfers and hence minimizes the risk of applicator positional changes from the surgical suite, to planning, through to the brachytherapy room for radiotherapy treatment.

24.5 MR SYSTEM CHARACTERIZATION

The MR system is an adaptation of the IMRIS Visius Surgical Theatre (IMRIS, USA), which includes an MR scanner and a ceiling-mounted MR transport system. The scanner is a 1.5 T Magnetom® Espree (Siemens Medical Systems, Erlangen, Germany) featuring an open bore (70 cm) design and a short system length (125 cm cover-to-cover) to maximize patient access and comfort during procedures. The key MRI transport system components are: (a) ceiling-mounted rails spanning the three procedure rooms (Figure 24.2), (b) a magnet mover carriage consisting of a slew ring rotator and translation drive, (c) a mobile quench line carrier and cable management system, (d) collision detection system, and (e) associated software/hardware interface to provide system control and real-time status. The scanner can be rotated 180° in the simulation suite to allow for two distinct imaging configurations in the brachytherapy room. Specifically, at 180° rotation, the MRI advances with its diagnostic table first into the brachytherapy room, and at 0° rotation, the MRI travels with its service end first, enabling the use of a third-party patient table. When the diagnostic table is not used, it can be retracted from the bore and undocked. For each imaging configuration in the brachytherapy room, the MR was actively shimmed and tuned up as per the manufacturer guidelines. MRI imaging performance was characterized following standard guidelines provided by the ACR, American Association of Physicists in Medicine (AAPM), and National Electrical Manufacturers Association (NEMA), including advanced 3D distortion mapping and quantitative diffusion-weighting imaging testing.

24.6 EXAMPLE CLINICAL WORKFLOW: MRI-GUIDED PROSTATE HDR BRACHYTHERAPY

For prostate brachytherapy, MRI offers state-of-the-art staging and geometric delineation capabilities using T2-weighted and diffusion-weighted imaging to accurately identify subglandular regions of tumor burden. Also, superior definition of anatomic boundaries is significantly improved with MRI. The purpose of an MRI-only workflow for prostate cancer is to use the MRI to assist the needle placement during the insertion and to facilitate registration to other preinsertion MRI, thus minimizing registration errors. MRI guidance during the implant is also useful to account for gland swelling during the insertion, and imaging just prior to treatment minimizes any catheter position uncertainties. Three prostate HDR brachytherapy protocols are active in this suite: a whole gland boost to external beam radiotherapy (EBRT) for intermediate and high-risk patients where patients receive 15 Gy in 1 fraction; salvage treatment to patients who have failed prior treatment where 2

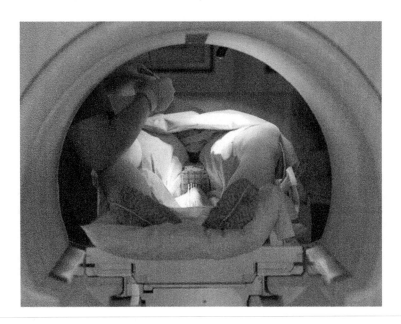

FIGURE 24.3 Image of the patient during a prostate HDR brachytherapy procedure as seen through the MRI bore. The template is also visible with some catheters inserted. Patient positioned on a raised support in frog-leg with pelvic tilt. The template is angled to manage pubic arch interference.

fractions of 13 Gy are focally administered to a tumor region defined by multiparametric MRI imaging (DCE-MRI and DWI); and a focal dose-escalation boost of 10 Gy to an intra-prostatic lesion (GTV) defined by multiparametric MRI combined with EBRT (76 Gy/38 fractions).

The workflow for all three protocols is similar. The patient arrives at the MR scanner and is prepared for treatment. Once placed on the MR scanner couch, the patient is anes-thetized using propofol sedation. The patient is positioned and immobilized in frog-leg on a raised patient support (Figure 24.3, Sentinel Endocoil Array, Siemens), which also anchors the template and endorectal coil. Navigation software (Aegis, Hologic, Inc.) is used to project template locations onto the targets and choose the needle depths. The template coordinates are selected manually, and 5F needles are stereotactically inserted two to four at a time. Needle positioning is verified by reimaging the gland only after each set of nee-dles that are inserted; this needle verification imaging is done between 4 and 15 times per insertion depending on the required number of needles and target size. The needle posi-tions appear as signal void areas under T2-weighted imaging. New template coordinates are then chosen until it is judged that sufficient needles have been inserted to cover the target area, either the whole gland for external beam boost or the subglandular targeted area for salvage or focal boost therapy.

After sufficient needles have been inserted and satisfactorily positioned, a high-resolu-tion 2-mm slice thickness final image set is acquired for planning purposes. Images are transferred to the planning system and a normal planning process takes place. During the time when treatment planning is taking place, the patient table is undocked from the MRI, and the MRI is moved back in the middle room, and the shielded doors are closed, thus preparing the treatment room for the delivery of the brachytherapy treatment. The patient is kept under anesthesia during planning and treatment delivery; once the dose is delivered, the needles and template are removed and the patient is recovered.

REFERENCES

Dimopoulos JC, Petrow P, Tanderup K et al. 2012. Recommendations from Gynaecological (GYN) GEC-ESTRO Working Group (IV): Basic principles and parameters for MR imaging within the frame of image based adaptive cervix cancer brachytherapy. *Radiother Oncol.* 103(1):113–122.

Haie-Meder C, Pötter R, Van Limbergen E et al. 2005. Gynaecological (GYN) GEC-ESTRO Working Group. Recommendations from Gynaecological (GYN) GEC-ESTRO Working Group (I): Concepts and terms in 3D image based 3D treatment planning in cervix cancer brachytherapy with emphasis on MRI assessment of GTV and CTV. *Radiother Oncol.* 74(3):235–245.

Hellebust TP, Kirisits C, Berger D et al. 2010. Gynaecological (GYN) GEC-ESTRO Working Group. Recommendations from Gynaecological (GYN) GEC-ESTRO Working Group: Considerations and pitfalls in commissioning and applicator reconstruction in 3D image-based treatment planning of cervix cancer brachytherapy. *Radiother Oncol.* 96(2):153–160.

Kanal E, Barkovich AJ, Bell C et al. 2013. ACR guidance document on MR safe practices: 2013. *JMRI* 37:501–530.

Kanal E, Borgstede JP, Barkovich AJ et al. 2002. American College of Radiology white paper on MR safety. *Am J Roentgenol* 178:1335–1347.

Pötter R, Haie-Meder C, Van Limbergen E et al. 2006. GEC ESTRO Working Group. Recommendations from gynaecological (GYN) GEC ESTRO working group (II): Concepts and terms in 3D image-based treatment planning in cervix cancer brachytherapy-3D dose volume parameters and aspects of 3D image based anatomy, radiation physics, radiobiology. *Radiother Oncol.* 78(1):67–77.

CHAPTER 25

Tata Memorial Hospital, Mumbai, India

Umesh Mahantshetty, Lavanya Naidu, Sarbani (Ghosh) Laskar, Ashwini Budrukkar, Jamema Swamidas, Siddhartha Laskar, and Shyamkishore Shrivastava

CONTENTS

25.1 INTRODUCTION

The burden of cancer in India is immense with over one million people being newly diagnosed with cancer every year and approximately 700,000 deaths. The Tata Memorial Centre, founded in 1941, is a national comprehensive center for the prevention and treatment of, and education and research in cancer and is recognized as one of the leading cancer centers in Asia. Approximately 37,000 cancer patients are registered and over 5000 undergo radiation therapy annually. The most common cancers are oral cavity, lung, cervix, breast, and esophageal cancers. The Radiation Oncology department is equipped with modern sophisticated state-of-the-art teletherapy and brachytherapy units offering the entire range of radiation treatments from conformal to IMRT with image guided radiotherapy (IGRT) to brachytherapy for most of the sites. The department is involved in the evaluation of newer technology to develop low-cost indigenous cobalt/linear accelerators and brachytherapy equipment.

25.2 HISTORY AND EVOLUTION OF BRACHYTHERAPY AT TATA MEMORIAL HOSPITAL

Brachytherapy at Tata Memorial Hospital (TMH) dates back to 1941 with the introduction of "radon seeds." An Indian physicist, Dr. Ramaya Naidu, who had worked as a postdoctoral student under Madame Curie during the late 1930s was responsible for setting up the radon plant at TMH. With the discovery of newer radioactive isotopes, the hospital acquired pre-loaded ^{60}Co and ^{137}Cs capsules in 1960. Manual afterloading techniques were introduced in 1972 using ^{60}Co sources. In 1976, we acquired ^{137}Cs tubes from the Bhabha atomic research centre (BARC) for use for intracavitary and interstitial brachytherapy. Manual afterloading ^{137}Cs sources for gynecological applications were initiated in 1981. Dr. K. A. Dinshaw introduced manual afterloading ^{192}Ir interstitial brachytherapy for the first time in the country at TMH in 1981. The ^{192}Ir sources used for the manual afterloading interstitial brachytherapy procedures were produced in the division of the Board of Radiation and Isotope Technology (BRIT) at the Bhabha Atomic Research Centre. The low-dose rate (LDR) remote afterloading units using ^{137}Cs and ^{192}Ir were acquired in 1986 and 1987, respectively. A further advancement in the brachytherapy facility came with the induction of the high-dose rate (HDR) microSelectron® (Elekta, Stockholm) unit using ^{192}Ir source in 1994. Endovascular radiation therapy was also started in 1997. Brachytherapy forms an integral part of many treatment protocols for various sites including gynecology, head and neck, breast, soft-tissue sarcomas (STSs), esophagus, etc. In the field of brachytherapy, TMH has continuously evolved to current modern brachytherapy practice and participation in various international collaborative research studies. Over the past two decades, we have treated approximately 6500 patients with brachytherapy comprising of cervix (3600), breast (800), esophagus (360), STSs (250), head and neck (300), bronchus and chest wall (60), rectum and anal canal (70), bile duct (30), and miscellaneous sites (400) including vagina, vulva, prostate, urethra, penile, and surface molds (Banerjee et al., 2014).

25.3 BRACHYTHERAPY INFRASTRUCTURE AND IMAGING AND PLANNING

The current brachytherapy infrastructure consists of a dedicated operation theatre with an adjoining HDR treatment room and two ^{192}Ir HDR afterloaders. The brachytherapy operation theatre is equipped for all major and minor procedures under general/spinal anesthesia, for breast and STSs, followed by intra-operative brachytherapy if feasible. Fluoroscopy and portable ultrasound units for guidance and a recovery room for post-procedure monitoring are also available. A wide range of brachytherapy applicators including Fletcher Suit, tandem-ring, computerised tomography scan (CT)/magnetic resonance imaging (MR) compatible tandem-ovoids, Vienna applicator, vaginal cylinders, Martinez Universal Perineal Interstitial Template (MUPIT), Syed-Neblett and indigenous templates for gynecology, templates for breast, prostate, anal canal, Leipzig applicators, tubes for endobiliary, endobronchial, and esophageal brachytherapy and implant kits to perform interstitial implants for head and neck, STSs, etc. The department is also equipped with two dedicated CT simulators, two conventional simulators, three magnetic resonance imaging (MR) suites (available in the Radiology department), and three treatment planning systems (Oncentra® v4.3.1). The staffing includes trained dedicated nurses (three), physicists (three), dosimetrists (two), and radiotherapy technologists (three).

Historically, the major load in brachytherapy was essentially gynecology and was traditionally treated with an orthogonal x-ray-based approach. With the acquisition of CT simulator and MR modalities, there was a smooth transition to an image-based approach since 2005. The current workload involves 6–8 intracavitary, 1–2 intracavitary/interstitial procedures on a daily basis, and 1–2 gynecological interstitial templates, 1–2 head and neck, 1–2 interstitial breast, 1–2 STSs, and 1–2 miscellaneous procedures, planning and treatment delivery on a weekly basis.

25.4 BRACHYTHERAPY PLANNING, QUALITY ASSURANCE, AND CONTROL

The department has evolved from traditional Manchester/Fletcher and Paris loading LDR systems to single stepping source HDR systems and point dose optimization to geometric/volume optimization and evaluation of inverse planning (Jamema et al., 2010). The dose calculation is based on TG43. The QA program has also evolved from machine specific QA, which included source activity measurement, positioning accuracy, timer accuracy, and other related tests to a process specific QA based on AAPM TG100. The QA tests for the treatment planning system are carried out using the recommendations of IAEA TRS 430 annually.

25.5 BENCHMARKS

25.5.1 Gynecological Malignancies

The most commonly performed procedures are the intracavitary and interstitial procedures for cervical and endometrial cancers. We have systematically evolved from LDR to HDR systems and orthogonal x-ray based to three-dimensional (3D) CT/MR/ultrasound (US)-based approach for cervical cancers. The transition from two-dimensional (2D) to 3D image-based brachytherapy for various processes including simplification of brachytherapy techniques (Sharma et al., 2003), imaging (CT/MR/US) (Jamema et al., 2008; Mahantshetty et al., 2011b, 2012), challenging point A and conventional International Commission on Radiation Units and Measurements (ICRU) point doses (Deshpande et al., 1997; Mahantshetty et al., 2011a), initial clinical outcome with use of MR and implementation of Groupe Européen de Curiethérapie (GEC) - European SocieTy for Radiotherapy & Oncology (GEC-ESTRO) recommendations (Mahantshetty et al., 2011b), and successful participation in international prospective studies like EMBRACE (international study on MR-guided brachytherapy in cervix cancer) was smooth and rewarding. Today, with the increasing current case load, the workflow for patients has been refined for optimal utilization of resources and improved quality care which has translated into better clinical outcome. Our center also has one of the largest experiences with the use of interstitial templates (MUPIT) and relatively better outcome (i.e., disease free survival (DFS) of 60% at 3 years with grade 3 bowel and bladder toxicity rates between 4% and 10%) with strict QA program for gynecological cancers (Mahantshetty et al., 2014b). With this vast experience, we have also explored the feasibility of salvage re-irradiation with intracavitary/interstitial image-based brachytherapy for recurrent cervical cancers and reported a local control rate of 44% at 2 years (Mahantshetty et al., 2014a).

25.5.2 Breast Brachytherapy

Breast brachytherapy has evolved over the years from LDR to HDR multicatheter interstitial, boost to radical setting, and whole breast to partial breast brachytherapy since 1982. Interstitial brachytherapy as a boost to the tumor after external beam radiotherapy (EBRT) to the whole breast is being practiced at our institute since 1982. The frequency of women with early breast cancer opting for accelerated partial breast irradiation (APBI) has increased substantially. In our set up, this is a feasible and ideal situation since patients come from long distances with socioeconomic hindrances to stay for protracted treatment schedules. APBI with brachytherapy (multicatheter interstitial brachytherapy (MIB)) is performed either as an intra-operative procedure (open cavity) or in the postoperative setting (closed cavity) after chemotherapy. We have the largest experience with the open cavity technique, this requires cooperation with the surgeons and pathologists. In this technique, after obtaining the frozen section, the procedure is performed in the absence of adverse features, and the patient undergoes a planning CT scan on day 3 after the surgical procedure, tumor bed and clinical target volume (CTV) are delineated which is modified from skin. Planning is done with the objectives of at least 90% coverage of CTV and dose homogeneity index (DHI) >0.75. The final histopathology report is obtained before the fifth fraction, in the presence of unfavorable features the procedure is abandoned and converted to boost. We have replaced 2D planning with 3D volumetric planning, this being practiced for the past 10 years as we envisaged the dosimetric advantages of CT-based planning. We have reported 5-year local control rates of 93% with 3D-based planning with good to excellent cosmetic outcomes in 77% of patients (Budrukkar et al., 2015).

Novel approaches include internal mammary nodal radiation by brachytherapy in locally advanced (inner quadrant or heavy axillary burden) breast cancer patients. A single intraluminal flexible catheter is placed in the internal mammary vessel during surgery and a dose of 34 Gy/10 fractions is delivered to treat the internal mammary nodal region with skin sparing measured by metal–oxide–semiconductor field-effect transistor (MOSFET) (Kinhikar et al., 2006) and a 5-year DFS of 68% has been demonstrated in more than 500 patients (unpublished).

25.5.3 Head and Neck Brachytherapy

Our institution is one of leading centers for head and neck brachytherapy. Brachytherapy is utilized for various subsites including lip, buccal mucosa, oral tongue, base of tongue, tonsil, soft palate, and epiglottic lesions. It was a common practice to perform radical brachytherapy for early mobile tongue cancers in the 1980s to early 2000s. An innovative technique with the use of plastic tubes and modified bead technique using LDR ^{192}Ir wires reported a local control rate of 70% in T1 tumors treated with the LDR technique (Bhalavat et al., 2009). With the institutional protocol of wide local excision in early tongue cancers, brachytherapy is offered in an adjuvant setting now. Brachytherapy in oropharyngeal (base of tongue, tonsil, soft palate, and epiglottis) cancers is offered to T1-2N0 lesions as a boost after external radiation. The brachytherapy techniques include loop and non-loop techniques depending on the tumor volume and extensions. We reported a 2-year local control rate of 82% for early epiglottic carcinomas (lingual) treated with brachytherapy boost using thyrohyoid approach (Bhalavat et al., 2007). Currently, a randomized trial comparing brachytherapy boost versus intensity modulated radiotherapy (IMRT) in terms of local control rates, xerostomia, and quality of life in early oropharyngeal cancer is ongoing at our institution. In the recent past, ^{90}Sr brachytherapy for conjunctival lesions and radical HDR brachytherapy for eyelid tumors has been started and we reported improved local control rates and cosmetic outcomes (Laskar et al., 2014, 2015).

25.5.4 STS Brachytherapy

Brachytherapy with or without EBRT for STS has being practiced for the past 30 years both in adults and children across various sites such as limbs, chest wall, paravertebral, and pelvis regions. The brachytherapy procedure is most commonly performed intra-operatively during surgery. The principles include placement of radiopaque clips to identify the excision bed, placement of brachytherapy catheters (usually single plane) avoiding the neurovascular bundle and bone and treatment in the postoperative period (preferably after 2–5 days after surgery) with HDR brachytherapy. We have reported local control rates of 82% at 4 years in children and 71% in adults treated with brachytherapy with excellent functional outcomes (Laskar et al., 2006, 2007).

25.5.5 Esophagus

Intraluminal brachytherapy for esophageal cancers has been practiced either as a boost to EBRT or as a part of palliation. The intraluminal brachytherapy tube is inserted with/without endoscopic/fluoroscopic guidance and a dose of 12 Gy in two fractions and 16 Gy in two fractions (once weekly) using HDR ^{192}Ir is delivered in the boost and palliative setting, respectively. We have reported dysphagia-free survival of 10 months in approximately half of the patients having improvement in dysphagia with complication rates of 30% mainly strictures and ulcers (Sharma et al., 2002).

25.5.6 Mold Brachytherapy

Customized surface mold HDR brachytherapy is offered for various challenging sites including face (basal cell carcinoma), chest wall recurrences, scalp, pinna, and hard palate cancers.

25.5.7 Training and Education

The Department of Radiation Oncology and Medical Physics is recognized for postgraduate training programs, long- and short-term fellowship programs, and 3–6 month training programs in brachytherapy (for all sites) for physicians. Apart from the formal training programs, our center is also recognized by various international and national organizations including UICC, WHO, IAEA, AROI, IBS, AGOI AMPI, etc. for exchange programs and training of physicians, physicists, dosimetrists, and technologists. Our center is also involved in conducting live workshops including demos in brachytherapy with an aim to promote and disseminate brachytherapy in India and other neighboring countries.

25.5.8 Clinical Trials Infrastructure and Research

Clinical research in cancer is one of the avowed objectives of the TMH. Numerous clinical research projects including several randomized clinical trials are currently in progress. The research infrastructure includes the TMH research administrative council (TRAC), the clinical research secretariat (CRS), three independent ethics committees, and dedicated staff for training and conduct of the research. The CRS offers a wide range of services from collection, maintenance, quality control and analysis of data to design and execution of prospective trials of importance to the institute and nation. Currently, clinical studies related to brachytherapy are ongoing including APBI in breast cancers, brachytherapy in head and neck cancers, image-based brachytherapy in cervical cancers (EMBRACE I & II, conventional vs. 3D image-based randomized clinical studies), etc. The department is also currently working on indigenous development of brachytherapy machines and applicators, which would benefit the population at large in low and middle income countries.

REFERENCES

Banerjee S, Mahantshetty U, Shrivastava S. Brachytherapy in India—A long road ahead. *J Contemp Brachytherapy*. 2014;6(3):331–5.

Bhalavat R, Mahantshetty U, Tole S, Jamema S. Treatment outcome with low-dose-rate interstitial brachytherapy in early-stage oral tongue cancers. *J Can Res Ther*. 2009;5(3):192.

Bhalavat R, Pathak K, Mahantshetty U, Jamema S. Brachytherapy boost: A novel approach for epiglottic carcinoma. *Brachytherapy*. 2007;6(3):212–7.

Budrukkar A, Gurram L, Upreti R, Munshi A, Jalali R, Badwe R et al. Clinical outcomes of prospectively treated 140 women with early stage breast cancer using accelerated partial breast irradiation with 3 dimensional computerized tomography based brachytherapy. *Radiother Oncol*. 2015;115(3):349–54.

Deshpande DD, Shrivastava SK, Pradhan AS, Viswanathan PS, Dinshaw KA. Dosimetry of intracavitary applications in carcinoma of the cervix: Rectal dose analysis. *Radiother Oncol*. 1997;42:163–6.

Jamema S, Kirisits C, Mahantshetty U, Trnkova P, Deshpande D, Shrivastava S et al. Comparison of DVH parameters and loading patterns of standard loading, manual and inverse optimization for intracavitary brachytherapy on a subset of tandem/ovoid cases. *Radiother Oncol*. 2010;97(3):501–6

Jamema S, Saju S, Mahantshetty U, Palled S, Deshpande D, Shrivastava S et al. Dosimetric evaluation of rectum and bladder using image-based CT planning and orthogonal radiographs with ICRU 38 recommendations in intracavitary brachytherapy. *J Med Phys*. 2008;33(1):3.

Kinhikar R, Sharma P, Tambe C, Mahantshetty U, Sarin R, Deshpande D et al. Clinical application of a OneDose™ MOSFET for skin dose measurements during internal mammary chain irradiation with high dose rate brachytherapy in carcinoma of the breast. *Phys Med Biol*. 2006;51(14):N263–8.

Laskar S, Bahl G, Ann Muckaden M, Puri A, Agarwal M, Patil N et al. Interstitial brachytherapy for childhood soft tissue sarcoma. *Pediatr Blood Cancer.* 2007;49(5):649–55.

Laskar S, Bahl G, Puri A, Agarwal M, Muckaden M, Patil N et al. Perioperative interstitial brachytherapy for soft tissue sarcomas: Prognostic factors and long-term results of 155 patients. *Ann Surg Oncol.* 2006;14(2):560–7.

Laskar S, Basu T, Chaudhary S, Chaukar D, Nadkarni M, Gn M. Postoperative interstitial brachytherapy in eyelid cancer: Long term results and assessment of Cosmesis After Interstitial Brachytherapy scale. *J Contemp Brachytherapy.* 2014;4:350–5.

Laskar S, Gurram L, Laskar SG, Chaudhuri S, Khanna N, Upreti R et al. Superficial ocular malignancies treated with strontium-90 brachytherapy: Long term outcomes. *J Contemp Brachytherapy.* 2015;7(5):369–73.

Mahantshetty U, Jamema S, Engineer R, Shrivastava S, Tiwana M, Mishra S et al. Additional rectal and sigmoid mucosal points and doses in high dose rate intracavitary brachytherapy for carcinoma cervix: A dosimetric study. *J Can Res Ther.* 2011a;7(3):298.

Mahantshetty U, Kalyani N, Engineer R, Chopra S, Jamema S, Ghadi Y et al. Reirradiation using high-dose-rate brachytherapy in recurrent carcinoma of uterine cervix. *Brachytherapy.* 2014a;13(6):548–53.

Mahantshetty U, Khanna N, Swamidas J, Engineer R, Thakur M, Merchant N et al. Trans-abdominal ultrasound (US) and magnetic resonance imaging (MRI) correlation for conformal intracavitary brachytherapy in carcinoma of the uterine cervix. *Radiother Oncol.* 2012;102(1):130–4.

Mahantshetty U, Shrivastava S, Kalyani N, Banerjee S, Engineer R, Chopra S. Template-based high-dose-rate interstitial brachytherapy in gynecologic cancers: A single institutional experience. *Brachytherapy.* 2014b;13(4):337–42.

Mahantshetty U, Swamidas J, Khanna N, Engineer R, Merchant NH, Deshpande DD et al. Reporting and validation of gynaecologicalGroupe Euopeen de Curietherapie European Society for Therapeutic Radiology and Oncology (ESTRO) brachytherapy recommendations for MR image-based dose volume parameters and clinical outcome with high dose-rate brachytherapy in cervical cancers: A single institution initial experience. *Int J Gynecol Cancer.* 2011b;21(6):1110–6.

Sharma V, Mahantshetty U, Dinshaw K, Deshpande R, Sharma S. Palliation of advanced/recurrent esophageal carcinoma with high-dose-rate brachytherapy. *Int J Radiat Oncol Biol Phys.* 2002;52(2):310–5.

Sharma V, Mahantshetty U, Menon V, Sharma D. A modified technique for high-dose-rate intracavitary brachytherapy in advanced cancer of the cervix. *Brachytherapy.* 2003;2(4):246–8.

CHAPTER 26

Institut Joliot-Curie Cancer Center, Dakar, Senegal

Implementing a Brachytherapy Program in a Resource Limited Setting

Derek Brown, Moustafa Dieng, Macoumba Gaye, Magatte Diagne, Alana Hudson, Adam Shulman, and John Einck

CONTENTS

26.1 INTRODUCTION

The Institut Joliot-Curie Cancer Center in Dakar, Senegal, is the only radiation therapy facility in the country and is a regional referral center for cervical cancer for Senegal and for all of West Africa. Senegal has a population of nearly 13 million people and has one of the highest incidence rates of cervical cancer in the world (Globocan, 2008). The vast majority of women do not have access to screening or treatment for the disease, leading to presentation at advanced stages and to high mortality rates. Compounding this problem is the lack of radiation treatment facilities in Senegal and many other parts of the African continent.

26.2 BACKGROUND

Consisting of two radiation oncologists, three medical physicists, a nurse, and two radiation therapists, the staff at the cancer center in Dakar treats approximately 30–40 patients per day with external beam radiation therapy (EBRT) using a conventional simulator and a ^{60}Co teletherapy unit. See Figure 26.1 for clinic layout. It is the only treatment facility within a 500 km radius and also serves several neighboring countries. Staggeringly, approximately 30% of the patients treated at the center have carcinoma of the cervix. Prior to 2013, hypofractionated chemoradiotherapy was given as both neoadjuvant treatment prior to definitive surgery and palliative treatment but not definitively because no access to brachytherapy was available.

In 2012, Radiating Hope, a nonprofit organization whose mission is to provide radiation therapy equipment to countries in the developing world received a donation of a high-dose rate (HDR) afterloading unit (microSelectron®, Elekta Brachytherapy, Veenendaal, The Netherlands) including three magnet resonance imaging (MRI)-compatible tandem and ring applicators. They worked with the Institut Joliot-Curie Cancer Center to secure an optimal location for the donated equipment and to initiate a curative cervical cancer treatment program. This chapter presents a description of the implementation of the HDR brachytherapy in Senegal; requiring a nonstandard fractionation schedule and a novel treatment planning approach. A detailed implementation strategy is presented at the end of the chapter and it is hoped that this may serve as a possible blueprint to providing this technology to other developing countries.

26.3 PROGRAM DESIGN

Typically, we would start such a program by determining what type of equipment was most suitable and then seek to purchase that equipment. When working with donated equipment, this task rather becomes determining whether or not the equipment available will be usable. Much upfront legwork was put into ensuring that the equipment could be used in the specific environment at the Institut Joliot-Curie Cancer Center.

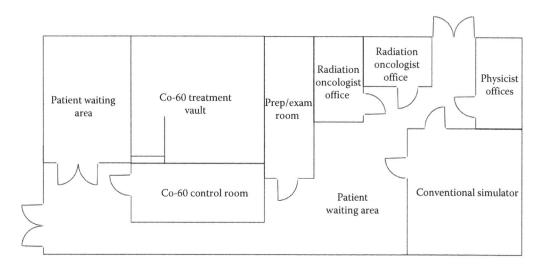

FIGURE 26.1 Blueprint showing the layout of the clinic.

There were a number of challenges to implementing HDR shielded brachytherapy at the center. First, the HDR afterloader would have to be located in the shielded ^{60}Co teletherapy vault, which is located at the opposite end of the center from the simulator where the applicators would be implanted. With a full schedule of ^{60}Co EBRT patients, the HDR procedures would have to be coordinated with this schedule. Second, moderate sedation which requires short acting medications and monitoring of the patients' oxygen saturation and heart rate during the procedure were not be available at the outset of the program which could potentially make the procedures longer and cause patient discomfort. Both of these limitations made it necessary to try to limit the number of brachytherapy fractions planned. Finally, although an Oncentra® treatment planning system (Elekta Brachytherapy, Veenendaal, The Netherlands) was supplied with the afterloader, there were connectivity issues that could not be immediately resolved making real-time treatment planning impossible. This lack of real-time treatment planning required a "work-around" involving the design of an innovative system of fixed geometry applicators (i.e., tandem and ring), preplanned dosimetry with a library of plans loaded directly into the afterloader control system and isodose overlays for use in the simulator to confirm that dose to International Commission on Radiation Units (ICRU) rectal and bladder points were within tolerance.

The radiation oncologists at the cancer center are well trained and experienced with low-dose rate (LDR) brachytherapy. Therefore, they were familiar with the general technique, placement of applicators, and proper geometry to reduce toxicity, but not with the idea of delivering multiple implants outside of the operating room and with the unique fractionation and radiobiological considerations required with HDR treatments. Therefore, our chief aims were to (1) design a treatment schema involving as few fractions of brachytherapy as possible to limit patient discomfort and staff workload without undue risk of rectal and bladder injury, (2) use a fixed geometry applicator and a library of treatment plans to safely deliver treatment without the requirement for computerized treatment planning, and (3) show that the planned system can effectively and safely be used for patient treatments in the facility.

26.4 TECHNICAL DESCRIPTION

Figure 26.2 illustrates an overview of the treatment workflow. The implementation was based on two-dimensional (2D) imaging using preplans and fixed geometry applicators. A series of tandem and ring applicators were imaged in North America using computed tomography (CT) scanners. Preplans were created for all tandem and ring applicators, treating point A to a prescription dose of 750 cGy. Dose distributions for anterior–posterior and lateral views of the applicator for all preplans were printed on transparencies prior to departure for Dakar. Note that the use of tandem and ring applicators enables the use of preplans as the geometry of tandem and ring applicators are fixed. The preplanned technique developed here would not be suitable for use with tandem and ovoid applicators, which do not have a fixed geometry.

Upon arrival in Dakar, preplans were entered into the treatment console of the HDR unit and saved as library plans. On the day of treatment, the applicator was implanted and 2D orthogonal images of the pelvis (anterior–posterior and lateral) were acquired

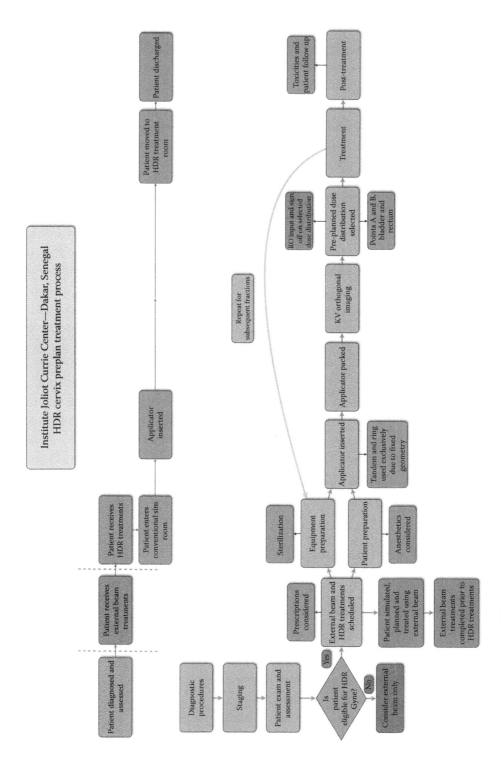

FIGURE 26.2 Treatment workflow implemented at the Institut Joliot-Curie Cancer Center, Dakar, Senegal.

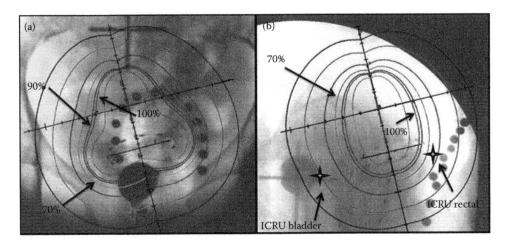

FIGURE 26.3 Transparency isodose overlays over anterior–posterior (a) and left-lateral (b) simulation films. Part (b) shows the ICRU bladder and rectal points falling outside of the 90% and 70 isodose lines, respectively. The tandem and ring applicators can be seen as dummy seeds with 1 cm spacing.

using the conventional kV simulator. Where necessary, the images were scaled using a metal ring of known diameter placed on the patient's skin during imaging. The images were then printed on regular paper. The dose distribution transparency corresponding to the implanted applicator was then overlaid on the printout and the dose distribution was assessed prior to treatment (Figure 26.3). If the dose distribution was deemed acceptable, the patient was then treated using the appropriate library plan.

26.5 FRACTIONATIONS AND COMBINED EBRT DOSE

After extensive literature review (Shigematsu et al., 1983; Okkan et al., 2003; Sood et al., 2003; Patel et al., 2005; Souhami et al., 2005; Jain et al., 2007; Tan et al., 2009; Verma et al., 2009; Das et al., 2011; Patidar et al., 2012; Tharavichitkul et al., 2012), and with an eye to keeping the total number of fractions as low as reasonably achievable, an HDR prescription of 7.5 Gy × 3 fractions was decided on. The center was already accustomed to treating the whole pelvis to 46 Gy in 23 fractions using a 4-field technique with concurrent 5-FU and cisplatinum chemotherapy. Three fractions of 7.5 Gy provide an equivalent dose in 2 Gy Fractions (EQD2) of 78.8 Gy for tumor for patients with non-bulky disease or good responses to chemoradiotherapy. An additional fraction of 7.5 Gy could be added to the poor responders for a total EQD2 of 89.75 Gy only if implant geometry and cumulative dose to rectal and bladder points were acceptable.

26.6 PATIENTS AND EQUIPMENT STATISTICS

A total of 61 patients have been treated using HDR brachytherapy since March of 2013 and the authors are currently working on a formal review of patient outcomes that will be reported elsewhere. Four source exchanges have taken place and Radiating Hope has confirmed funding for new sources through 2017.

26.7 FUTURE WORK

The major focus of ongoing efforts at the Institut Curie is focused on transitioning to image-guided HDR treatments. Progress has been made in terms of securing time on the CT scanner at the main hospital and vendor installation of the treatment planning system. A significant amount of commissioning and training remains to be done but there are plans to begin this work in the fall of 2016. We anticipate the first image-guided treatments to take place in early 2017.

26.7.1 A Detailed Implementation Strategy

Throughout the course of the implementation, we attempted as much as possible to follow a standardized implementation strategy. This strategy, in part developed during our work in Senegal, is described in detail elsewhere (Brown et al., 2014) but is summarized briefly below.

> *Step 1—Site and resource assessment.* The first step in the development of any new technology or technique should be the creation of an implementation team generally consisting of radiation oncologists, medical physicists, and radiation therapists (minimum one of each with more as needed depending on the scope of the project and staff availability). The importance of the site and resource assessment cannot be overstated. The International Atomic Energy Agency (IAEA) has created a very useful spreadsheet for calculating the estimated required staff resources given various workload inputs (International Atomic Energy Agency, 2013b, 2014).
>
> *Step 2—Evaluation of equipment and funding.* Based on the results of the assessment, the adequacy and appropriateness of equipment and funding must be evaluated. Efforts to implement inappropriate equipment, or appropriate equipment with insufficient support and funding, are wasteful at best and potentially dangerous for patients at worst. There must be some balance here as often equipment will have been donated and as such may not be ideal for the desired location or purpose. Care must be taken to ensure that the equipment will be usable in the environment in which it is placed. Realistic appraisals of capacity, expertise, infrastructure, and funding increase the likelihood that the right equipment is installed in the right environment and that patients will benefit from the new technology/treatment technique (Chalkidou et al., 2014).
>
> *Step 3—Establishing timelines.* Realistic timelines should be established at the outset of the project. Timelines should include the acquisition of the required radiation licenses as well as any renovations or construction that must occur, and must include sufficient time for commissioning, training, and system testing (Loudon, 2012).
>
> *Step 4—Defining the treatment process.* Significant gains in safety and efficiency can be realized through proper design and testing (see *Step 8—System testing*) of the treatment process. For a practical guide to process mapping, see Thomadsen et al. (2013), Trebble et al. (2010), and Savory et al. (2001).

Step 5—Equipment commissioning. Comprehensive equipment commissioning must be performed. The extent of this work will depend on the variety and complexity of the new technology/treatment technique being implemented. Sufficient commissioning measurements must be performed to ensure that equipment, under all clinically relevant circumstances, performs within tolerance. Wherever possible, the treatment process and equipment commissioning documents, as well as the quality control and preventive maintenance schedules, should be reviewed by someone external to the project. The IAEA provides guidance on equipment commissioning (International Atomic Energy Agency, 2009), as does the American Association of Physicists in Medicine (Kubo et al., 1998). All these documents plus others can be found in a new online chapter by Van Dyk (2013).

Step 6—Training and competency assessment. This is an absolutely crucial and often neglected step. Training is time consuming and the adequacy of training is difficult to assess, making it easy to short change the user on this step of implementation. This must be avoided. A useful resource to guide training and competency assessment has been produced Smajo et al. (2012).

    a. Where possible, training should be provided in the user's native language

    b. Competency assessment can be achieved through quiz taking or by peer observation of task completion

    c. Presentations and video or telephone conferences provide good introductory information, but the value and necessity of visual, step-by-step training cannot be overstated

    d. Staff should be completely comfortable performing system testing—from initial patient assessment, through treatment, to post-treatment follow-up—prior to initiating actual patient treatments

Step 7—Prospective risk assessment. A multidisciplinary, prospective risk assessment should be undertaken by the team involved in commissioning, training, and using the equipment. The tool used to perform this assessment is far less important than the act of performing the assessment (Ford et al., 2009).

Step 8—System testing. The purpose of system testing is to ensure that the entire process is functioning as expected—from patient assessment through treatment to post-treatment follow-up. Going through the entire process using a mock patient (e.g., phantom, preferably with dosimeters inserted) can be very useful for determining where the process might fail and for improving the efficiency of the process.

Step 9—External dosimetric audits and incident learning systems. An external dosimetric audit is an excellent way to ensure that, at least under specific circumstances, the implemented technology/treatment technique is performing within expected tolerances. IAEA (http://www-naweb.iaea.org/nahu/DMRP/tld.html) and the IROC

Houston quality assurance (QA) Center (formerly known as the Radiological Physics Center (RPC)) (http://rpc.mdanderson.org/RPC/home.htm) offer thermo luminescent dosimeter (TLD) audit programs (Izewska and Andreo, 2000; Izewska et al., 2003; International Atomic Energy Agency, 2007, 2013a; MD Anderson—IROC Houston Quality Assurance Center; 2013).

Step 10—Support and follow-up. Once established, the implemented technology/treatment technique must be supported and maintained if it is to remain safe and effective. Initial support should take the form of reviewing treatment plans, fielding questions that arise from staff, and facilitating patient and machine specific QA activities. Longer-term support should include patient follow-up, assistance with required post-repair measurements, and general advice.

REFERENCES

Azad SK, Choudhary V. Treatment results of radical radiotherapy of carcinoma uterine cervix using external beam radiotherapy and high dose rate intracavitary radiotherapy. *J Cancer Res Ther* 2010, 6(4):482–486.

Brown DW, Shulman A, Hudson A, Smith W, Fisher B, Hollon J, Pipman Y, Van Dyk J, Einck J. A framework for the implementation of new radiation therapy technologies and treatment techniques in low-income countries. *Phys Med* 2014, 30(7):791–798.

Chalkidou K, Marquez P, Dhillon PK, Teerawattananon Y, Anothaisintawee T, Gadelha CA, Sullivan R. Evidence-informed frameworks for cost-effective cancer care and prevention in low, middle, and high-income countries. *Lancet Oncol* 2014, 15(3):e119–131.

Das D, Chaudhuri S, Deb AR, Aich RK, Gangopadhyay S, Ray A. Treatment of cervical carcinoma with high-dose rate intracavitary brachytherapy: Two years follow-up study. *Asian Pac J Cancer Prev* 2011, 12(3):807–810.

Ford EC, Gaudette R, Myers L, Vanderver B, Engineer L, Zellars R, Song DY, Wong J, Deweese TL. Evaluation of safety in a radiation oncology setting using failure mode and effects analysis. *Int J Radiat Oncol Biol Phys* 2009, 74(3):852–858.

Globocan 2008. International Agency for Research on Cancer (IARC). http://globocan.iarc.fr/

International Atomic Energy Agency, Quality Assurance Team for Radiation Oncology: *Comprehensive Audits of radiotherapy Practices: A Tool for Quality Improvement.* Quality Assurance Team for Radiation Oncology (QUATRO). Vienna, Austria: International Atomic Energy Agency; 2007.

International Atomic Energy Agency. *Clinical Training of Medical Physicists Specializing in Radiation Oncology.* Training Course Series 37. Vienna, Austria: International Atomic Energy Agency; 2009.

International Atomic Energy Agency. Dosimetry and Medical Radiation Physics. 2013a. http://www-naweb.iaea.org/nahu/DMRP/about.html

International Atomic Energy Agency. Human Health Campus, Medical Physics, Radiotherapy. International Atomic Energy Agency, Vienna, 2013b.

International Atomic Energy Agency. *Staffing and Cost Calculation.* International Atomic Energy Agency; 2014. ISBN:978-92-0-156715-4. http://www-pub.iaea.org/books/IAEABooks/10800/Staffing-in-Radiotherapy-An-Activity-Based-Approach

Izewska J, Andreo P. The IAEA/WHO TLD postal programme for radiotherapy hospitals. *Radiother Oncol J Eur Soc Ther Radiol Oncol* 2000, 54(1):65–72.

Izewska J, Andreo P, Vatnitsky S, Shortt K. The IAEA/WHO TLD postal dose quality audits for radiotherapy: A perspective of dosimetry practices at hospitals in developing countries. *Radiother Oncol: J Eur Soc Ther Radiol Oncol* 2003, 69:91–97.

Jain VS, Singh KK, Shrivastava R, Saumsundaram KV, Sarje MB, Jain SM. Radical radiotherapy treatment (EBRT + HDR-ICRT) of carcinoma of the uterine cervix: Outcome in patients treated at a rural center in India. *J Cancer Res Ther* 2007, 3(4):211–217.

Kubo H. D. et al. High dose-rate brachytherapy treatment delivery: Report of the AAPM Radiation Therapy Committee Task Group No. 59. *Med Phys* 1998, 25(4):375–403.

Loudon J. Applying project management processes to successfully complete projects in radiation medicine. *J Med Imaging Radiat Sci* 2012, 43(4):253–258.

MD Anderson—IROC Houston Quality Assurance Center; 2013. http://rpc.mdanderson.org/RPC/home.htm

Okkan S, Atkovar G, Sahinler I, Oner Dincbas F, Koca A, Koksal S, Turkan S, Uzel R. Results and complications of high dose rate an'd low dose rate brachytherapy in carcinoma of the cervix: Cerrahpasa experience. *Radiother Oncol* 2003, 67(1):97–105.

Patel FD, Rai B, Mallick I, Sharma SC. High-dose-rate brachytherapy in uterine cervical carcinoma. *Int J Radiat Oncol Biol Phys* 2005, 62(1):125–130.

Patidar AK, Kumar HS, Walke RV, Hirapara PH, Jakhar SL, Bardia MR. Evaluation of the response of concurrent high dose rate intracavitary brachytherapy with external beam radiotherapy in management of early stage carcinoma cervix. *J Obstet Gynaecol India* 2012, 62(5):562–565.

Savory P, Olson J. Guidelines for using process mapping to aid improvement efforts. *Hosp Manage Q* 2001, 22(3):10–16.

Shigematsu Y, Nishiyama K, Masaki N, Inoue T, Miyata Y, Ikeda H, Ozeki S, Kawamura Y, Kurachi K. Treatment of carcinoma of the uterine cervix by remotely controlled afterloading intracavitary radiotherapy with high-dose rate: A comparative study with a low-dose rate system. *Int J Radiat Oncol Biol Phys* 1983, 9(3):351–356.

Smajo M: *Medical Physics Clinical Skills Workbook for Therapy Physics.* Chicago, Illinois. Rosalind Franklin University of Medicine and Science College of Health Professionals; 2012.

Sood BM, Gorla GR, Garg M, Anderson PS, Fields AL, Runowicz CD, Goldberg GL, Vikram B. Extended-field radiotherapy and high-dose-rate brachytherapy in carcinoma of the uterine cervix: Clinical experience with and without concomitant chemotherapy. *Cancer* 2003, 97(7):1781–1788.

Souhami L, Corns R, Duclos M, Portelance L, Bahoric B, Stanimir G. Long-term results of high-dose rate brachytherapy in cervix cancer using a small number of fractions. *Gynecol Oncol* 2005, 97(2):508–513.

Tan LT, Coles CE, Hart C, Tait E. Clinical impact of computed tomography-based image-guided brachytherapy for cervix cancer using the tandem-ring applicator—The Addenbrooke's experience. *Clin Oncol (R Coll Radiol)* 2009, 21(3):175–182.

Tharavichitkul E, Klunkin P, Lorvidhaya V, Sukthomya V, Chakrabhandu S, Pukanhaphan N, Chitapanarux I, Galalae R. The effects of two HDR brachytherapy schedules in locally advanced cervical cancer treated with concurrent chemoradiation: A study from Chiang Mai, Thailand. *J Radiat Res* 2012, 53(2):281–287.

Thomadsen B, Brown D, Ford E, Huq S, Rath F. Risk assessment using the TG-100 methodology. In: *Quality and Safety in Radiotherapy.* Thomadsen B (ed.). Madison, Wisconsin: Medical Physics Publishing; 2013: pp. 98–102.

Trebble TM, Hansi N, Hydes T, Smith MA, Baker M. Process mapping the patient journey: An introduction. *BMJ* 2010, 341:c4078.

Van Dyk J. The modern technology of radiation oncology: A compendium for medical physicists and radiation oncologists. In: *Radiation Oncology Medical Physics Resources for Working, Teaching and Learning.* Van Dyk J (ed.). Madison, Wisconsin: Medical Physics Publishing; 2013, Vol. 3: pp. 695–752.

Verma AK, Arya AK, Kumar M, Kumar A, Gupta S, Sharma D, Rath G. Weekly cisplatin or gemcitabine concomitant with radiation in the management of locally advanced carcinoma cervix: Results from an observational study. *J Gynecol Oncol* 2009, 20(4):221–226.

IV

Is Brachytherapy a Competitive Modality?

CHAPTER 27

EBRT or Brachytherapy?

Kathy Han, Eve-Lyne Marchand,
Jennifer Croke, and Té Vuong

CONTENTS

27.1 INTRODUCTION

This chapter outlines competing techniques for the treatment of cervical, endometrial, prostate, breast, and rectal cancer, and discusses the recent utilization and trend of brachytherapy compared to external beam radiotherapy (EBRT).

27.2 CERVICAL CANCER

The standard of care for the curative management of locally advanced cervical cancer consists of EBRT, concurrent chemotherapy, and brachytherapy. Brachytherapy is an essential component of the treatment that enables the most conformal dose escalation to the primary tumor. A planning study comparing brachytherapy versus intensity-modulated radiotherapy (IMRT) and intensity-modulated proton therapy (IMPT) showed that brachytherapy was superior to IMRT and IMPT in target doses and organ at risk (OAR) sparing (Georg et al. 2008).

Three recent studies reported a disturbing decline in brachytherapy utilization for locally advanced cervical cancer. One study using the Surveillance, Epidemiology, and End Results (SEER) database found a 25% decrease in brachytherapy utilization rate between 1998 and 2009 (though the decline may be partially related to the revision of the SEER coding manual) (Han et al. 2013). Using propensity-score matching to adjust for differences in baseline characteristics, brachytherapy treatment was independently associated with better cause-specific and overall survival. Similarly, a study using the National Cancer

Data Base (NCDB) also showed a decline in brachytherapy utilization from 97% to 86% from 2004 to 2011, contrasted with an increase in IMRT and stereotactic body radiation therapy (SBRT) from 3% to 14% (Gill et al. 2014). IMRT or SBRT boost was associated with inferior overall survival (hazard ratio [HR] 1.86%, 95% confidence interval [CI] 1.35–2.55), and a survival detriment stronger than that associated with omitting chemotherapy (HR 1.61%, 95% CI 1.27–2.04). Finally, the Quality Research in Radiation Oncology (formerly Patterns of Care) study reviewed records of 261 randomly selected patients treated between 2005 and 2007 from 45 institutions and found that 12.5% were treated with EBRT instead of brachytherapy (Eifel et al. 2014). This was almost double the rate (6.4%) reported in the 1996–1999 survey.

Possible reasons for this undesirable decline in brachytherapy utilization include: (1) the adoption of less invasive techniques such as IMRT and SBRT, (2) reimbursement policies favoring these techniques over brachytherapy, (3) inadequate training of radiation oncology residents given that only 8% of US facilities treat more than three patients with locally advanced cervical cancer per year, and (4) insufficient maintenance of brachytherapy skills among practicing radiation oncologists (Han et al. 2013, Eifel et al. 2014, Gill et al. 2014, Petereit et al. 2015).

27.3 ENDOMETRIAL CANCER

Endometrial cancer is surgically staged. Intermediate, high-intermediate, and high-risk disease are treated with adjuvant radiotherapy (Klopp et al. 2014). The Norwegian, PORTEC1, GOG99, and ASTEC/EN5 trials reported a reduction in pelvic relapse rate with EBRT to the pelvis (Aalders et al. 1980, Creutzberg et al. 2000, Keys et al. 2004, Scholten et al. 2005, Blake et al. 2009). However, EBRT causes gastrointestinal toxicities and most of the relapses occur in the vagina. Hence, the PORTEC2 trial randomized women with high-intermediate risk endometrial cancer (age >60 with deeply invasive grade 1 or 2 disease or minimally invasive grade 3 disease) to adjuvant pelvic EBRT versus vaginal brachytherapy (Nout et al. 2010). There were no significant differences in vaginal, locoregional, or distant relapse between the two arms. Patients who received vaginal brachytherapy reported lower levels of bowel symptoms and better social functioning 5 years after treatment compared to EBRT (Nout et al. 2012).

Even prior to the publication of the PORTEC2 study results, a SEER study demonstrated an increasing trend in the use of adjuvant vaginal brachytherapy for women with stage I/II endometrial cancer (13% in 1995 compared to 33% in 2005), with a corresponding decrease in the use of EBRT (56% in 1995 vs. 46% in 2005) (Patel et al. 2012). The trend in vaginal brachytherapy utilization after PORTEC2 has not been examined, but is expected to be increasing given its favorable toxicity profile.

27.4 PROSTATE CANCER

For low- and intermediate-risk prostate cancer, definitive treatment options include radical prostatectomy (RP), EBRT, or brachytherapy, all of which have comparable outcomes (D'Amico et al. 1998). Brachytherapy is more cost effective compared to RP and EBRT (Hayes et al. 2013). Furthermore, a comparison of brachytherapy, IMRT, and proton therapy found that brachytherapy yielded better overall "value" with respect to sexual function, urinary incontinence, urinary bother, bowel function, biochemical relapse-free survival, and cost (Frank 2013).

Prostate brachytherapy can be delivered using low-dose rate (LDR) or high-dose rate (HDR) either as a monotherapy or a boost to EBRT. For localized prostate cancer, brachytherapy has favorable long-term outcomes with 15-year biochemical recurrence free survival rates of 86%, 80%, and 62% for patients with low-, intermediate-, and high-risk disease, respectively (Sylvester et al. 2011). There are no published randomized studies comparing brachytherapy monotherapy to EBRT or RP but retrospective studies have shown the superiority of brachytherapy (Spratt et al. 2014, Crook 2015). Dosimetric studies also show that brachytherapy is superior with respect to bladder/rectal wall sparing compared to IMRT and proton therapy (Georg et al. 2014). The incorporation of brachytherapy to EBRT has been shown to provide better rates of local control, biochemical relapse-free survival, and distant metastases-free survival compared to EBRT alone for intermediate-risk prostate cancer (Spratt et al. 2014). In the recently presented ASCENDE-RT trial, men with intermediate- and high-risk prostate cancer who were randomized to receive LDR brachytherapy boost (after androgen deprivation and 46 Gy/32 fractions of EBRT) had better disease-free survival than those randomized to receive an additional 32 Gy/16 fractions of EBRT (Morris et al. 2015). Furthermore, EBRT with brachytherapy and androgen therapy was shown to offer better long-term functional outcomes and freedom from biochemical failure for patients with intermediate/high-risk disease compared to RP (Crook 2015).

The trend in the utilization of prostate brachytherapy has decreased over the past decade while EBRT use has increased (Mahmood et al. 2014, Martin et al. 2014). In the NCDB, there has been a steady decline in the use of brachytherapy since 2003. It reached a peak use of 16.7% in 2002 and declined to a low of 8% in 2010 (Martin et al. 2014).

There are several hypotheses for the disturbing decline in the use of prostate brachytherapy (Mahmood et al. 2014, Martin et al. 2014, Petereit et al. 2015). First, there has been an increase in the number of RP since the introduction of robotic surgery (44% in 2002 vs. 60% in 2012) (Mahmood et al. 2014). Second, newer advanced EBRT techniques such as IMRT, image-guided radiotherapy (IGRT), and protons, have been rapidly adopted. In addition, there is more financial incentive in the United States to treat patients with EBRT over brachytherapy. Finally, brachytherapy is a technically demanding and complex skill that requires specialized training and upkeep in order to maintain proficiency.

27.5 BREAST CANCER

Whole breast radiation therapy is routinely offered following breast-conserving surgery to reduce local recurrence (Vinh-Hung and Verschraegen 2004). In light of data showing that the rate of ipsilateral breast recurrences occurring in areas other than the tumor bed is similar to the recurrence of contralateral second primary breast cancer (Fisher et al. 2001), whole breast irradiation has been criticized as an overtreatment.

Accelerated partial breast irradiation (APBI) is a novel technique that treats only the lumpectomy bed plus a 1–2 cm margin, rather than the whole breast. Advantages of this approach include lower radiation dose to uninvolved portions of the breast and normal tissue by delivering radiation locally and conformably to breast tissue at high risk of recurrence following breast-conserving surgery. APBI also uses higher radiation fraction size, therefore allowing treatment delivery over a shorter period of time.

To date, the strongest evidence for APBI comes from both retrospective and prospective studies showing that the vast majority of local recurrence occurs within or close to the tumor bed (Fisher and Anderson 1994, Holli et al. 2001, Veronesi et al. 2001, Malmstrom et al. 2003). There is, however, a possibility of occult foci of cancer elsewhere in the breast that will not be treated leading to a potentially increased recurrence rate in the treated breast when using APBI. Therefore, in the absence of definitive evidence establishing long-term efficacy and safety of APBI, the American Society for Radiation Oncology (ASTRO) and the Groupe Européen de Curiethérapie—the European SocieTy for Radiotherapy & Oncology (GEC-ESTRO) independently convened a task force of breast cancer experts to elaborate guidelines on patient selection for APBI application outside of a clinical trial (Smith et al. 2009, Polgar et al. 2010).

APBI can be delivered using EBRT, LDR, or HDR brachytherapy through multiple interstitial catheters implant or intracavitarily using a balloon catheter. Compared to EBRT, all brachytherapy approaches for APBI result in lower dose to uninvolved portions of the breast, normal tissue, and integral dose.

Multi-catheter interstitial brachytherapy is the APBI technique that has been utilized the longest and with the longest extensive follow-up (Arthur and Vicini 2005, Sanders et al. 2007, Antonucci et al. 2009, Offersen et al. 2009, Polgar et al. 2010). Catheter configuration and their relation to the tumor target volume are crucial for effective treatment; therefore, a high level of experience is necessary to obtain a good quality implant. Balloon-based brachytherapy devices include the Mammosite®, Axxent® electronic brachytherapy, and Contura®. To date, there is limited published data regarding the long-term tumor control and cosmesis associated with Mammosite but the available results are promising (Niehoff et al. 2006, Benitez et al. 2007, Chen et al. 2007, Dragun et al. 2007, Patel et al. 2008, Vicini et al. 2008, 2011, Belkacemi et al. 2009, Jeruss et al. 2011), though in one study, the 5-year actuarial incidence of telangiectasia was 33% (Vargo et al. 2014).

Limitations associated with APBI include availability of data regarding tumor control with this approach as well as technical limitations such as the experience of the brachytherapist being essential in obtaining a good quality implant particularly with the multiple catheter interstitial approach. Cost effectiveness also remains an issue, with reduced cost to patients for the HDR-based APBI approaches being overshadowed by substantial increases in cost to payers, and therefore resulting in higher total societal costs (Suh et al. 2005a, b, Sher et al. 2009). However, APBI using applicator-based brachytherapy has been shown to be a cost-effective option with regard to cosmesis and toxicity in appropriately selected patients (Shah et al. 2014). Cost effectiveness of brachytherapy is discussed in Chapter 29.

Results of the NSABP B-39/RTOG 0413 and RAPID trials are still pending. These randomized phase III trials will answer whether APBI following lumpectomy provides equivalent local tumor control in the breast compared to conventional whole breast irradiation for early stage breast cancer (Hiatt et al. 2006, Sanghani and Wazer 2007). The recently published GEC-ESTRO phase III, non-inferiority trial of low-risk invasive and ductal carcinoma *in situ* showed no significant difference between adjuvant whole breast radiotherapy versus APBI using multi-catheter brachytherapy (Strnad et al. 2016). Medicare and SEER studies have shown a trend of rapid adoption of breast brachytherapy (Abbott et al. 2011, Smith et al. 2011).

27.6 RECTAL CANCER

Endocavitary brachytherapy for rectal cancer was developed as a highly conformal radiation modality (Vuong et al. 2005) in the era of modern imaging with magnetic resonant imaging (MRI) in conjunction with three-dimensional (3D) treatment planning in 1998 at McGill University, and the development of quality rectal surgery with total mesorectal resection. It is the best possible targeted form of radiation due to the inverse square law but it requires quality imaging to allow for target volume identification and dedicated teamwork. The treatment is delivered by the ^{192}Ir source HDR brachytherapy afterloader with a flexible silicon rubber intracavitary mold applicator (Elekta Brachytherapy, The Netherlands) of cylindrical shape (27 cm long and 2 cm in diameter), designed with eight channels distributed in equal angular increments over the circumference of the applicator, and a central cavity that allows for the use of a central shield (either lead or tungsten). Figure 27.1 illustrates the endorectal applicator.

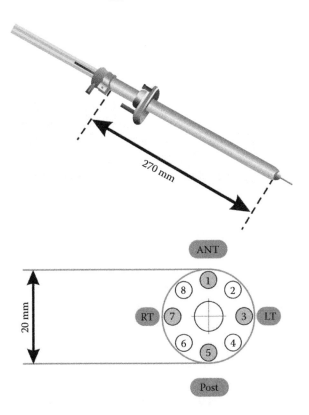

FIGURE 27.1 Intracavitary mold applicator. The bottom schematics represent the catheter convention (1–8) as well as the catheters engaged with x-ray markers (1, 3, 4, 5, and 7; gray-shaded circles). (Reprinted with permission from Vuong T. 2015. *J Contemp Brachytherapy* 7(2):183–8.)

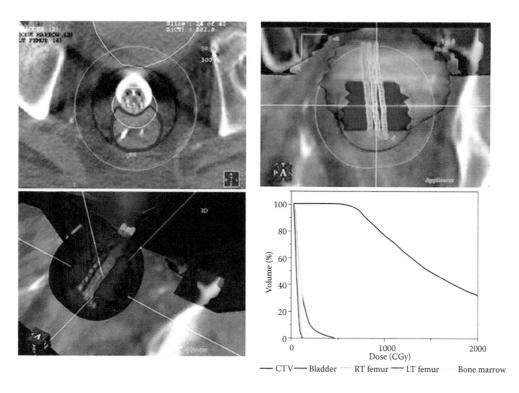

FIGURE 27.2   Dose distribution obtained by treatment planning system and illustrates the difference between dose coverage to the CTV (neoadjuvant setting) and sparing of the surrounding critical structures in terms of the corresponding dose-volume histograms. (Reprinted with permission from Vuong T. 2015. *J Contemp Brachytherapy* 7(2):183–8.)

Initially, the modality was tested as a neoadjuvant radiation modality to allow for tumor down staging. This modality was validated by more than 670 patients with operable low T2 and T3 rectal tumors and resulted in an excellent local control with a 4.7% local recurrence rate at 5 years and median follow up of 62 months (Devic et al. 2007) and allows for significant sparing of normal tissues exposure to radiation when compared to EBRT. Figure 27.2 illustrates the dose distribution of a treatment plan and corresponding dose-volume histogram of the clinical target volume (CTV) and OARs.

As imaging technologies evolved, the introduction of daily imaging permits adaptive radiation treatment (Devic et al. 2007, Appelt et al. 2013). HDR endorectal brachytherapy (HDREBT) was explored as a means to allow for dose escalation as a boost treatment in elderly unfit patients, patients refusing surgery, and nowadays for those patients desiring organ preservation. To address the risks of microscopic disease with intra-mesorectal nodes and deposits, the dose of 45–50 Gy has been shown to be adequate in clinical trials. However, to achieve local control in macroscopic disease for adenocarcinoma, a dose as high as 92 Gy (Appelt et al. 2013) is suggested. There are two challenges to overcome in such a quest: first, there is a high dose-volume correlation for long-term toxicities for rectal mucosa from the experience of prostate brachytherapy

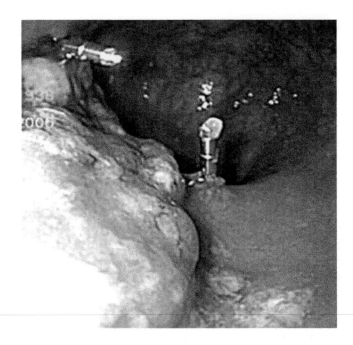

FIGURE 27.3 Marking proximal and distal limit of a tumor using radiopaque clips under direct rectoscopy. (Reprinted with permission from Vuong T. 2015. *J Contemp Brachytherapy* 7(2):183–8.)

and, second, the rectum is a mobile organ with gas interference. Thus, keys to achieving this goal are primary accurate tumor bed identification and, second, optimal sparing of the adjacent rectal wall. Boosting the tumor bed after EBRT as identification of the target volume is challenging in the presence of post radiation changes. Therefore, quality-planning steps include: direct visualization with rectoscopy, radio-opaque clips positioning in addition to a post-EBRT pelvic MRI. An example of placed radio-opaque clips is seen in Figure 27.3. Optimal protection of the rectal mucosa is best achieved during HDREBT for the following reasons: the applicator has a central cavity allowing for the use of central shielding and it can accommodate two balloons with water or iodine fillings. The latter apparatus offers three purposes: (1) to supplement shielding, that is, to move the uninvolved parts of the rectal wall further from the source positions, (2) to increase the distance from the source to the tumor itself; this makes the dose gradients across the tumor shallower due to the nature of the dose falloff, and thus prescription to deeper points is possible without an unacceptable increase in dose to the surface of the rectal mucosa, and (3) the combination of the applicator and the double balloons will move along with the rectum, minimizing intra-treatment motion and permitting accurate dose delivery. Using this technique, a 2-year local control rate of 72.8% (Vuong et al. 2015) was achieved in a cohort of 79 medically unfit patients with median age of 82 years, stages T2-4, and without surgery.

Many centers across the world have adopted brachytherapy for rectal cancer recently including: Canada, United States, Hong Kong, Denmark, United Kingdom, and The Netherlands. An ongoing phase II randomized trial to test whether the preoperative

HDREBT improves pathologic complete response rates compared to pelvic IMRT and capecitabine is being currently carried out (NCT02017704).

27.7 CONCLUSION

In summary, brachytherapy plays an important role in the management of many cancers because of superior results but there is a disturbing decline in its use in cervical and prostate cancers partly due to the adoption of newer EBRT techniques such as IMRT, proton therapy, and SBRT. Breast brachytherapy will likely be adopted more widely. Ongoing trials should further establish the role of brachytherapy compared to EBRT in rectal cancer.

REFERENCES

Aalders, J., V. Abeler, P. Kolstad, and M. Onsrud. 1980. Postoperative external irradiation and prognostic parameters in stage I endometrial carcinoma: Clinical and histopathologic study of 540 patients. *Obstet Gynecol* 56:419–27.

Abbott, A. M., E. B. Habermann, and T. M. Tuttle. 2011. Trends in the use of implantable accelerated partial breast irradiation therapy for early stage breast cancer in the United States. *Cancer* 117:3305–10.

Antonucci, J. V., M. Wallace, N. S. Goldstein et al. 2009. Differences in patterns of failure in patients treated with accelerated partial breast irradiation versus whole-breast irradiation: A matched-pair analysis with 10-year follow-up. *Int J Radiat Oncol Biol Phys* 74:447–52.

Appelt, A. L., J. Ploen, I. R. Vogelius, S. M. Bentzen, and A. Jakobsen. 2013. Radiation dose-response model for locally advanced rectal cancer after preoperative chemoradiation therapy. *Int J Radiat Oncol Biol Phys* 85:74–80.

Arthur, D. W. and F. A. Vicini. 2005. Accelerated partial breast irradiation as a part of breast conservation therapy. *J Clin Oncol* 23:1726–35.

Belkacemi, Y., M. P. Chauvet, S. Giard et al. 2009. Partial breast irradiation as sole therapy for low risk breast carcinoma: Early toxicity, cosmesis and quality of life results of a mammosite brachytherapy phase II study. *Radiother Oncol* 90:23–9.

Benitez, P. R., M. E. Keisch, F. Vicini et al. 2007. Five-year results: The initial clinical trial of mammosite balloon brachytherapy for partial breast irradiation in early-stage breast cancer. *Am J Surg* 194:456–62.

Blake, P., A. M. Swart, J. Orton et al. 2009. Adjuvant external beam radiotherapy in the treatment of endometrial cancer (MRC ASTEC and NCIC CTG EN.5 randomised trials): Pooled trial results, systematic review, and meta-analysis. *Lancet* 373:137–46.

Chen, S., A. Dickler, M. Kirk et al. 2007. Patterns of failure after mammosite brachytherapy partial breast irradiation: A detailed analysis. *Int J Radiat Oncol Biol Phys* 69:25–31.

Creutzberg, C. L., W. L. van Putten, P. C. Koper et al. 2000. Surgery and postoperative radiotherapy versus surgery alone for patients with stage-1 endometrial carcinoma: Multicentre randomised trial. PORTEC Study Group. Post Operative Radiation Therapy in Endometrial Carcinoma. *Lancet* 355:1404–11.

Crook, J. 2015. Long-term oncologic outcomes of radical prostatectomy compared with brachytherapy-based approaches for intermediate- and high-risk prostate cancer. *Brachytherapy* 14:142–7.

D'Amico, A. V., R. Whittington, S. B. Malkowicz et al. 1998. Biochemical outcome after radical prostatectomy, external beam radiation therapy, or interstitial radiation therapy for clinically localized prostate cancer. *JAMA* 280:969–74.

Devic, S., T. Vuong, B. Moftah et al. 2007. Image-guided high dose rate endorectal brachytherapy. *Med Phys* 34:4451–8.

Dragun, A. E., J. L. Harper, J. M. Jenrette, D. Sinha, and D. J. Cole. 2007. Predictors of cosmetic outcome following mammosite breast brachytherapy: A single-institution experience of 100 patients with two years of follow-up. *Int J Radiat Oncol Biol Phys* 68:354–8.

Eifel, P. J., A. Ho, N. Khalid, B. Erickson, and J. Owen. 2014. Patterns of radiation therapy practice for patients treated for intact cervical cancer in 2005 to 2007: A quality research in radiation oncology study. *Int J Radiat Oncol Biol Phys* 89:249–56.

Fisher, B. and S. Anderson. 1994. Conservative surgery for the management of invasive and non-invasive carcinoma of the breast: NSABP trials. National Surgical Adjuvant Breast and Bowel Project. *World J Surg* 18:63–9.

Fisher, E. R., S. Anderson, E. Tan-Chiu, B. Fisher, L. Eaton, and N. Wolmark. 2001. Fifteen-year prognostic discriminants for invasive breast carcinoma: National Surgical Adjuvant Breast and Bowel Project Protocol-06. *Cancer* 91:1679–87.

Frank, S. J. 2013. Defining value in prostate brachytherapy. *Paper presented at the University of Texas MD Anderson Cancer Center 2nd International Prostate Brachytherapy Conference: The Value and Future of Prostate Brachytherapy in a Changing Healthcare Environment*, Houston, Texas.

Georg, D., J. Hopfgartner, J. Gora et al. 2014. Dosimetric considerations to determine the optimal technique for localized prostate cancer among external photon, proton, or carbon-ion therapy and high-dose-rate or low-dose-rate brachytherapy. *Int J Radiat Oncol Biol Phys* 88:715–22.

Georg, D., C. Kirisits, M. Hillbrand, J. Dimopoulos, and R. Potter. 2008. Image-guided radiotherapy for cervix cancer: High-tech external beam therapy versus high-tech brachytherapy. *Int J Radiat Oncol Biol Phys* 71:1272–8.

Gill, B. S., J. F. Lin, T. C. Krivak et al. 2014. National Cancer Data Base analysis of radiation therapy consolidation modality for cervical cancer: The impact of new technological advancements. *Int J Radiat Oncol Biol Phys* 90:1083–90.

Han, K., M. Milosevic, A. Fyles, M. Pintilie, and A. N. Viswanathan. 2013. Trends in the utilization of brachytherapy in cervical cancer in the United States. *Int J Radiat Oncol Biol Phys* 87:111–9.

Hayes, J. H., D. A. Ollendorf, S. D. Pearson et al. 2013. Observation versus initial treatment for men with localized, low-risk prostate cancer: A cost-effectiveness analysis. *Ann Intern Med* 158:853–60.

Hiatt, J. R., S. B. Evans, L. L. Price, G. A. Cardarelli, T. A. Dipetrillo, and D. E. Wazer. 2006. Dose-modeling study to compare external beam techniques from protocol NSABP B-39/RTOG 0413 for patients with highly unfavorable cardiac anatomy. *Int J Radiat Oncol Biol Phys* 65:1368–74.

Holli, K., R. Saaristo, J. Isola, H. Joensuu, and M. Hakama. 2001. Lumpectomy with or without postoperative radiotherapy for breast cancer with favourable prognostic features: Results of a randomized study. *Br J Cancer* 84:164–9.

Jeruss, J. S., H. M. Kuerer, P. D. Beitsch, F. A. Vicini, and M. Keisch. 2011. Update on DCIS outcomes from the American Society of Breast Surgeons accelerated partial breast irradiation registry trial. *Ann Surg Oncol* 18:65–71.

Keys, H. M., J. A. Roberts, V. L. Brunetto et al. 2004. A phase III trial of surgery with or without adjunctive external pelvic radiation therapy in intermediate risk endometrial adenocarcinoma: A Gynecologic Oncology Group study. *Gynecol Oncol* 92:744–51.

Klopp, A., B. D. Smith, K. Alektiar et al. 2014. The role of postoperative radiation therapy for endometrial cancer: Executive summary of an American Society for Radiation Oncology Evidence-Based Guideline. *Pract Radiat Oncol* 4:137–44.

Mahmood, U., T. Pugh, S. Frank et al. 2014. Declining use of brachytherapy for the treatment of prostate cancer. *Brachytherapy* 13:157–62.

Malmstrom, P., L. Holmberg, H. Anderson et al. 2003. Breast conservation surgery, with and without radiotherapy, in women with lymph node-negative breast cancer: A randomised clinical trial in a population with access to public mammography screening. *Eur J Cancer* 39:1690–7.

Martin, J. M., E. A. Handorf, A. Kutikov et al. 2014. The rise and fall of prostate brachytherapy: Use of brachytherapy for the treatment of localized prostate cancer in the National Cancer Data Base. *Cancer* 120:2114–21.

Morris, W., S. Tyldesley, S. Rodda et al. 2015. LDR brachytherapy is superior to 78 Gy of EBRT for unfavourable risk prostate cancer: The results of a randomized trial. *Radiother Oncol* 115:S239.

Niehoff, P., C. Polgar, H. Ostertag et al. 2006. Clinical experience with the mammosite radiation therapy system for brachytherapy of breast cancer: Results from an international phase II trial. *Radiother Oncol* 79:316–20.

Nout, R. A., H. Putter, I. M. Jurgenliemk-Schulz et al. 2012. Five-year quality of life of endometrial cancer patients treated in the randomised Post Operative Radiation Therapy in Endometrial Cancer (PORTEC-2) trial and comparison with norm data. *Eur J Cancer* 48:1638–48.

Nout, R. A., V. T. Smit, H. Putter et al. 2010. Vaginal brachytherapy versus pelvic external beam radiotherapy for patients with endometrial cancer of high-intermediate risk (PORTEC-2): An open-label, non-inferiority, randomised trial. *Lancet* 375:816–23.

Offersen, B. V., M. Overgaard, N. Kroman, and J. Overgaard. 2009. Accelerated partial breast irradiation as part of breast conserving therapy of early breast carcinoma: A systematic review. *Radiother Oncol* 90:1–13.

Patel, M. K., M. L. Cote, R. Ali-Fehmi, T. Buekers, A. R. Munkarah, and M. A. Elshaikh. 2012. Trends in the utilization of adjuvant vaginal cuff brachytherapy and/or external beam radiation treatment in stage I and II endometrial cancer: A surveillance, epidemiology, and end-results study. *Int J Radiat Oncol Biol Phys* 83:178–84.

Patel, R. R., M. E. Christensen, C. W. Hodge, J. B. Adkison, and R. K. Das. 2008. Clinical outcome analysis in "high-risk" versus "low-risk" patients eligible for national surgical adjuvant breast and bowel b-39/Radiation Therapy Oncology Group 0413 trial: Five-year results. *Int J Radiat Oncol Biol Phys* 70:970–3.

Petereit, D. G., S. J. Frank, A. N. Viswanathan et al. 2015. Brachytherapy: Where has it gone? *J Clin Oncol* 33:980–2.

Polgar, C., E. Van Limbergen, R. Potter et al. 2010. Patient selection for accelerated partial-breast irradiation (APBI) after breast-conserving surgery: Recommendations of the Groupe Europeen De Curietherapie-European Society for Therapeutic Radiology and Oncology (GEC-ESTRO) breast cancer working group based on clinical evidence (2009). *Radiother Oncol* 94:264–73.

Sanders, M. E., T. Scroggins, F. L. Ampil, and B. D. Li. 2007. Accelerated partial breast irradiation in early-stage breast cancer. *J Clin Oncol* 25:996–1002.

Sanghani, M. and D. E. Wazer. 2007. Patient selection for NSABP B-39/RTOG 0413: Have we posed the right questions in the right way? *Brachytherapy* 6:119–22.

Scholten, A. N., W. L. van Putten, H. Beerman et al. 2005. Postoperative radiotherapy for stage 1 endometrial carcinoma: Long-term outcome of the randomized PORTEC trial with central pathology review. *Int J Radiat Oncol Biol Phys* 63:834–8.

Shah, C., T. Lanni, J. B. Wilkinson et al. 2014. Cost-effectiveness of 3-dimensional conformal radiotherapy and applicator-based brachytherapy in the delivery of accelerated partial breast irradiation. *Am J Clin Oncol* 37:172–6.

Sher, D. J., E. Wittenberg, W. W. Suh, A. G. Taghian, and R. S. Punglia. 2009. Partial-breast irradiation versus whole-breast irradiation for early-stage breast cancer: A cost-effectiveness analysis. *Int J Radiat Oncol Biol Phys* 74:440–6.

Smith, B. D., D. W. Arthur, T. A. Buchholz et al. 2009. Accelerated partial breast irradiation consensus statement from the American Society for Radiation Oncology (ASTRO). *Int J Radiat Oncol Biol Phys* 74:987–1001.

Smith, G. L., Y. Xu, T. A. Buchholz et al. 2011. Brachytherapy for accelerated partial-breast irradiation: A rapidly emerging technology in breast cancer care. *J Clin Oncol* 29:157–65.

Spratt, D. E., Z. S. Zumsteg, P. Ghadjar et al. 2014. Comparison of high-dose (86.4 Gy) IMRT vs combined brachytherapy plus IMRT for intermediate-risk prostate cancer. *BJU Int* 114:360–7.

Strnad, V., O. J. Ott, G. Hildebrandt et al. 2016. 5-year results of accelerated partial breast irradiation using sole interstitial multicatheter brachytherapy versus whole-breast irradiation with boost after breast-conserving surgery for low-risk invasive and in-situ carcinoma of the female breast: A randomised, phase 3, non-inferiority trial. *Lancet* 387:229–38.

Suh, W. W., B. E. Hillner, L. J. Pierce, and J. A. Hayman. 2005a. Cost-effectiveness of radiation therapy following conservative surgery for ductal carcinoma *in situ* of the breast. *Int J Radiat Oncol Biol Phys* 61:1054–61.

Suh, W. W., L. J. Pierce, F. A. Vicini, and J. A. Hayman. 2005b. A cost comparison analysis of partial versus whole-breast irradiation after breast-conserving surgery for early-stage breast cancer. *Int J Radiat Oncol Biol Phys* 62:790–6.

Sylvester, J. E., P. D. Grimm, J. Wong, R. W. Galbreath, G. Merrick, and J. C. Blasko. 2011. Fifteen-year biochemical relapse-free survival, cause-specific survival, and overall survival following I(125) prostate brachytherapy in clinically localized prostate cancer: Seattle experience. *Int J Radiat Oncol Biol Phys* 81:376–81.

Vargo, J. A., V. Verma, H. Kim et al. 2014. Extended (5-year) outcomes of accelerated partial breast irradiation using mammosite balloon brachytherapy: Patterns of failure, patient selection, and dosimetric correlates for late toxicity. *Int J Radiat Oncol Biol Phys* 88:285–91.

Veronesi, U., E. Marubini, L. Mariani et al. 2001. Radiotherapy after breast-conserving surgery in small breast carcinoma: Long-term results of a randomized trial. *Ann Oncol* 12:997–1003.

Vicini, F., P. D. Beitsch, C. A. Quiet et al. 2008. Three-year analysis of treatment efficacy, cosmesis, and toxicity by the American Society of Breast Surgeons MammoSite Breast Brachytherapy Registry Trial in patients treated with accelerated partial breast irradiation (APBI). *Cancer* 112:758–66.

Vicini, F., P. Beitsch, C. Quiet et al. 2011. Five-year analysis of treatment efficacy and cosmesis by the American Society of Breast Surgeons MammoSite Breast Brachytherapy Registry Trial in patients treated with accelerated partial breast irradiation. *Int J Radiat Oncol Biol Phys* 79:808–17.

Vinh-Hung, V. and C. Verschraegen. 2004. Breast-conserving surgery with or without radiotherapy: Pooled-analysis for risks of ipsilateral breast tumor recurrence and mortality. *J Natl Cancer Inst* 96:115–21.

Vuong, T., S. Devic, B. Moftah, M. Evans, and E. B. Podgorsak. 2005. High-dose-rate endorectal brachytherapy in the treatment of locally advanced rectal carcinoma: Technical aspects. *Brachytherapy* 4:230–5.

Vuong, T., T. Niazi, and S. Devic. 2015. SP-0222: Role of Endoluminal brachytherapy for rectal cancer: Current status and challenges. *Radiother Oncol* 115:S111–2.

CHAPTER 28

Particle Therapy or Brachytherapy?

Dietmar Georg and Richard Pötter

CONTENTS

28.1 GENERAL ASPECTS OF BRACHYTHERAPY AND PARTICLE THERAPY

Brachytherapy and particle therapy are often compared and considered as rival techniques in radiation oncology. On the other hand, when compared to advanced photon beam therapy, both techniques have several common features. The similarities and the major differences are briefly reviewed in the following.

Brachytherapy and particle therapy allow achieving highly conformal dose distributions, with dose gradients (from targets toward organs at risk [OARs]) being much steeper than the ones typical for high-energy photon beam therapy. Furthermore, low-dose volumes are smaller than in photon beam therapy. In particle beam therapy, a range of target volumes can be applied from very small to large, whereas in brachytherapy target volumes are usually very small to intermediate size (up to 100–150 cm^3). While dose distributions in particle beam therapy benefit from the selective ballistics of protons or ions, in brachytherapy the radial dose distributions of the generally small sources are the backbone of their advantageous dose distribution. A major difference is the application of radiation, with particles coming from outside the body and with brachytherapy coming from a radioactive source (within an applicator) inside the body, or even inside the target. Another major difference, basically associated with these different application techniques, is related to the treatment planning goal in the two modalities. In particle beam therapy, the paradigm of a homogeneous dose in the target has so far been followed, which is typically for all external beam radiotherapy (EBRT) techniques. On the other hand, brachytherapy has been based on inhomogeneous dose distributions allowing for limited size high dose volumes and areas within the target. Brachytherapy always utilized the benefit of these dose gradients inside the target and of a controlled dose inhomogeneity (e.g., ICRU 58). As far as margins in treatment planning are concerned to compensate for uncertainties, both proton therapy and brachytherapy have their own peculiarities with distinct differences from photon beam therapy (ICRU Reports 62, 83). This has led to specific elaborations within dedicated ICRU reports for proton therapy (ICRU Report 78) and for cervical cancer brachytherapy (ICRU Report 89). In brachytherapy, if there is a rigid link between applicator and the clinical target volume (CTV) (e.g., in many interstitial and intracavitary techniques), such a link reduces the need for a margin for intra-fraction effects orthogonal to the longitudinal source axis (Tanderup et al. 2010). In particle therapy, the pronounced sensitivity of the particle range on density variations and the interaction mechanisms of particles leading to different longitudinal and lateral beam spread resulted in recommendations for beam specific margins, for example, along the beam axis and in the orthogonal direction (ICRU Report 78).

Finally, there are important radiobiological aspects due to effects from fractionation and overall treatment time when comparing external beam photon therapy, particle therapy, and brachytherapy. The dose per fraction may vary considerably in treatments with different planning aims, for example, one schedule using (near) conventional fractionation (1.8–2.2 Gy per fraction) and another hypo-fractionation (3–15 Gy per fraction). Such differences may be even observed within one patient, in particular for inhomogeneous doses as applied in brachytherapy or between target and OAR doses. In addition, there are planning aims when using pulsed-dose rate (PDR) brachytherapy techniques, where the dose in the target may be only 0.5–0.8 Gy per fraction or even lower in adjacent OARs. In addition, overall treatment time differs considerably when comparing hypo-fractionated accelerated and conventional treatment schedules. Hypo-fractionated schedules are frequently used in particle therapy and high-dose rate (HDR) brachytherapy, whereas hyper-fractionated accelerated schedules are typical for PDR brachytherapy. Such alternative fractionation schedules have so far been rarely used in photon EBRT, except for palliative or stereotactic radiotherapy. These differences become even more complex when taking into account the effects from differences in linear energy transfer (LET), pronounced in carbon ion radiotherapy. Consequently, treatment planning and/or clinical practice of brachytherapy and particle therapy implies systematic radiobiological modeling when comparing or combining with photon beam therapy. Also, there is a need for clinical and translational research to better understand the underlying mechanisms and effects of variations in fractionation and overall treatment time.

28.2 REVIEW OF CURRENT CLINICAL PRACTICE AND RESEARCH

Prostate carcinoma and uveal melanoma are the two major primary indications where brachytherapy and particle therapy are recognized at present as competing techniques with rather comparable clinical outcome (Colaco et al. 2015, Shields and Shields 2015, Wang et al. 2013). Another indication in clinical practice for both modalities is pediatric radiation oncology, mainly for minimizing long-term adverse effects in the radiosensitive growing tissues. However, experience so far is very limited to a few institutions worldwide (Habrand et al. 2009).

While treatment planning and delivery techniques for uveal melanoma have not changed much for brachytherapy and proton therapy during the last decades , prostate treatments have evolved significantly. In particle therapy, scanning techniques eclipsed the traditional passive scattering techniques, and in brachytherapy image-guided mono-therapy (low-dose rate [LDR] and HDR), as well as boost and focal therapy and treatment plan optimization have gained more attention and importance. Historical and recent outcome reports of monotherapy with ^{125}I sources on large patient cohorts underline its efficiency, with a recent trend also to HDR monotherapy (Hoskin et al. 2014, Jawad et al. 2015, Martinez et al. 2010, Yoshioka et al. 2016). Furthermore, there is an evident and important role of HDR brachytherapy for intermediate and high-risk prostate as a boost modality after external beam photon therapy (Grimm et al. 2012). A common trend in prostate cancer radiotherapy, both in EBRT and HDR brachytherapy is hypo-fractionation, also for particle therapy (Habl et al. 2014), which is motivated by the relatively low alpha-beta ratio. For both uveal melanoma and prostate cancer, there is a lack of comparative clinical studies on the effects of photons, particles, or brachytherapy addressing disease outcome and treatment associated morbidity and impairments in quality of life.

Besides for uveal melanoma and prostate cancer, the advantages of brachytherapy and particle therapy have been explored in numerous treatment planning studies for other malignancies, in particular for particle therapy with a focus on brain, head and neck, abdomen, and lung (Kamada et al. 2015). In this context, one needs to be critical in the assessment of such in silico studies, since any comparison between external beam therapy and brachytherapy is influenced by the choice of dose and dose volume constraints for relevant structures. Moreover, these studies often do not include robustness aspects and uncertainties related to anatomic inter- and intra-fraction variations for external beam therapy. The major drawback of these in silico studies is the often missing estimate of the impact on clinical outcome.

For example, for treating cervical cancer, particle beam therapy has large clinical potential for elective irradiation of the para-aortic (and pelvic) region or lymph node recurrences with optimal kidney, bowel (bladder), and spinal cord sparing. As such, the combination of particle therapy with brachytherapy in the overall management of cervical cancer needs to be further investigated. The role of particle as a boost modality instead of brachytherapy is limited (Georg et al. 2008) and should be taken with caution taking into account the clinical experience with photon EBRT, even with advanced techniques (Tanderup et al. 2014).

28.3 PAST, PRESENT, AND FUTURE OF PARTICLE THERAPY AND BRACHYTHERAPY

Treatment plan optimization in today's advanced radiotherapy setting is based on current clinical evidence for dose volume constraints predicting clinical outcome and relying on "inverse" planning principles and cost functions that drive the computerized optimization process. The mainstream in technology development was driven by intensity modulated photon beam therapy and has certainly influenced both brachytherapy and particle therapy (e.g., Dinkla et al. 2015, Safai et al. 2013, van de Water et al. 2013). Today, the automated treatment plan optimization has become an integral part for any advanced treatment technique in radiation oncology

Besides the paradigm change in treatment planning from forward toward inverse planning, the role of imaging in radiation oncology has increased tremendously during the last decades. Improved target definition, advanced treatment planning, and dose delivery—as introduced in a wide range for photon EBRT worldwide—is now also increasingly used for brachytherapy and particle therapy. Utilizing advanced volumetric imaging combined with treatment plan optimization and improved technology for dose delivery (such as applicator developments in brachytherapy and pencil beam scanning for particles) including also online verification enhance the basic physics advantages inherent in brachytherapy and particle therapy with respect to their ability to create highly conformal and individually tailored dose distributions with optimal OAR sparing. In general, however, when applying highly conformal doses, (geometric) uncertainties are less forgiving. In both brachytherapy and particle therapy, this fact is even more important compared to conventional fractionated photon beam therapy due to the characteristics of steep dose gradients. Therefore, the role of image guidance and adaptive techniques will increase in the future in these fields. In this respect, ultrasound guidance has been clinically proven for prostate brachytherapy during the last decades (Ash et al. 2006, Grimm et al. 2001, Kovács et al. 1999) and MR guidance has been recently pioneered for cervical cancer and the clinical benefit has been demonstrated (Pötter et al. 2011).

In addition, the exploration and utilization of functional imaging methods such as positron emission tomography (PET) or multi-parametric magnetic resonance imaging (mpMRI) for tumor characterization will continue and will have an impact on photon EBRT development but also on the future clinical practice of both brachytherapy and particle beam therapy. This holds especially true for dose escalation strategies for sub-volumes indicating a high risk for recurrent disease (e.g., hypoxic). Both particle therapy and brachytherapy are ideal candidates for dose painting strategies introducing prescribed heterogeneous dose distributions within the traditional CTV, directed at sub-volumes that can be visualized by advanced (functional) imaging techniques. Figures 28.1 and 28.2 illustrate the advantages of brachytherapy and proton therapy compared to photon beam therapy in terms of dose distributions and dose volume histogram (DVH) for dose painting, respectively. This is in contrast to the currently still valid radiotherapy practice in EBRT where the dogma of homogenous dose delivery still dominates clinical practice (e.g., ICRU 62, 78, 83). Such dose painting approaches are intensively discussed and investigated in the radiation oncology community at present, but mainly with high-energy photon beams. Brachytherapy and particle beam therapy are, however, advantageous based on

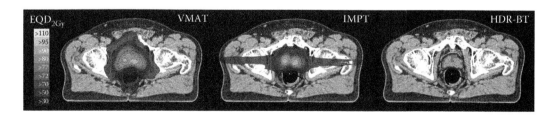

FIGURE 28.1 Representative dose distributions on axial slices.

their inherent physics characteristics and hence more ideal "brushes" for such dose painting strategies (Andrzejewski et al. 2015). The utilization of MRI and PET for brachytherapy is also discussed in Chapters 16 and 17, respectively.

Instead of comparing particles and brachytherapy, future studies should focus on combining external beam therapy including particle therapy with advanced brachytherapy by using the inherent advantages of both options, for example, photon or particle therapy for large targets and brachytherapy for limited size accessible targets (including sub-volumes of interest). Such activities are compromised by the lack of availability of comprehensive technological infrastructure and professional expertise in the different fields. This also holds true for the lack of comprehensive research and insight across these fields. Also, commercial

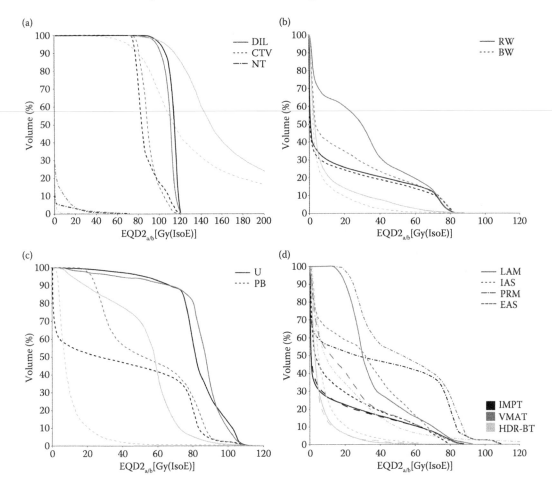

FIGURE 28.2 Representative DVHs for (a) CTV, DIL, and NT, (b) RW and BW, (c) PB and U, (d) IAS, EAS, LAM, and PRM for the three treatment techniques depicted in Figure 28.1. All doses were converted to EQD2. BW, bladder wall; CTV, clinical target volume; DIL, dominant intraprostatic lesion; DVH, dose volume histogram; EAS, external anal sphincter muscle; EQD2α/β, 2Gy fractions dose equivalent; HDR-BT, high-dose rate brachytherapy; IAS, internal anal sphincter muscle; IMPT, intensity modulated proton therapy; LAM, levator ani muscle; NT, normal tissue; PB, penile bulb; PRM, puborectalis muscle; RW, rectal wall; U, urethra; VMAT, volumetric modulated arc therapy;

treatment planning systems (TPSs) in which composite treatment plans can be fully integrated are missing. Such systems and deformable image registration with dose accumulation are in the research and development focus, but so far no practical solution is available for daily clinical use. However, comprehensive clinical, physics, and biological research, clinical professional expertise, and information technology solutions (software) are needed to better understand the potential and the limitations for combination treatment in regard to toxicity and tumor control with an estimate of the "true" dose picture as backbone. Such links between clinical radiobiology, medical physics, IT technology, and radiation oncology are a prerequisite to further optimize radiation oncology, with a special focus on normal tissue which drives the optimization in state-of-the-art treatment planning.

During the last decades, proton and other particle beam therapies took place mainly in dedicated treatment and/or large research centers, with activities ranging from continuous technological developments toward clinical implementation and exploration. This situation has already changed and will certainly continue to change in the near future. The reasons are multilayered. The main reasons are the more pronounced interest and investment in particle therapy from the radiation oncology community in general, and the adapted solutions from particle therapy vendors now also offering single room facilities. In other words, proton therapy is rapidly becoming a standard treatment option in several large cancer treatment centers besides advanced photon beam therapy. Consequently, the parallel nonintegrated clinical research, technology, and/or conceptual developments in photon and particle therapy are likely to change to an integrated approach. Advanced brachytherapy is also very much linked with and dependent on local infrastructure, such as operation theatres, volumetric imaging and treatment planning facilities, as well as manual skills and training, and comprehensive professional expertise to fully utilize its potential. At present, dynamic developments in the field of advanced image-guided adaptive gynecologic brachytherapy are driving forces in the field of brachytherapy which need complementary developments in the other traditional fields (e.g., prostate, breast) or new areas (e.g., liver).

The setting in radiation oncology is likely to change and in large centers the future "radiotherapy toolbox" will cover photons, electrons, particles, and should also comprise advanced brachytherapy. Each patient is then ideally treated with the optimal technique or combinations, respectively, corresponding to the individual need. In such a setting, the clinical outcome of radiotherapy can be optimized in regard to widening the therapeutic window. The technological and methodological differences of the various treatment options are mainly discussed within the radiotherapy community but are far beyond the interests of the clients of radiotherapy, that is, the patients and other medical and oncological disciplines. The maximized clinical outcome of radiotherapy is obviously their interest, irrespective of technique. The question whether brachytherapy or particle therapy is the better treatment option is in principle irrelevant, since both techniques are and will become essential complementary pillars in comprehensive radiotherapy. The question to be answered is under which clinical circumstances one or the other or any combinations represent the optimal treatment method for the individual needs of a specific patient. Such comprehensive radiotherapy will finally lead to highly individualized and high precision radiotherapy.

REFERENCES

Andrzejewski, P., Kuess, P., Knäusl, B., Pinker, K., Georg, P., Knoth, J., Berger, D. et al. 2015. Feasibility of dominant intraprostatic lesion boosting using advanced photon-, proton- or brachytherapy. *Radiother Oncol.* 117(3):509–14.

Ash, D., Al-Qaisieh, B., Bottomley, D., Carey, B., Joseph, J. 2006. The correlation between D90 and outcome for I-125 seed implant monotherapy for localised prostate cancer. *Radiother Oncol.* 79(2):185–9.

Colaco, R.J., Hoppe, B.S., Flampouri, S., McKibben, B. T., Henderson, R. H., Bryant, C., Nichols, R. C. et al. 2015. Rectal toxicity after proton therapy for prostate cancer: An analysis of outcomes of prospective studies conducted at the University of Florida Proton Therapy Institute. *Int J Radiat Oncol Biol Phys.* 91(1):172–81.

Dinkla, A.M., van der Laarse, R., Koedooder, K., Kok, P.H., van Wieringen, N., Pieters, B.R., Bel, A. 2015. Novel tools for stepping source brachytherapy treatment planning: Enhanced geometrical optimization and interactive inverse planning. *Med Phys.* 42(1):348–53.

Georg, D., Kirisits, C., Hillbrand, M., Dimopoulos, J., Pötter, R. 2008. Image-guided radiotherapy for cervix cancer: High-tech external beam therapy versus high-tech brachytherapy. *Int J Radiat Oncol Biol Phys.* 71(4):1272–8.

Grimm, P.D., Blasko, J.C., Sylvester, J.E., Meier, R.M., Cavanagh, W. 2001. 10-year biochemical (prostate-specific antigen) control of prostate cancer with (125)I brachytherapy. *Int J Radiat Oncol Biol Phys.* 51(1):31–40.

Grimm, P.D., Billiet, I., Bostwick, D., Dicker, A.P., Frank, S., Immerzeel, J., Keyes, M. et al. 2012. Comparative analysis of prostate-specific antigen free survival outcomes for patients with low, intermediate and high risk prostate cancer treatment by radical therapy. Results from the Prostate Cancer Results Study Group. *BJU Int* 109, Suppl 1:22–9.

Habl, G., Hatiboglu, G., Edler, L., Uhl, M., Krause, S., Roethke, M., Schlemmer, H.P., Hadaschik, B., Debus, J., Herfarth, K. 2014. Ion Prostate Irradiation (IPI)—A pilot study to establish the safety and feasibility of primary hypofractionated irradiation of the prostate with protons and carbon ions in a raster scan technique. *BMC Cancer.* 14:202.

Habrand, J.L., Demarzi, L., Haie-Meder, C., Dumas, I., Belaide, A., Koedooder, K., Pieters, B. 2009. Proton therapy (PT) is the best radiotherapeutical modality for pediatric tumors. For the proposition. *Radiother Oncol.* 91, Suppl 1:S26.

Hoskin, P., Rojas, A., Ostler, P., Hughes, R., Alonzi, R., Lowe, G., Bryant, L. 2014. High-dose-rate brachytherapy with two or three fractions as monotherapy in the treatment of locally advanced prostate cancer. *Radiother Oncol.* 112(1):63–7.

ICRU Report 58. *Dose and Volume Specification for Reporting Interstitial Therapy*, 1997.

ICRU Report 62. *Prescribing, Recording, and Reporting Photon-Beam Therapy*, 1999.

ICRU Report 78. *Prescribing, Recording, and Reporting Proton-Beam Therapy*, 2007.

ICRU Report 83. *Prescribing, Recording, and Reporting Intensity-Modulated Photon-Beam Therapy (IMRT)*, 2010.

ICRU Report 89. *Prescribing, Recording, and Reporting Brachytherapy for Cancer of the Cervix*, 2016.

Jawad, M.S., Dilworth, J.T., Gustafson, G.S., Ye, H., Wallace, M., Martinez, A., Chen, P.Y., Krauss, D.J. 2015. Outcomes associated with 3 treatment schedules of high-dose-rate brachytherapy monotherapy for favorable-risk prostate cancer. *Int J Radiat Oncol Biol Phys.* pii: S0360-3016(15)26576-5

Kamada, T., Tsujii, H., Blakely, E.A., Debus, J., De Neve, W., Durante, M., Jäkel, O. et al. 2015. Carbon ion radiotherapy in Japan: An assessment of 20 years of clinical experience. *Lancet Oncol. 2015* 1n6(2):e93–e100.

Kovács, G., Galalae, R., Loch, T., Bertermann, H., Kohr, P., Schneider, R., Kimming, B. 1999. Prostate preservation by combined external beam and HDR brachytherapy in nodal negative prostate cancer. *Strahlenther Onkol.* 175, Suppl 2:87–8.

Martinez, A.A., Demanes, J., Vargas, C., Schour, L., Ghilezan, M., Gustafson, G.S. 2010. High-dose-rate prostate brachytherapy: An excellent accelerated-hypofractionated treatment for favorable prostate cancer. *Am J Clin Oncol.* 33(5):481–8.

Pötter, R., Georg, P., Dimopoulos, J. C., Grimm, M., Berger, D., Nesvacil, N., Georg, D. et al. 2011. Clinical outcome of protocol based image (MRI) guided adaptive brachytherapy combined with 3D conformal radiotherapy with or without chemotherapy in patients with locally advanced cervical cancer. *Radiother Oncol.* 100(1):116–23.

Safai, S., Trofimov, A., Adams, J.A., Engelsman, M., Bortfeld, T. 2013. The rationale for intensity-modulated proton therapy in geometrically challenging cases. *Phys Med Biol.* 58(18):6337–53.

Shields, J.A., Shields, C.L. 2015. Management of posterior uveal melanoma: Past, present, and future: The 2014 Charles L. Schepens lecture. *Ophthalmology.* 122(2):414–28.

Tanderup, K., Eifel, P.J., Yashar, C.M., Pötter, R., Grigsby, P.W. 2014. Curative radiation therapy for locally advanced cervical cancer: Brachytherapy is NOT optional. *Int J Radiat Oncol Biol Phys.* 88(3):537–9.

Tanderup, K., Pötter, R., Lindegaard, J.C., Berger, D., Wambersie, A., Kirisits, C. 2010. PTV margins should not be used to compensate for uncertainties in 3D image guided intracavitary brachytherapy. *Radiother Oncol.* 97(3):495–500.

van de Water, S., Kraan, A.C., Breedveld, S., Schillemans, W., Teguh, D.N., Kooy, H.M., Madden, T.M., Heijmen, B.J., Hoogeman, M.S. 2013. Improved efficiency of multi-criteria IMPT treatment planning using iterative resampling of randomly placed pencil beams. *Phys Med Biol.* 58(19):6969–83.

Wang, Z., Nabhan, M., Schild, S.E., Stafford, S.L., Petersen, I.A., Foote, R.L., Murad, M.H. 2013. Charged particle radiation therapy for uveal melanoma: A systematic review and meta-analysis. *Int J Radiat Oncol Biol Phys.* 86(1):18–26.

Yoshioka, Y., Suzuki, O., Isohashi, F. Seo, Y., Okubo, H., Yamaguchi, H., Oda, M. et al. 2016. High-dose-rate brachytherapy as monotherapy for intermediate- and high-risk prostate cancer: Clinical results for a median 8-year follow-up. *Int J Radiat Oncol Biol Phys.* 94(4):675–82.

CHAPTER 29

Is Brachytherapy Cost Effective?

Peter Orio, Benjamin Durkee, Thomas Lanni, Yolande Lievens, and Daniel Petereit

CONTENTS

29.1 UTILIZATION OF BRACHYTHERAPY

Brachytherapy is the ultimate form of conformal radiotherapy (RT) with an ability to deliver high doses to the tumor while sparing the adjacent normal structures due to the steep dose gradient. There is a disturbing trend in the United States whereby the use of brachytherapy is in rapid decline for prostate, cervical, vaginal, and inoperable endometrial cancer (Rajagopalan et al., 2015).

Even though prostate brachytherapy is the least costly alternative, with outcomes as good if not superior to other modalities, two recent studies detail a significant decrease in utilization. Martin et al. found that brachytherapy utilization reached a peak of 17% in 2002 and steadily declined to a low of 8% using the National Cancer Database (NCDB) (Martin et al., 2014). Similarly, Mahmood et al. reported prostate brachytherapy procedures decreased from 44% in 2004 to 38% in 2009 using the SEER (Surveillance, Epidemiology, and End Results) database (Mahmood et al., 2014). The most dramatic decline in brachytherapy procedures was seen at academic centers (48%), though it was also significant at comprehensive community (41%) and community cancer centers (30%) (Martin et al., 2014).

While brachytherapy is a standard component of potentially curative treatment in the management of locally advanced cervical cancer, a surprising number of patients do not receive it. Han et al. reported a 25% reduction in brachytherapy utilization and a 13% reduction in the cause-specific survival rate during the time span of 1988–2009 using the SEER data (Han et al., 2013). Similarly, Gill et al. reported a brachytherapy reduction from 97% to 86% with a concomitant increase of intensity modulated radiation therapy (IMRT) and stereotactic body radiation therapy (SBRT) from 3.3% to 14% using the NCDB (2004–2011) ($p < 0.01$). The median survival time was 71 months for patients who received brachytherapy compared to 47 months for those treated with either IMRT or SBRT as an alternative to brachytherapy (Gill et al., 2014).

With the expansion of health care under the Affordable Healthcare Act (ACA) in the United States, it is possible that patients may present with earlier stages of cancer, and therefore, increase the population base eligible for brachytherapy. The Congressional Budget Office projects that the ACA will cover 19 million Americans in 2016 and 23 million by 2025 (Congressional Budget Office, 2015). The newly insured are mostly covered through insurance exchanges or the Medicaid expansion (Congressional Budget Office, 2015). Men and women covered under the ACA are more likely to receive brachytherapy (Grant et al., 2015) and see improved outcomes (Mahal et al., 2014). On the other hand, the overall impact on brachytherapy is likely to be small since the ACA primarily extends coverage to Americans under the age of 65. This younger population has a relatively low rate of cancer compared to the Medicare cohort. In women under the age of 65, the incidences of cervical and endometrial cancers are 7 and 16 per 100,000, compared to 12 and 84 per 100,000 for women aged 65 and over (Mahal et al., 2014). In men under 65, the incidence of prostate cancer is 58 per 100,000, compared to 769 per 100,000 for men aged 65 and over (US Mortality Files).

Appreciating that there is a difference in both the utilization trends and number of centers with brachytherapy in the United States, a recent survey of the International Atomic Energy Agency (IAEA) identified a total of 945 brachytherapy centers in Europe, helping to highlight these differences. The majority used high-dose rate (HDR) sources (n = 546), followed by 328, 40, and 31 using low-dose rate (LDR), pulsed-dose rate (PDR), and medium-dose rate (MDR), respectively. Overall, 52% of all RT centers in Europe had brachytherapy facilities, but ranging from less than 40% in France, Italy, and Spain, to 60% or more in northern, eastern, and southeastern European countries (Rosenblatt et al., 2013). High-income countries, typically with a high level of RT use per head of population, treat markedly less patients annually with brachytherapy compared to countries with a lower income. A possible explanation is the easier access to more sophisticated external beam equipment in the case of higher economic status. Another hypothesis is that there is a higher tendency to centralize brachytherapy in less-affluent countries in order to avoid multiplying the cost of establishing new brachytherapy facilities.

Some countries in Europe, such as Denmark and the United Kingdom have policies to centralize brachytherapy and RT services which translates into larger centers but this is certainly not the case in the majority of countries (Guedea et al., 2010; Dunscombe et al., 2014; Grau et al., 2014). Centralization provides advantages in terms of economies of scale with reduced costs and increased physician expertise (Remonnay et al., 2010; Hulstaert et al., 2013), but must be balanced against the potential disadvantages of increased waiting times and reduced access (Guedea et al., 2010). Another observation ensuing from economic incentives is the increased uptake of HDR compared to LDR in case of lower welfare conditions, potentially related to the fact that HDR equipment allows a significantly larger population of patients to be treated for the same costs compared to traditional LDR (Guedea et al., 2010).

29.2 COSTING ANALYSIS OF BRACHYTHERAPY

There are many challenges to assessing cost effectiveness by classic economic modeling techniques such as Markov models. The most obvious is lack of high-quality data directly comparing brachytherapy to other modalities. The best models are validated against well-designed prospective randomized clinical trials. Lesser challenges include limited quality of life data using instruments that have been validated against utility weights used to estimate quality-adjusted life years. Functional outcomes across all modalities are widely variable due to different patient selection and other confounding factors. Such variability would contribute substantial uncertainty to the model, which would have to be carefully addressed in deterministic and probabilistic sensitivity analyses. Finally, cost-effectiveness models are typically designed from the perspective of a single government payer. Such models would be limited in their ability to address many of the most important costs and benefits to the patient and treating institution.

Cost of therapy to the treating institution is difficult to estimate, even with access to internal data. Accountants must consider not only direct costs but also cost of space, equipment, maintenance, and personnel. Shah et al. attempted to account for these variables and found that the institutional costs for HDR ($5467) and LDR ($2395) are substantially lower than IMRT ($23,665) (Shah et al., 2012). Their cost analysis showed economic efficiencies for both payer and provider when brachytherapy was chosen over IMRT (Shah et al., 2012). The authors acknowledge the shortcomings of using Medicare's Current Procedural Terminology codes or other charge-based accounting methodologies, and specifically site time-driven activity-based costing as a potentially superior method. Time-driven activity-based costing is a simpler and more powerful accounting tool developed by investigators at Harvard (Kaplan et al., 2003, 2014). Costs cover an entire cycle of care and the analysis is based on time used by each resource, including the physician's time spent planning and executing the procedure. Radiation oncologists at the MD Anderson Cancer Center have experimented with this method for costing in proton therapy (Thaker et al., 2015) and are applying this model to LDR prostate brachytherapy. The cost calculation performed by the Belgian Health Care Knowledge Centre also used this methodology to compute real-life costs of standard and innovative RT techniques, including external beam radiation therapy (EBRT) as well as brachytherapy (Hulstaert et al., 2013).

The cost of therapy from the payer's perspective is more straightforward. Insurers contract privately with the treating institution, usually paying a pre-negotiated premium on the Medicare rate. For example, for an HDR monotherapy treatment of 4 fractions, the 2015 Medicare reimbursement would be $6629 at the national payment amount (Table 29.1). As an exercise, we estimated the Medicare cost of treating low and intermediate-risk prostate cancers with HDR brachytherapy using the previously described method (Durkee and Buyyounouski, 2015) and projections from the Congressional Budget Office (June 2015) report. The total estimated cost of treating all patients with brachytherapy from 2016 to

TABLE 29.1 Sample Medicare Reimbursement for Treatment of Prostate Cancer with Interstitial HDR Brachytherapy in Four Fractions Based on 2015 CPT

CPT	Description	# of Units	Prices	Total
77263	Treatment planning	1	$165.90	$165.90
77470	Special radiation treatment	1	$155.89	$155.89
77290	Complex simulation	1	$510.93	$510.93
76873	Prostate tumor volume study	4	$167.69	$670.76
76872	Transrectal ultrasound	1	$94.39	$94.39
76965	Ultrasound guidance for interstitial application	4	$90.82	$363.28
55758	Placements of needles	4		$0.00
C1728	Brachytherapy catheters	4		$0.00
77280	Simple simulation	1	$271.38	$271.38
77332	Simple device	4	$82.59	$330.36
77300	MU calculation	4	$63.29	$253.16
77328	Brachytherapy isodose plan	1		$0.00
C1717	Brachytherapy source, HDR ^{192}Ir	4		$0.00
77295	3D isodose plan with DVH for brachytherapy	1	$489.12	$489.12
77787	HDR brachytherapy application over 12 channels	4	$782.67	$3130.68
77336	Physics consult and quality assurance	1	$76.87	$76.87
77370	Special physics consult	1	$116.92	$116.92
		Total HDR × 4	$6629.64	

Note: CPT, current procedural terminology; DVH, dose-volume histogram; HDR, high-dose rate; MU, monitor units.

2025 is $28.7 billion over 10 years (2015 dollars). This represents a $19.4 billion savings over IMRT, which we estimate at $48.2 billion over the same time period.

The comparison between brachytherapy and other treatment strategies will often boil down to weighing the costs with the differences in local control and quality of life. In the context of accelerated partial breast irradiation (APBI) after breast conserving surgery, balloon brachytherapy, providing slightly inferior results in terms of in-breast local control, was found unlikely to become cost effective in comparison to external beam techniques, unless the quality of life would be proven to be superior (Sher et al., 2009). In another study, looking at the entire spectrum of breast-sparing treatments—whole breast RT and APBI—all accelerated approaches were found cost effective compared to IMRT due to the lower costs. When the comparison was made to 3D-conformal radiotherapy (CRT), external beam techniques represented a more cost-effective approach based on the lower cost, but brachytherapy became cost effective when quality of life was taken into account (Shah et al., 2013).

Amin et al. determined that brachytherapy was consistently found to be more cost effective when compared with surgery and EBRT options for prostate cancer due to its lower costs, low side-effect profile, and good quality of life for patients after treatment (Amin et al., 2014). A similar conclusion is reached in the work of the Institute for Clinical and Economic Review (ICER) using their evidence-based rating matrix assessment, which combines a rating for comparative clinical effectiveness and a rating for comparative value. In this approach, IMRT and brachytherapy were identified to be of comparable clinical effectiveness but with an enhanced comparative value of brachytherapy versus IMRT. Results from this decision analytic model suggest that brachytherapy is likely to be less expensive and result in a slightly improved quality of life for a general population of patients compared with IMRT or proton beam therapy (Ollendorf et al., 2009).

29.3 IMPLEMENTATION OF BRACHYTHERAPY PROGRAMS

The cost to implement either a brachytherapy or EBRT program can vary based on the technology and technique used to deliver the specific treatment. In the past 15 years, there has been a significant change in the way patients are treated with the advancement of image guidance and dose escalation. These advancements have allowed physicians to shorten the treatment course with comparable clinical outcomes. In the area of prostate cancer, there are a number of treatment options available with radiation therapy. These options range from LDR, HDR, to EBRT. Within HDR and EBRT, the fractionation schemes can be different based on the disease stage, institutional adoption of treatment techniques, and patient preference. Each of these treatment options have a different level of commitment from an organization, from capital equipment and staffing to physical space. There is also a difference in the cost to construct the physical space to accommodate each treatment modality. In Table 29.2, we breakdown the actual capital and operational expenses to start a program in the United States.

TABLE 29.2 Sample List of Equipment, Supplies, and Maintenance Expense to Initiating a Brachytherapy and/or EBRT Program

U.S. Dollars	LDR	HDR	EBRT
Capital Equipment			
Unit	44,000	485,068	3,000,000
Treatment planning System	109,335	109,335	125,000
CT simulator	0	0	750,000
C-arm	148,770	148,770	0
Ultrasound	120,000	120,000	0
Anesthesia cart	130,000	130,000	0
Room equipment	310,186	310,186	
Physics equipment	6500	6500	500,785
Shielding	0	1,000,000	1,000,000
Total Capital Expense	$868,791	$2,309,859	$5,375,785
Service Maintenance			
Annual contract	$46,500	$46,500	$200,000
Supplies			
^{103}Pd seeds	2650	0	0
Operating room	628	628	0
Hospitalization	1129	1129	0
Immobilization device	0	0	3.25
Total Supply Expense Per Case	$4407	$1757	$3

TABLE 29.3   Sample Labor Cost to Treat Prostate Cancer by LDR, HDR, or EBRT

Labor	Unit Cost ($/h)[a]	Labor Cost per Procedure			Labor Cost per Entire Treatment		
		LDR	HDR	EBRT	LDR 1 Fraction	HDR 2 Fractions	EBRT 25 Fractions
Radiation therapist[a]	38.51	0.00	0.00	28.88	0.00	0.00	722.06
Nursing[a]	32.04	64.08	32.04	1.60	64.08	64.08	40.05
Nursing assistant[a]	12.07	6.04	6.04	0.00	6.04	12.07	0.00
Medical physicist[b]	89.80	269.40	224.50	8.98	269.40	449.00	224.50
Dosimetrist	43.60	21.80	21.80	109.00	21.80	43.60	2725.00
Radiation oncologist[c]	243.28	486.56	182.46	48.66	486.56	364.92	1216.40
Anesthetist[c]	204.83	204.83	102.42	0.00	204.83	204.83	0.00
		$1052.71	$569.25	$197.12	$1052.71	$1138.50	$4928.01

[a] Unit costs were hourly pay rates based on the U.S. dollar.

LDR represents the lowest capital expenditure due to lower equipment costs and shielding requirements for treatment; however, it does have the highest supply expense per case. Although LDR is the least expensive option, it has limitations to the types of cancer it can treat based on disease site. Even though the most expensive in terms of capital dollars and service maintenance costs, EBRT has the ability to treat all cancers in a variety of different techniques. With the continual improvements in the technology for EBRT, physicians have the ability to offer their patients a variety of options from standard fractionation to hypofractionation with comparable outcomes. Even with these advancements, EBRT still has the largest labor costs for treatment. When reviewing the average fractionation schedule for a typical department today, the labor cost to perform EBRT is more than both LDR and HDR, while the difference between LDR and HDR is negligible (Table 29.3).

In order to justify the expenditure for any of these programs, a financial analysis is conducted in order to review the rate of return for this investment. Based on the U.S. Medicare reimbursement fee schedule and a published study by Shah et al. (2012), a *pro forma* for each modality was developed to evaluate the number of patients that are needed to break-even on the capital investment. In year 1 of operation, the number of patients needed to be treated is 105, 68, and 136 for LDR, HDR, and EBRT, respectively. The main drivers are reimbursement per fraction and fractionation schedules. The 5-year projection for overall net income will be highest for EBRT due to the ability to treat more patients and increased fractionation schedule versus the other modalities. Even with the adoption of new treatment modalities and fractionation schedules, financial pressures to reduce cost and reimbursement will continue to weigh on both hospitals and free standing centers in the investment of these technologies.

29.4 CONCLUSION

Although prostate brachytherapy is the least costly alternative as compared to other forms of RT, with outcomes as good if not superior to other modalities, international utilization has seen a decline at academic centers, comprehensive community centers, and community cancer centers with degrees of utilization varying across economic and geographic landscapes. Given the increasing pressures facing RT centers across the globe, consideration needs to be given to the utilization of brachytherapy as a form of conformal RT for its ability to safely deliver high doses of radiation for local tumor control, cost effectiveness as an efficacious treatment modality, and low cost of implementation. Although challenges arise in assessing the cost effectiveness of brachytherapy across economic markets, consideration of direct costs, space, equipment, maintenance, and personnel show that HDR and LDR can be utilized in the setting of multiple disease site diagnoses with substantially lower costs than EBRT for payers as well as providers.

REFERENCES

Amin NP, Sher DJ, Konski AA. Systematic review of the cost effectiveness of radiation therapy for prostate cancer from 2003 to 2013. *Appl Health Econ Health Policy*, 2014;12:391–408.

Congressional Budget Office. Budgetary and Economic Effects of Repealing the Affordable Care Act. June 2015:9. http://www.cbo.gov/publication/50252.

Dunscombe P, Grau C, Defourny N. et al. for the HERO Consortium. Guidelines for equipment and staffing of radiotherapy facilities in the European countries: Final results of the ESTRO-HERO survey. *Radiother Oncol*, 2014;112:165–77.

Durkee BY, Buyyounouski MK. The case for prostate brachytherapy in the Affordable Care Act era. *Int J Radiat Oncol Biol Phys*, 2015;91(3):465–7. doi: 10.1016/j.ijrobp.2014.10.014

Gill BS, Lin JF, Krivak TC et al. National Cancer Data Base analysis of radiation therapy consolidation modality for cervical cancer: The impact of new technological advancements. *Int J Radiat Oncol Phys*, 2014;90:1083–90.

Grant SR, Walker GV, Koshy M et al. Impact of insurance status on radiation treatment modality selection among potential candidates for prostate, breast, or gynecologic brachytherapy. *Int J Radiat Oncol Biol Phys*, 2015;93(5). doi: 10.1016/j.ijrobp.2015.08.036.

Grau C, Defourny N, Malicki J et al. for the HERO consortium. Radiotherapy equipment and departments in the European countries: Final results from the ESTRO-HERO survey. *Radiother Oncol*, 2014;112:155–64.

Guedea F, Venselaar J, Hoskin P et al. Patterns of care for brachytherapy in Europe: Updated results. *Radiother Oncol*, 2010;97:514–20.

Han, K., Milosevic, M., Fyles A et al. Trends in the utilization of brachytherapy in cervical cancer in the United States. *Int J Radiat Oncol Biol Phys*, 2013;87(1):111–9.

Hulstaert F, Mertens A-S, Obyn C et al. *Innovative radiotherapy techniques: A multicentre time-driven activity-based costing study.* Health Technology Assessment (HTA) Brussels: Belgian Health Care Knowledge Centre (KCE) KCE Reports. Brussels: Belgian Health Care Knowledge Centre (KCE); 2013 Available at https://kce.fgov.be/sites/default/files/page_documents/kce_198c_innovative_highrisk_medical_devices_0.pdf.

Kaplan SA. Re: Measuring the cost of care in benign prostatic hyperplasia using time-driven activity-based costing (TDABC). *J Urol*, 2015;194(5):1355–6. doi: 10.1016/j.juro.2015.08.037.

Kaplan RS, Anderson SR. Time-driven activity-based costing. *SSRN J*, 2003. doi: 10.2139/ssrn.485443.

Mahal BA, Aizer AA, Ziehr DR et al. The association between insurance status and prostate cancer outcomes: Implications for the Affordable Care Act. *Prost Cancer Prost Dis*, 2014;17(3):273–9. doi: 10.1038/pcan.2014.23.

Mahmood U, Pugh T, Frank S et al. Declining use of brachytherapy for the treatment of prostate cancer. *Brachytherapy*, 2014;13(2):157–62.

Martin JM, Handorf EA, KutiKov A et al. The rise and fall of prostate brachytherapy: Use of brachytherapy for the treatment of localized prostate cancer in the National Cancer Data Base. *Cancer*, 2014;120(14):2114–21.

Ollendorf DA, Hayes J, McMahon P, Kuba M, Pearson SD. *Management Options for Low-risk Prostate Cancer: A Report on Comparative Effectiveness and Value.* Boston, MA: Institute for Clinical and Economic Review, December 2009: Available at: http://www.icer-review.org/index.php/mgmtoptionlrpc.html.

Rajagopalan MS, Xu KM, Lin J et al. Patterns of care and brachytherapy boost utilization for vaginal cancer in the United States. *Pract Radiat Oncol*, 2015;5(1):56–61.

Remonnay R, Morelle M, Pommier P et al. Economic assessment of pulsed dose-rate (PDR) brachytherapy with optimized dose distribution for cervix carcinoma. *Cancer Radiothér*, 2010;14:161–8.

Rosenblatt E, Izewska J, Anacak Y et al. Radiotherapy capacity in European countries: An analysis of the Directory of Radiotherapy Centres (DIRAC) database. *Lancet Oncol*, 2013;14:79–86.

Shah C, Lanni TB, Ghilezan MI et al. Brachytherapy provides comparable outcomes and improved cost-effectiveness in the treatment of low/intermediate prostate cancer. *Brachytherapy*, 2012;11(6):441–5. doi:10.1016/j.brachy.2012.04.002.

Shah C, Lanni TB, Saini H et al. Cost-efficacy of acceleration partial-breast irradiation compared with whole-breast irradiation. *Breast Cancer Res Treat*, 2013;138:127–35.

Sher DJ, Wittenberg E, Suh WW et al. Partial-breast irradiation versus whole-breast irradiation for early-stage breast cancer: A cost-effectiveness analysis. *Int J Radiat Oncol Biol Phys*, 2009;74:440–6.

Thaker NG, Frank SJ, Feeley TW. Comparative costs of advanced proton and photon radiation therapies: Lessons from time-driven activity-based costing in head and neck cancer. *J Comp Eff Res*, 2015;4(4):297–301. doi: 10.2217/cer.15.32.

US Mortality Files, National Center for Health Statistics, Centers for Disease Control, Prevention. SEER Cancer Statistics Review 1975–2011: Cancer of the Prostate. http://seer.cancer.gov/archive/csr/1975_2011/browse_csr.php?sectionSEL=23&pageSEL=sect_23_table.07.html#table2. Accessed November 1, 2015.

V

Vision 20/20: Industry Perspective

CHAPTER 30

Elekta Brachytherapy

Mehmet Üzümcü

CONTENTS

30.1 INTRODUCTION—ELEKTA'S VIEW: VISION 2020—EYE ON THE FUTURE OF BRACHYTHERAPY

Since the discovery of radium in 1898, radiation therapy has been applied to treat various cancers. Brachytherapy was the first modality of radiation therapy to perform such treatments by placing applicators pre-loaded with radium in contact with the skin to treat skin tumors and later by inserting pre-loaded needles and applicators in cavities in the body to treat, for example, gynecological cancers.

Since then, numerous innovations have led to the growth of the utilization of brachytherapy. Most of these innovations—such as the selectronLDR® (low-dose rate), microSelectron® HDR/PDR (high-dose rate/pulsed-dose rate), Flexitron®, numerous applicators and the state-of-the-art treatment planning software Oncentra® Brachy—were brought to the community by Elekta Brachytherapy Solutions (formerly Nucletron) in the past decades. Figure 30.1 shows the evolution of our treatment delivery platforms throughout the past decades and Figure 30.2 shows examples of innovations in the area of applicators. Through these innovations, brachytherapy has become more efficient, more effective, and more precise over the years.

Brachytherapy has numerous advantages over other radiation therapy modalities. Because irradiation takes place from within the body with a small radioactive source, a very high dose can be delivered to target structures. Due to the high gradient in dose fall-off,

microSelectron v1 classic   microSelectron digital   Flexitron

FIGURE 30.1   Evolution of the elekta brachytherapy treatment delivery platforms in the past decades.

healthy tissue close to the target structure can be optimally spared. These benefits enable organ sparing treatments that in many cases reduce the need for hysterectomy, prostatectomy, cystectomy, etc. and have a high and positive impact on the quality of life of patients after treatment. Furthermore, brachytherapy has been shown to be a cost-effective treatment and is appealing to patients due to the inherent hypo-fractionated nature of the therapy.

The innovations in and advantages of brachytherapy have led to increased utilization and adoption of brachytherapy over the past decades. However, to date, still less than 10% of all radiotherapy procedures are done with brachytherapy, and not all eligible patients have access to or are offered the therapy. Innovations in alternative therapy options have made those attractive from a workflow and efficacy perspective and therefore the need for innovating brachytherapy is greater than ever.

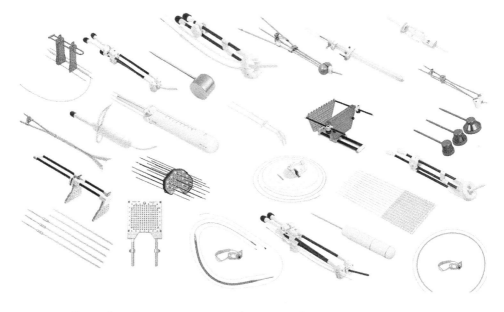

FIGURE 30.2   Examples of innovations on applicators in the past years.

With Vision 2020, we identify future improvement areas for brachytherapy. The objective is to make brachytherapy break out from its current limited/niche-utilization and make it accessible to the wide range of patients that can benefit from brachytherapy as part of their cancer treatment. Rather than focusing on specific solutions, our Vision 2020 points out a number of themes in which future innovations are necessary to increase the utilization of brachytherapy.

30.2 VISION 2020 THEMES

Within our Vision 2020, we have identified six themes in which future innovations are needed in our view. Per theme, multiple simultaneous and subsequent solutions and developments are foreseen. The identified themes are as shown in Figure 30.3: *System Integration, Access to Brachytherapy, Adaptive Combined Therapy, Treatment Certainty, Patient Experience*, and *Workflow Efficiency*. In the sections below these themes are further explained and examples are given of our recent solutions and future ideas and concepts to further improve on these themes.

30.2.1 Theme I: System Integration

Our definition of the *System Integration* theme is as follows:

> Provide real-time adaptive solutions for application in various cancer types by further integrating treatment planning, imaging, and treatment delivery.

Recent studies have shown that when a patient is moved after applicator and/or needle insertion, for example, from the brachytherapy treatment room to the radiology department and back, significant displacements and shifts in the position of the applicator and/or needles may occur (Holly et al., 2011). This may lead to a deviation between the planned and the actual position of the implant, leading to suboptimal dosimetry in the actual treatment delivery.

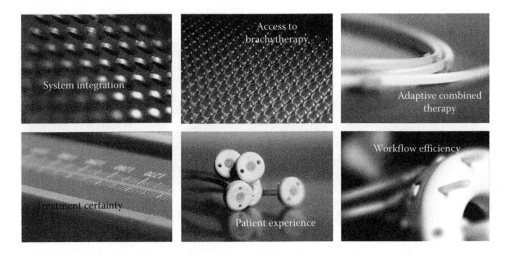

FIGURE 30.3　The six main innovation themes of the Vision 2020.

By integrating image acquisition, treatment planning, and treatment delivery, they can all be performed in the same room removing the need for patient movement, thereby keeping displacements in the applicator and/or needles to a minimum. Furthermore, having such an integrated solution, images providing real-time information can be acquired at any time during treatment, allowing for quick checks and adjustments to be performed prior to delivery of the treatment. In this way the most conformal dose plan can be made, based on the actual location of implant versus target and organ at risk (OAR).

An example of one of our existing solutions related to this theme is our fully integrated real-time adaptive prostate solution for HDR (Oncentra Prostate) and LDR (Oncentra Seeds) treatments. Here, imaging, treatment planning, and treatment delivery are fully integrated, eliminating the need for patient movement and therewith improving treatment results (Hinnen et al., 2010). These results encourage us to investigate similar real-time adaptive image-guided solutions to be used in various other body sites, such as GYN treatments. Figure 30.4 gives an impression of a concept for real-time adaptive GYN treatments.

30.2.2 Theme II: Access to Brachytherapy

Our definition of the *Access to Brachytherapy* theme is as follows:

> Reduce physical, financial, and organizational hurdles. Enable the application of brachytherapy outside the bunker by qualified individuals.

Today, the high energy of ^{192}Ir poses requirements on the treatment room with respect to shielding (Lymperopoulou et al., 2006). Clinics that consider starting an HDR brachytherapy practice are faced with high investments for the construction of a bunker. Due to their dedicated set-up these bunkers can often only be used for brachytherapy, thus occupying

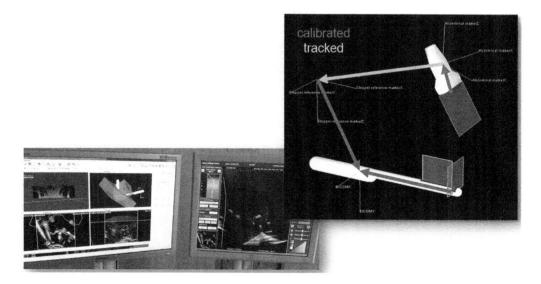

FIGURE 30.4 Real-time ultrasound image-guided GYN treatment concept. (Images courtesy of AKH Vienna, Austria and MedCom GmbH, Germany.)

space in the clinic regardless of its utilization. The high initial investment in infrastructure and the limited use of space can be barriers for starting with HDR brachytherapy.

Furthermore, as HDR brachytherapy needs to take place in a bunker, this brings along organizational challenges. The anesthesiologists for instance might not be willing or able to bring their equipment over to the brachytherapy treatment rooms.

Reducing the need for shielding requirements enables application of brachytherapy in other parts of the clinic from the bunkers in the radiation oncology department, while qualified individuals (radiation oncologists and physicists) are still involved in handling of radiation emitting equipment.

A recent innovation that we have commercialized is our Esteya® solution for treatment of non-melanoma skin cancers (NMSCs) (Figure 30.5). This solution encompasses a low energy x-ray source that enables irradiation of NMSCs in a minimally shielded treatment room (Ballester-Sánchez et al., 2016). Furthermore, we are investigating the possibility to use a low energy isotope that has a sufficiently high-dose rate to enable HDR brachytherapy treatments without the need for a heavily shielded bunker.

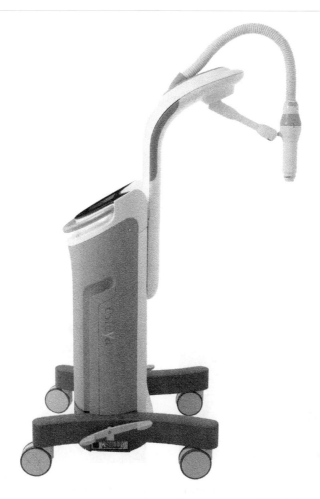

FIGURE 30.5 The Esteya low-energy x-ray solution for treatment of NMSCs.

30.2.3 Theme III: Adaptive Combined Therapy

Our definition of the *Adaptive Combined Therapy* theme is as follows:

> Make brachytherapy a full member of the total cancer care continuum with tools to manage combined therapies and integrated informatics in the Oncology Information System environment.

Brachytherapy is often given as a boost on top of an external beam irradiation (Deutsch et al., 2010; Hoskin et al., 2012; Han et al., 2013). Today, the boost dose is determined empirically based on prior clinical experience and outcomes, and the same boost dose is used for all patients in clinic. Summation of external beam and brachytherapy physical doses does not provide clinically relevant results since external beam is typically delivered in many fractions with a low dose per fraction and brachytherapy is typically delivered in few fractions with high dose per fraction. Both types of treatment have a different radiobiological effect on tumors and healthy tissues (Bianchi et al., 2008).

To overcome this, several clinics use their own in-house developed spreadsheets, in which radiobiological conversions can be made and the biological dose from external beam and brachytherapy can be summed. However, these spreadsheets require manual input of many parameters, making this method prone to errors.

Besides the need for radiobiological conversions of the physical dose, also the image datasets used in external beam and brachytherapy treatment planning must be co-registered (Moulton et al., 2015; Vásquez Osorio et al., 2015). Due to large anatomical changes, a deformable image registration method is needed to map the voxels from one image data set onto the other. Various deformable image registration methods are commercially available today. However, existing solutions have primarily been developed for use in external beam treatments only, to map the image data sets between subsequent EBRT fractions. However, deformations between external beam and brachytherapy image datasets are large due to insertion of applicators and/or needles (Figure 30.6) and existing methods do not perform adequately in these situations.

To achieve the abovementioned goals, integration of brachytherapy in the Oncology Information Systems of the radiation therapy (RT) department is a must. If all relevant

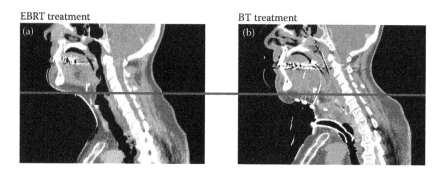

FIGURE 30.6 Large deformations between EBRT (a) and brachytherapy (b). (Images courtesy of Erasmus Medical Center, Rotterdam, The Netherlands.)

data for each patient, such as image data sets for external beam and brachytherapy treatments, treatment plans and records for both modalities are stored in one central location, they can be made accessible for dose summation purposes and be used to optimize patient treatments. An example of a recently introduced solution in this aspect is the connectivity of our afterloaders to the various OIS systems in the field, especially with MOSAIQ®. This enables storage of patient and plan data for brachytherapy and external beam treatments in the same location, making brachytherapy a full member of the radiation oncology department and enabling a better and paperless workflow.

Another recent release is ACE: the Advanced Collapsed cone Engine in Oncentra Brachy v4.4. ACE enables model-based dose calculations, taking into account tissue heterogeneities, lack of backscattering tissue, and high-Z shielding materials corrections (Tedgren and Ahnesjö 2003; Ballester et al., 2015). This brings the accuracy level of brachytherapy dose calculations to the same level as EBRT dose calculations, which is a requirement for accurate dose summation.

We are currently investigating dedicated deformable image registration methods that can account for large deformations as described above. Our vision is that in the future accurate dose summation (Sun et al., 2015) will be possible and the brachytherapy boost dose can be more accurately determined per case, thereby looking at the desired biological effects rather than only at the physical dose.

30.2.4 Theme IV: Treatment Certainty

Our definition of the *Treatment Certainty* theme is as follows:

> Add certainty to safety to ensure accurate delivery of the planned dose to both the target and the organs at risk by providing tools for determining the actual delivered dose.

Over the past decades, brachytherapy has proven to be a precise and effective modality for treatment of cancers in certain body sites. In patients eligible for treatment with brachytherapy, cumulated clinical evidence shows beneficial results in both local control and toxicity (Hinnen et al., 2010; Hoskin et al., 2012; Han et al., 2013).

We believe that determining the actual delivered dose will provide users valuable insight into what is truly going on inside the patient during treatment delivery (Nesvacil et al., 2013; Kertzscher et al., 2014; Kertzscher et al., 2014). For instance, changes such as shifts in applicator position and anatomical changes may occur inside the patient both intra- and inter-fraction (Holly et al., 2011). Insight into the actual delivered dose will enable the tailoring of treatments and provide direction for adaptations to subsequent fractions to compensate for fractions that have not been delivered as they had been planned (Kirchheiner et al., 2016). Also in case of possible use errors, the user will immediately be made aware and can make adjustments in a timely manner to prevent these cases from occurring. Insights in the actual delivered dose will elevate the focus in dose delivery from safety to a level of certainty where the clinical efficacy of brachytherapy over time can be raised to an even higher level.

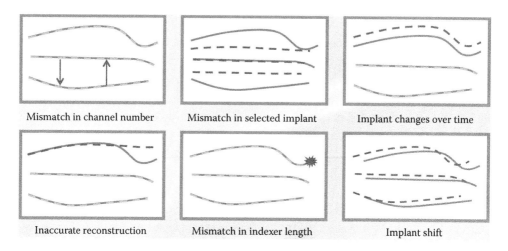

FIGURE 30.7   Possible implant changes that may be undetected at present.

An existing solution in this direction is the Flexitron afterloader, which is designed to eliminate possible use errors by having a standard transfer tube length, the reference "zero" dwell position in the connector instead of the tip and metric dwell position numbering. The Flexitron is our afterloading platform for the future and both the electronics and the mechanics have been designed with possible future expansions in mind. For example, sensors can be integrated into the Flexitron that can accurately measure positions or dose.

Figure 30.7 shows what kind of possible changes and errors can be detected and prevented if a position measuring sensor is integrated in the afterloader. Examples are wrong connections of transfer tubes, wrong implant, shift of one catheter, change in shape of a catheter, mismatch in indexer lengths, and shifts of entire implants.

30.2.5 Theme V: Patient Experience

Our definition of the *Patient Experience* theme is as follows:

> Make the application of brachytherapy more comfortable and less of a burden for the patient.

A patient's perception of a treatment modality and information regarding experiences from other patients may be determining factors in their choice to accept or ask for a certain treatment. Shorter treatment regimens, whereby the patient and their relatives do not have to travel to a hospital many times, may be more attractive from a patient's perspective as this minimizes the impact on their daily lives. The physical environment in which a treatment takes place can influence the perception of the total treatment. Less invasive treatments often are more attractive as they facilitate a faster recovery. Shorter hospital stays and faster recovery enable patients to get back to their normal lives much quicker. Figure 30.8 illustrates possible improvement areas from a patient's perspective.

Explaining to patients that brachytherapy enables organ preserving treatments that in many cases can help prevent prostatectomy (Zelefsky et al., 2016), cystectomy (van der

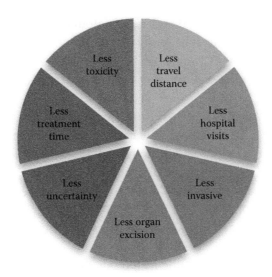

FIGURE 30.8   Possible improvement areas from a patient's perspective.

Steen-Banasik et al., 2002, 2009; Bos et al., 2014), hysterectomy (Mazeron et al., 2013), and mastectomy (Polgár et al., 2013) leading to significantly improved quality of life after treatment will help to increase the adoption and utilization of brachytherapy worldwide.

Recently, we have commercialized the Luneray™ applicator. This solution enables robot-assisted minimally invasive implantation of catheters in the bladder wall to accommodate a brachytherapy boost to the bladder and makes bladder preservation for patients possible (Van der Steen-Banasik et al., 2002; Nap-van Klinken et al., 2014). Furthermore, we have supported a study on brachytherapy-based accelerated partial breast irradiation (APBI) for breast cancers (Kamrava et al., 2015). This study shows that brachytherapy APBI is an attractive option for a large group of patients as it can provide similar outcomes to EBRT treatments in a much shorter timeframe.

Areas of focus for our future innovation include the following topics. Enabling further hypo-fractionation with brachytherapy (Mavroidis et al., 2014), which will further decrease the number of required hospital visits. Less invasive applicators that can be inserted more easily and are less uncomfortable for the patient due to, for example, possibly reducing the need for additional interstitial needles (Han et al., 2014). Focal therapy for prostate cancers (Banerjee et al., 2015), that is, irradiating only the lesion instead of the whole prostate gland, which will further decrease toxicities related to treatments.

30.2.6 Theme VI: Workflow Efficiency

Our definition of the *Workflow Efficiency* theme is as follows:

> Provide solutions that improve the overall workflow efficiency, while still providing tools to increase accuracy and conformity.

In recent studies the benefits of performing 3D imaging and image-guided adaptive brachytherapy were demonstrated (Tanderup et al., 2010; Kirchheiner et al., 2016).

However, acquiring 3D image sets and outlining contours for the target and OARs in these image datasets can be very time consuming—especially if this needs to be repeated for each fraction. Thus, although the benefits of 3D image-guided adaptive brachytherapy are clear, the penalty in terms of procedure time and required resources is at present too high for it to be used on many patients.

Also in certain steps of the treatment, such as treatment planning, many repetitive elements are present that are therefore time consuming. We believe that by focusing on improvement of the overall workflow, we can take away such hurdles and enable the use of advanced techniques to treat many more patient treatments.

In recent years we have implemented functionality into Oncentra Brachy to speed up the treatment planning workflow, such as Applicator Modeling and Library Plans. These functionalities automate many manual steps in the treatment planning process and speed up the workflow substantially.

Future areas of interest in this theme include the following: automation of applicator reconstruction, automation of delineation of target structures and organs at risk, and protocol-based workflow to further remove repetitive manual steps in the process.

30.3 CONCLUSION

Elekta brachytherapy solutions have historically been the driver for innovation in brachytherapy. We remain committed to brachytherapy and will continue to drive the further development of this modality. With our large customer base, our network of key opinion leaders, and the knowledge and experience within the company, we have all the right ingredients to further elevate brachytherapy to a higher level and increase its adoption and utilization.

With a clear vision based on ideas for future developments, we can bring new innovations to our customers. Delivering on our Vision 2020 will further increase attractiveness of brachytherapy as a treatment option for many patients who currently do not have access to this therapy. Above all, we believe that discussions and partnerships with our customers are key success factors in the innovation of brachytherapy. For further information about our current solutions and future innovations, please visit http://www.elekta.com/brachytherapy.

REFERENCES

Ballester F, Carlsson Tedgren Å, Granero D, Haworth A, Mourtada F, Fonseca GP et al. 2015. A generic high-dose rate (192)Ir brachytherapy source for evaluation of model-based dose calculations beyond the TG-43 formalism. *Med Phys*, 42(6), 3048–3061.

Ballester-Sánchez R, Pons-Llanas O, Candela-Juan C, Celada-Álvarez FJ, Barker CA, Tormo-Micó A, Pérez-Calatayud J, Botella-Estrada R. 2016. Electronic brachytherapy for superficial and nodular basal cell carcinoma: A report of two prospective pilot trials using different doses. *J Contemp Brachytherapy*, 8(1), 1–8.

Banerjee R, Park SJ, Anderson E, Demanes DJ, Wang J, Kamrava M. 2015. Brachytherapy, From whole gland to hemigland to ultra-focal high-dose-rate prostate brachytherapy: A dosimetric analysis. *Brachytherapy* 14(3), 366–372.

Bianchi C, Botta F, Conte L, Vanoli P, Cerizza L. 2008. Biological effective dose evaluation in gynaecological brachytherapy: LDR and HDR treatments, dependence on radiobiological parameters, and treatment optimisation. *Radiol Med*, 113(7), 1068–1078.

Bos MK, Marmolejo RO, Rasch CRN, Pieters BR. 2014. Bladder preservation with brachytherapy compared to cystectomy for T1-T3 muscle-invasive bladder cancer: A systematic review. *J Contemp Brachytherapy*, 6(2), 191–199.

Deutsch I, Zelefsky MJ, Zhang Z, Mo Q, Zaider M, Cohen G, Cahlon O, Yamada Y. 2010. Comparison of PSA relapse-free survival in patients treated with ultra-high-dose IMRT versus combination HDR brachytherapy and IMRT. *Brachytherapy*, 9(4), 313–318.

Han DY, Webster MJ, Scanderbeg DJ, Yashar C, Choi D, Song B, Devic S, Ravi A, Song WY. 2014. Direction-modulated brachytherapy for high-dose-rate treatment of cervical cancer. I: Theoretical design. *Int J Radiat Oncol Biol Phys* 89(3), 666–673.

Han K, Milosevic M, Fyles A, Pintilie M, Viswanathan AN. 2013. *Int J Radiat Oncol Biol Phys*, 87(1), 111–119.

Hinnen KA, Battermann JJ, van Roermund JG, Moerland MA, Jürgenliemk-Schulz IM, Frank SJ, van Vulpen M. 2010. Long term biochemical and survival outcome of 921 patients treated with I-125 permanent prostate brachytherapy. *Int J Radiat Oncol Biol Phys*, 76(5), 1433–1438.

Holly R, Morton GC, Sankreacha R, Law N, Cisecki T, Loblaw DA, Chung HT. 2011. Use of cone-beam imaging to correct for catheter displacement in high dose-rate prostate brachytherapy. *Brachytherapy*, 10(4), 299–305.

Hoskin PJ, Rojas AM, Bownes PJ, Lowe GJ, Ostler PJ, Bryant L. 2012. Randomised trial of external beam radiotherapy alone or combined with high-dose-rate brachytherapy boost for localised prostate cancer. *Radiother Oncol*, 103(2), 217–222.

Kamrava M, Kuske RR, Anderson B, Chen P, Hayes J, Quiet C, Wang PC, Veruttipong D, Snyder M, Jeffrey Demanes D. 2015. Outcomes of breast cancer patients treated with accelerated partial breast irradiation via multicatheter interstitial brachytherapy: The pooled registry of multicatheter interstitial sites (PROMIS) experience. *Ann Surg Oncol.*, 22(Suppl. 3), S404–11.

Kertzscher G, Rosenfeld A, Beddar S, Tanderup K, Cygler JE. 2014. In vivo dosimetry: Trends and prospects for brachytherapy. *Br J Radiol*, 87(1041).

Kirchheiner K, Pötter R, Tanderup K, Lindegaard JC, Haie-Meder C, Petrič P, Mahantshetty U et al. and EMBRACE Collaborative Group. 2016, Health-related quality of life in locally advanced cervical cancer patients after definitive chemoradiation therapy including image guided adaptive brachytherapy: An analysis from the EMBRACE Study. *Int J Radiat Oncol Biol Phys*. 94(5), 1088–1098.

Kirisits C, Rivard MJ, Baltas D, Ballester F, De Brabandere M, van der Laarse R, Niatsetski Y et al. 2014. Review of clinical brachytherapy uncertainties: Analysis guidelines of GEC-ESTRO and the AAPM. *Radiother Oncol*, 110(1), 199–212.

Lymperopoulou G, Papagiannis P, Sakelliou L, Georgiou E, Hourdakis CJ, Baltas D. 2006. Comparison of radiation shielding requirements for HDR brachytherapy using 169Yb and 192Ir sources. *Med Phys*, 33(7), 2541–2547.

Mavroidis P, Milickovic N, Cruz WF, Tselis N, Karabis A, Stathakis S, Papanikolaou N, Zamboglou N, Baltas D, 2014. Comparison of different fractionation schedules toward a single fraction in high-dose-rate brachytherapy as monotherapy for low-risk prostate cancer using 3-dimensional radiobiological models. *Int J Radiat Oncol Biol Phys*, 88(1), 216–223.

Mazeron R, Gilmore J, Dumas I, Champoudry J, Goulart J, Vanneste B, Tailleur A, Morice P, Haie-Meder C 2013. Adaptive 3D image-guided brachytherapy: A strong argument in the debate on systematic radical hysterectomy for locally advanced cervical cancer. *Oncologist*, 18(4), 415–422.

Moulton CR, House MJ, Lye V, Tang CI, Krawiec M, Joseph DJ, Denham JW, Ebert MA. 2015. Registering prostate external beam radiotherapy with a boost from high-dose-rate brachytherapy: A comparative evaluation of deformable registration algorithms. *Radiat Oncol*, 10, 254.

Nap-van Klinken A, Bus SJEA, Janssen TG, Van Gellekom MPR, Smits G, Van der Steen-Banasik E. 2014. Interstitial brachytherapy for bladder cancer with the aid of laparoscopy. *J Contemp Brachytherapy*, 6(3), 313–317.

Nesvacil N, Tanderup K, Hellebust TP, De Leeuw A, Lang S, Mohamed S, Jamema SV, Anderson C, Pötter R, Kirisits C. 2013. A multicentre comparison of the dosimetric impact of inter- and intra-fractional anatomical variations in fractionated cervix cancer brachytherapy. *Radiother Oncol*, 107(1), 20–25.

Polgár C, Fodor J, Major T, Sulyok Z, Kásler M. 2013, Breast-conserving therapy with partial or whole Breast Irradiation: Ten-year results of the Budapest randomized trial. *Radiother Oncol*, 108(2), 197–202.

Sun B, Yang D, Esthappan J, Garcia-Ramirez J, Price S, Mutic S, Schwarz JK, Grigsby PW, Tanderup K. 2015. Three-dimensional dose accumulation in pseudo-split-field IMRT and brachytherapy for locally advanced cervical cancer. *Brachytherapy*, 14(4), 481–489.

Tanderup K, Nielsen SK, Nyvang GB, Pedersen EM, Røhl L, Aagaard T, Fokdal L, Lindegaard JC. 2010. From point A to the sculpted pear: MR image guidance significantly improves tumour dose and sparing of organs at risk in brachytherapy of cervical cancer. *Radiother Oncol.*, 94(2), 173–180.

Tedgren AK, Ahnesjö A. 2003. Accounting for high Z shields in brachytherapy using collapsed cone superposition for scatter dose calculation. *Med Phys*, 30(8), 2206–2217.

van der Steen-Banasik E, Ploeg M, Witjes JA, van Rey FS, Idema JG, Heijbroek RP, Karthaus HF, Reinders JG, Viddeleer A, Visser AG. 2009. Brachytherapy versus cystectomy in solitary bladder cancer: A case control, multicentre, East-Netherlands study. *Radiother Oncol*, 93(2), 352–357.

Van der Steen-Banasik EM, Visser AG, Reinders JG, Heijbroek RP, Idema JG, Janssen TG, Leer JW. 2002. Saving bladders with brachytherapy: Implantation technique and results. *Int J Radiat Oncol Biol Phys*, 53(3), 622–629.

Vásquez Osorio EM, Kolkman-Deurloo IK, Schuring-Pereira M, Zolnay A, Heijmen BJ, Hoogeman MS. 2015. Improving anatomical mapping of complexly deformed anatomy for external beam radiotherapy and brachytherapy dose accumulation in cervical cancer. *Med Phys*, 42(1), 206–220.

Zelefsky MJ, Poon BY, Eastham J, Vickers A, Pei X, Scardino PT. 2016. Longitudinal assessment of quality of life after surgery, conformal brachytherapy, and intensity-modulated radiation therapy for prostate cancer. *Radiother Oncol.* 118(1), 85–91.

CHAPTER 31

Eckert & Ziegler BEBIG

Carmen Schulz, Irina Fotina, Michael Andrassy,
Thomas Osche, and Sven Beerheide

CONTENTS

31.1 INTRODUCTION: ECKERT & ZIEGLER BEBIG'S VISION

Since its first use over a century ago, brachytherapy has allowed high radiation doses to be delivered to clinical targets while sparing adjacent healthy tissues. Many technological developments have enabled the range and complexity of treatments to enhance, from simple skin treatments and manual gynecological insertions, to sophisticated image-guided implants. The pace of change has never been greater than it is today. Brachytherapy plays an important role both as a monotherapy and in multimodal settings and a range of radionuclides are available to provide low-dose rate (LDR) or high-dose rate (HDR) treatment. Eckert & Ziegler BEBIG were the first to introduce ^{60}Co miniaturized HDR sources in 2003, and continue to innovate and contribute to the further advancement of brachytherapy (e.g., see Figure 31.1), with promising new technologies emerging in the near future. Faster computing, smaller detectors, and individual 3D printing solutions will update imaging and treatment planning, refine dose delivery verification, and bring individualization of therapy to a higher level. The introduction of nanoparticles might have the potential to further improve the clinical effectiveness of brachytherapy.

31.2 ADVANCED IMAGING TECHNOLOGIES

The importance of 3D imaging in modern radiation therapy techniques was recognized long ago, and brachytherapy is no exception. Contemporary 3D image-guided techniques

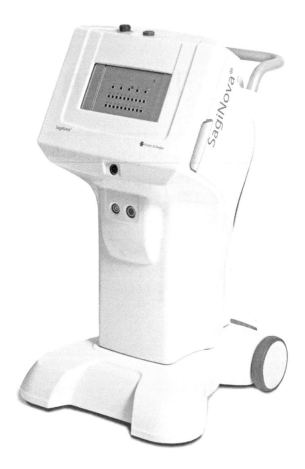

FIGURE 31.1   SagiNova afterloader: Commercial launch in 2015.

include the use of various imaging modalities at each stage of treatment, through diagnosis, planning, and clinical follow-up. Verification imaging at the point of treatment has been widely adopted for external beam radiotherapy, in 2D, 3D, and 4D, both for target and organs at risk localization and dosimetry assessments, but such imaging has not yet been utilized in brachytherapy, other than simple verification of applicator position. Future directions in this area might include development of online integrated 3D ultrasound (US) imaging systems or cone beam CT or cone beam computed tomography (CBCT)-integrated brachytherapy equipment for gynecological patients. Possible benefits of imaging during treatment for brachytherapy would not be limited at image-guidance but could include imaging for quality control of dose delivery and online monitoring of the deposited dose. Image processing and storage demands for brachytherapy may be significant in the future, as has been seen for external beam therapy over the last decade. Modern image processing software will be required, enabling fusion, and deformable image registration (DIR) of the multiple image modalities, including CT, MRI, and ultrasound. Currently, implementation of DIR methods in clinical brachytherapy is limited by a need for further research on uncertainties, commissioning, and quality control methodologies (Pukala et al., 2013;

Teo et al., 2015) (for in-depth discussion, see Chapter 7), but it is expected that DIR will be integrated into routine clinical treatment planning systems for brachytherapy in the near future.

Safe and effective brachytherapy workflow based on multiple image modalities at all stages of the treatment process would be dependent on the accurate definition of the target volume and organs at risk. Despite significant research and clinical interest in automated segmentation techniques, and use of hybrid segmentation approaches (model and atlas-based), there are currently still very limited data available on their acceptance and use for contouring in brachytherapy. A promising new approach to automatic image segmentation is "machine-learning," in which software would be "trained" for automatic detection of abnormal tissue across various imaging modalities. Rapid implementation of these techniques may be facilitated if models and atlases for commercially available software were shared in the community via educational platforms or open source software.

Sharing of knowledge and aiding to the use of advanced brachytherapy technologies via shared libraries of data for automatic organ segmentation, planning templates, and results based on models for DVH estimates, and biological evaluation should be supported both from educational organizations and equipment vendors. It is anticipated that "cloud based" technology to share such information will increase in popularity and importance over the next 5 years.

31.3 MC SIMULATION AS STANDARD FOR TREATMENT PLANNING

Monte Carlo techniques are widely used to generate data describing the variation in dose-rate around specific brachytherapy source models (Perez-Calatayud et al., 2012). Among established calculation algorithms, Monte Carlo (MC) is recognized as the gold standard because it avoids any significant approximations for radiation transport, and can offer comparatively low uncertainties. Although experimental techniques cannot match MC calculations in terms of uncertainty, measurements are useful and necessary for validation of MC data.

The majority of commercial clinical treatment planning systems subsequently use the MC-generated information as the basis for brachytherapy dose calculations, using simple summation algorithms, based on the formalism introduced by AAPM TG-43 (Nath et al., 1995). While the TG-43 dosimetry formalism has provided robust and consistent clinical dosimetry planning for many years, it does not take account of patient-specific scatter and radiological differences from the source-specific homogeneous water geometry. We may soon see a significant clinical uptake of new brachytherapy treatment calculation algorithms, already known from external beam therapy, that consider heterogeneous structures in the treatment volume and provide improved accuracy of dose calculation. The clinical relevance of this progress is to be evaluated, and is likely to be more significant for low-energy sources than for high energy, and for sites with significant heterogeneity such as breast and esophagus, or for shielded applicators.

The introduction of accelerated algorithms and faster computers, parallel processing in local, cluster-, and cloud-based configurations intended for other technologies, are encouraging MC techniques to become a reality in commercial planning software for brachytherapy as well.

Healthcare will, without doubt, benefit from translational developments of information technology (IT)-technologies, driven by a multitude of needs from modern global society. MC-codes running on graphical processor units (GPUs) have already been adopted in treatment planning systems to reduce calculation times from tens of minutes to several seconds (Hissoiny et al., 2012). Other research studies on fast calculation techniques are based on cloud configurations to increase calculation capacity.

Although further development is required before MC systems for brachytherapy become commercially available, we can anticipate that significant improvements in computer performance will replace present algorithms and bring a new quality into adaptive, near-real time treatment planning and optimization.

31.4 ONLINE DOSE DELIVERY VERIFICATION

Although in vivo dosimetry (IVD) has been available for brachytherapy applications for many years, its clinical adoption has been limited. Technical solutions have mainly focused on dose monitoring near organs at risk and detection of gross errors during treatment delivery. There is now a renewed interest to develop online in vivo dose verification methods to improve treatment delivery quality reporting and actual dose delivery for clinical trials and patient follow-up. However, substantial progress in methodology and infrastructure is needed to realize the full potential of IVD. It is expected that new imaging systems will be developed to monitor the position of brachytherapy sources, applicators, and the dose delivered to clinical target and organs at risk in real time. Consequently, new innovations will be required to handle and visualize 4D-information of this kind in state-of-the-art software products. New detector applications, characterized for the specific conditions in brachytherapy fields including high gradients and changing radiation quality, will become the basis for IVD over the coming years.

The balance of treatment quality and cost effectiveness in the clinic, as well as commercial success in the medical device industry are challenges with a common goal—a new level of clinical dosimetry to further advance the quality of brachytherapy.

31.5 PREVENTIVE MAINTENANCE PREDICTION

Current afterloader maintenance practice is mainly driven by fixed time schedules based on previous risk estimates. Maintenance schedules for different system components are usually sequenced into a yearly schedule. This may create unnecessary costs for healthcare providers and manufacturers or not be aligned with a specific local uptime demand. A new approach to maintenance that takes into account the actual, rather than predicted, use in a specific clinic, combined with integrated sensors for equipment performance monitoring could increase the overall system uptime and change maintenance from a common planned preventative schedule to one of customized interventions based on actual need.

As current afterloader systems are completely digitally controlled, they can gather and analyze direct status information from different electronic components, such as motor movement counters, battery status, online/offline time, last exchange of components, component production dates, etc. If all these data are analyzed and used to adjust service intervention to the actual usage of the system, it would allow a more flexible scheduling of maintenance dates without compromising the safety and efficacy of the system.

31.6 DOSE DELIVERY INNOVATIONS

3D printing technology has numerous potential applications (also see Chapter 11). A promising opportunity in brachytherapy is to create customized, truly patient-specific HDR applicators.

Currently, individual patient anatomy is usually accommodated by means of selecting the closest sized applicators, or with flexible applicators, which is not optimum in all cases. For nonstandard anatomies such as irregular vault shapes or strongly asymmetric tumors, designing a treatment plan for appropriate clinical target volume coverage may not be possible with standard applicators, even with supplementary interstitial needles.

The fabrication of customized HDR applicators with 3D print technology could be created using patient anatomy data from CT, US, or MRI. Hence, customized applicators can be designed with high accuracy and with comparatively little effort by CAD systems, and manufactured by 3D printing directly. The implementation of algorithms to support the user in the design process is conceivable (Garg et al., 2013).

One crucial factor for the success of this approach is the availability of appropriate polymer materials (Cunha et al., 2014, 2015). Biocompatibility, sterilizability, and homogeneous density are specific requirements to be addressed, as well as methods for appropriate quality control checking of the resulting custom-applicator.

The usage of 3D printing technology appears to be a promising approach and a realistic scenario for many brachytherapy departments. First practical experiences in this field have already been made with this or similar technologies as stereolithography and have predominantly shown good results (Cumming et al., 2014; Wiebe et al., 2015).

31.7 DOSE ENHANCEMENT BY NANOTECHNOLOGIES

For over a decade, nanoparticles have been widely investigated in different fields of application for tumor therapy. These range from support of imaging to tumor labeling, from drug delivery to hyperthermia techniques, and, moreover, are considered as a method of providing radiation dose enhancement. The use of nanoparticles in radiotherapy combines the expertise of both nuclear medicine and brachytherapy. Although the origin of radiation emission is not a classical brachytherapy sealed source of specified dimensions, it differs as well from radiopharmaceuticals because the radiation of interest is emitted by inactive particles in a secondary process induced by an external radiation field. The increased dose in the vicinity of such particles can be understood via photo-absorption due to the emission of photons with lower energy and thus significantly higher mass–energy absorption coefficient, or by Auger-electrons.

Low-energy photons have a higher reaction probability and thus brachytherapy sources may exhibit a greater benefit with nanoparticles than for megavoltage photons delivered by external beam radiotherapy. The distribution of dose depends on the spatial nanoparticle distribution, their concentration, and contained chemical elements, as well as the relative position and dose distribution of the radiation sources themselves. Research efforts are concentrated on designing these particles with an optimized structure for radiation emission, particle deposition, time of detention, as well as visibility by means of different imaging modalities (Nawroth et al., 2006). Understanding cell survival, controlling the macro- and in-cell particle distributions and micro-dosimetrical calculations are examples of research requiring new interdisciplinary co-operations.

Much work is needed including feasibility studies and clinical trials to fulfill regulatory requirements, but this promising field of nanotechnology will mature and contribute to progress in radiation therapy. It has the potential to enhance the relevance of brachytherapy as an important option for the treatment of cancer.

REFERENCES

Cumming, I., Joshi C, Lasso A, Rankin A, Falkson C, Schreiner J & Fichtinger G. 3D printed patient-specific surface mould applicators for brachytherapy treatment of superficial lesions (abstract). *Med Phys* 2014;41:222.

Cunha J, Sethi R, Mellis K et al. Commissioning and clinical use of PC-ISO for customized, 3D printed, gynecological brachytherapy applicators (abstract). *Med Phys* 2014;41(6):514.

Cunha JAM, Mellis K, Sethi R, Siauw T, Sudhyadhom A, Garg A, Goldberg K, Hsu I-C, and Pouliot J. Evaluation of PC-ISO for customized, 3D printed, gynecologic 192Ir HDR brachytherapy applicators. *J Appl Clin Med Phys* 2015;16(1):5168.

Garg A, Patil S, Siauw T, Cunha JAM, Hsu I-C, Abbeel P, Pouliot J, Goldberg K. An algorithm for computing customized 3D printed implants with curvature constrained channels for enhancing intracavitary brachytherapy radiation delivery. *2013 IEEE International Conference on Automation Science and Engineering (CASE)*, Madison WI, USA.

Hissoiny S, D'Amours M, Ozell B, Després, P, Beaulieu L. Sub-second high dose rate brachytherapy Monte Carlo calculations with bGPUMCD. *Med Phys* 2012:39(7);4559–4567.

Nath R, Anderson LL, Luxton G, Weaver KA, Williamson JF, Meigooni AS. Dosimetry of interstitial brachytherapy sources. *Med Phys* 1995;22(2):209–234.

Nawroth T, Le Duc G, Meesters C, Decker H, Corde S, Requardt H, Bravin A. Therapeutic imaging and indirect radiation treatment IRT with target nanoparticles for cancer therapy at ESRF-ID17. *ESRF User Meeting*, Grenoble, 2006.

Perez-Calatayud J, Ballester F, Das RK, DeWerd LA, Ibbott GS, Meigooni AS, Ouhib Z, Rivard MJ, Sloboda RS, Williamson JF. Dose calculation for photon-emitting brachytherapy sources with average energy higher than 50 keV. Report of AAPM and ESTRO. *Med Phys* 2012;39(5):2904–2929.

Pukala J, Meeks SL, Staton RJ, Bova FJ, Mañon RR, Langen KM. A virtual phantom library for the quantification of deformable image registration uncertainties in patients with cancers of the head and neck. *Med Phys.* 2013;40(11):111703.

Teo BK, Bonner Millar LP, Ding X, Lin LL. Assessment of cumulative external beam and intracavitary brachytherapy organ doses in gynecologic cancers using deformable dose summation. *Radiother Oncol.* 2015;115(2):195–202.

Wiebe E, Easton H, Thomas G, Barbera L, D'Alimonte L, Ravi A. Customized vaginal vault brachytherapy with computed tomography imaging-derived applicator prototyping. *Brachytherapy* 2015;14(3):380–384.

Index

Printed and bound by CPI Group (UK) Ltd, Croydon, CR0 4YY

01/11/2024

01782601-0008